THE LONG PURPLE LINE

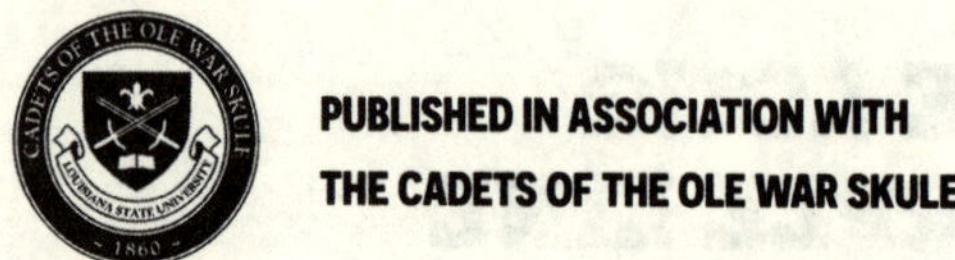
CADETS OF THE OLE WAR SKULE
LOUISIANA STATE UNIVERSITY
1860
PUBLISHED IN ASSOCIATION WITH
THE CADETS OF THE OLE WAR SKULE

THE LONG PURPLE LINE

THE MILITARY HISTORY OF LSU AND THE HEROISM OF ITS CADETS AND ASSOCIATES

RONALD J. DREZ

LOUISIANA STATE UNIVERSITY PRESS
BATON ROUGE

Published with the assistance of the V. Ray Cardozier Fund

Published by Louisiana State University Press
lsupress.org

Manufactured in the United States of America
First printing

Designer: Kaelin Chappell Broaddus
Typefaces: Miller Text, text; Novecento Slab, Novecento Sans, display
Printer and binder: Sheridan Books, Inc.

Jacket photograph courtesy Eddy Perez, LSU.

Cataloging-in-Publication Data are available from the Library of Congress.

ISBN 978-0-8071-8451-6 (cloth: alk. paper)

To all who have honorably served

A nation without a past is a lost nation,
and a people without a past is a people without a soul.

—SERETSE KHAMA,
Botswana's first President, 1970

CONTENTS

PREFACE

This book has been a five-year project of the Cadets of the Ole War Skule, an alumni association of former members of the LSU Corps of Cadets. It is the story of a school and its history, an important part of which is its military history—a Long Purple Line. The story is told through the lives of its participants. But first a few facts as a backdrop to those histories.

The institution opened as a military school in 1860, largely founded by General George Mason Graham, a wealthy Louisiana planter from Virginia who attended West Point and served in the Mexican War. His vision was a military school following the Virginia Military Institute model. Some say he is the father of LSU. He championed William Tecumseh Sherman to be its first President. A graduate of West Point, Sherman later became a famous general in the War Between the States and Chief of Staff of the U.S. Army under President Ulysses Grant. This is a decidedly military origin. After the war, Sherman donated to LSU two brass cannons that, under the command of Louisiana General P. G. T. Beauregard, fired on Fort Sumter. They are still on the campus today in front of the Military Science Building. LSU has had three generals serve as its President, as Major General Campbell B. Hodges and Lieutenant General Troy H. Middleton followed Sherman in the 20th century.

Of the first faculty of five professors under Sherman, three joined the Confederate Army, with David F. Boyd serving as a colonel and second President of LSU after the war. The fourth faculty member tried to join but was not accepted. All of the cadets in that first class save one joined the Confederate Army. One source credits the school's "Fighting Tigers" moniker and its tiger mascot to General Robert E. Lee's remark on observing a successful movement of a Louisiana regiment: "Those Louisianians fight like tigers."

After the war, the school reopened, moved to Baton Rouge, and occupied the Pentagon Barracks, a historic military structure on the banks of the Mississippi River that still exists as a historic landmark. It was still a military school when World War I came about, and LSU formed one of the first federal officer training units, which later became the Reserve Officer Training Corps. LSU's ROTC is still operating on campus today. During World War II the program produced the fourth most officers of any school, behind West Point, Annapolis, and Texas A&M. Lieutenant General Troy H. Middleton, the defender of Bastogne, served at LSU before the war and then as its President afterward.

While no longer exclusively a military school, LSU's long military history—over 160 years and counting—is still recognizable today and is visible to any visitor to the campus. Its most striking building is the magnificent Memorial Tower at the center of the campus constructed by the American Legion as a memorial to Louisiana's dead in World War I. It houses a splendid LSU military museum open at no charge to the public. Across from it is the honored Parade Grounds, where military formations and parades are still performed. The two polished brass cannons from Sherman, a U.S. Air Force jet fighter, and other military hardware are displayed on campus. The school newspaper, the *Reveille,* got its name from a military bugle call. The traditional class ring depicts the Memorial Tower with sabers on either side. The impressive World War II Memorial on the Parade Grounds is inscribed with the names of alumni who gave their last full measure in that war and all since, and nearby is the Memorial Oak Grove, remembering LSU's lost in World War I. Hodges, Johnston, and Lockett Halls on campus are named after prominent military personnel connected to LSU. The Pentagon Barracks military student dorms on campus are named Zachary Taylor, P. G. T. Beauregard, John A. Lejeune, and Andrew Jackson Halls after famous Louisiana military leaders, and East and West Germaine Laville Halls are named after a Marine heroine. Lastly, President Thomas Boyd in the early 1900s affectionately and respectfully named LSU the "Ole War Skule." Its military alumni association bears that name today.

But the real military history of LSU was and is being written by the soldiers, sailors, Marines, and airmen. From 1860 to now and hopefully into the distant future the Long Purple Line continues. Its history is told in the succeeding chapters through the LSU alumni, associates, and faculty who lived it.

W. Henson Moore
President, Cadets of the Ole War Skule

ACKNOWLEDGMENTS

This is my third book published by LSU Press. All have sought to illuminate U.S. military history during climactic times of war. In 1994, the Eisenhower Center at the University of New Orleans, where I served as assistant director under the late Dr. Stephen Ambrose, partnered with the press to publish previously silent voices of the men who were there during the invasion of Normandy on June 6, 1944. Our research had uncomfortably discovered that there were significant errors in the reports and writings of that day that not only ignored some extraordinary heroism but even proclaimed that those heroic actions had not even happened. My production editor from LSU Press at the time was Julie Schorfheide, who worked tirelessly with me on *Voices of D-Day* to see that new research was verified and presented to put an end to fifty years of stolen valor. Our joint effort was rewarded with the 1995 George Washington Honor Medal from the Freedoms Foundation at Valley Forge in the field of Communications.

Then, in 2014, as the nation approached the bicentennial celebration of the War of 1812, the Honorable Mr. W. Henson Moore, a former Louisiana congressman, chaired the Louisiana Battle of New Orleans Bicentennial Commission. It sought to correct long-standing errors and revisionism that had crept into the fascinating history of the Battle of New Orleans. He invited me to join him and research all the details to refute the revisionists' proclamations. Our children were taught that the battle was "useless" and had been fought after a peace treaty had been signed. Schoolbooks and teachers nationwide had taught this bad history for a hundred years.

Again, it was LSU Press that offered its publishing and editing skills to get our new primary research into print and correct errors once and for all. Mr. Moore pressed to get the correcting documents into print because, if it were not done for the bicentennial, there was little doubt that it would ever be corrected. Director MaryKatherine Callaway made available the press's most capable people to bring this venture to fruition, and *The War of 1812* became the singular work that published the secret British documents that refuted the long-standing revisionism. Without Henson Moore and the LSU Press, this would never have happened.

Now LSU Press has again stepped up to publish a new work detailing the long history of the military at LSU. That history is illuminated by presenting the heroism and leadership of its cadets, soldiers, sailors, Marines, and airmen. Included in those numbers are its associates and teachers. All have formed *The Long Purple Line* that began in 1860 and remains unbroken to this day. This five-year historic project was initiated by W. Henson Moore, Member of the Board, who convinced the Cadets of the Ole War Skule to undertake the project. He, as its President, chaired the lengthy effort that resulted in its printing.

LSU's President William F. Tate IV enthusiastically approved the project as a necessary attempt to add a long-missing story to the historiography of LSU. It would be a daunting task to conduct the proper research and unearth long-hidden documents and memoirs, but as Mr. Moore stated, "If it was easy, someone else would already have done it."

Providing vital assistance and support to navigate the pitfalls inherent in the completion of this work has been Ms. Jane Cassidy, Senior Vice Provost. I cannot thank her enough. Also greatly assisting has been Mr. James Gregory, the executive director of the Tower Military History Museum. While serving in this capacity, Mr. Gregory also graciously edited the manuscript, wrote a valuable review, and worked closely with Ms. Cassidy. He also pursues his own Ph.D. in military history. Mr. Randy Gure, the Executive Director of the Cadets of the Ole War Skule, opened the archives to me for research and provided valuable assistance.

The vital work of copy editing fell under the guiding hand of Ms. Catherine L. Kadair, the press's Managing Editor and Assistant Director, who designated the excellent freelance line editor Mr. Derik Shelor to finalize the work. In addition to his impressive credentials, Mr. Shelor brought an often-missing talent, melding military elements of style with the traditional elements to perfect the manuscript.

Finally, heartfelt thanks to my longtime friend and associate Dr. Douglas Brinkley, Katherine Tsanoff Brown Chair in Humanities and Professor of History at Rice University. For thirty years, Dr. Brinkley and I have worked together and collaborated on research and writing. His reputation as a significant historian is well known. He read and edited *The Long Purple Line* and wrote an important review. He encouraged and delighted in the entire story.

THE LONG PURPLE LINE

Introduction

The writing and publishing of this important work has been long overdue. The history of Louisiana State University, originally named the Louisiana State Seminary of Learning, began in 1860 and has continued unbroken for over 180 years. During that time, LSU alumni, authors, freelance writers, and essayists have filled millions of pages of print with words that detail and honor the scholarly, sociological, athletic, military, and technological prowess of the university. It is indeed a rich collection of writing.

The Louisiana State University Press has long been the source of publications on all subjects of Louisiana life, and much more. But conspicuously missing is any work on individual heroic exploits, and there were many from the veterans who passed through LSU's portals or who were associated with the Ole War Skule.

I have been privileged to have had two books on military history published by LSU Press—one on the Invasion of Normandy, and the other on the War of 1812.

LSU came into being at the end of a long line of established southern universities. If it was a little late to the party, a quick look at its geographical position might explain why. To modern-day students and alumni of LSU, the school's ancient title—the Louisiana State Seminary of Learning—dating back to the 1850s, might seem to indicate a religious connection. Today's definition of the word *seminary* is almost exclusively taken to mean a theological college. But in 1806, and again in 1811, the fledgling U.S. government created federal land grants for each of the states; and Louisiana, upon admission to the Union in 1812, qualified for over forty-six thousand acres of land. The grant authorization languished somewhere until it was rediscovered in 1843, and the sale of the land to support a "seminary of learning" began in 1844.[1]

Perhaps it was not so much a case of "rediscovering" the federal land grant as it was a lack of interest in constructing an institution of higher learning in Louisiana. A quick look at the founding dates of the southern state universities reveals that those states located in the east were quick to take advantage of the revenues generated by the sale of their federal lands to build state universities. Daily life in Louisiana was agricultural, and not conducive for young men to seek a higher education. Wealthy Creole plantation owners (first-generation French born in America) might send their sons back to the old country for their higher education. Others might send their children to a university along the East Coast, or in New England. But those numbers were minuscule when compared to the nation's educational statistics of the time. Less than 1.7 percent of people between eighteen and twenty-five years old were in college, and 20 percent of the entire population was illiterate.[2] Higher education was the least concern of the general population.

Unlike the eastern universities, LSU found itself on the extreme western boundary of what could be called southern society. That remote area would later become known as the Trans-Mississippi Department of the South. The Seminary's original location was close to Alexandria, and adjacent to the Red River. The Mississippi River was seventy miles to the east.

The University of Georgia had begun in 1785, followed by the University of Tennessee in 1794. Then the University of Virginia was founded in 1819. As the westerly expansion of the population continued, Alabama opened its doors in 1831. Mississippi followed in 1848. But in that Trans Mississippi Department, Louisiana (1860) and Arkansas (1871) lagged.

In fairness, its western remoteness was not the main culprit in the state's tardiness in developing a school for higher education. As in all ventures, jealousies and politics played a big part. Historian Walter Fleming wrote, "The East was alarmed by the swift growth of the West and Southwest; it feared the diminution of its political power in the union, and . . . the loss of its population to emigration."[3] Even so, Louisiana might have lagged even farther behind had it not been for the imagination, vision, forethought, and drive of one man: General G. Mason Graham, a wealthy planter and resident of Rapides Parish.

Graham was not native to Louisiana. He had been born in Virginia in 1807, into the wealth of a Virginia plantation. That birthright afforded him the opportunity to attend the U.S. Military Academy at West Point and, later, the University of Virginia. For whatever reasons, he never received a degree from either school, but learning from his enrolment experiences, he concluded that education was best served in an institution that also had military discipline.

In 1828, as a twenty-one-year-old, Mason Graham set out from the family plantation in Virginia to carve out his own niche in the world, and that new world was to the west. He moved into a sparsely populated area of Rapides Parish in central Louisiana and successfully developed his own plantation holdings. Thirty years later, just before the outbreak of the Civil War, those holdings were worth over $250,000.[4]

In 1846, Graham had organized an infantry company of volunteers that he called the Rapides Horse Guards to serve in the Mexican War. They returned home having never seem action, but Graham stayed in Mexico and fought in the Battle of Monterrey and was cited for bravery.[5]

Over time, Graham had become a tireless advocate for the development of a state university that would be located in his own Rapides Parish. He had continued to study how universities were run and had observed that many of them had failed. And those failures seemed to have happened at institutions that presented a strictly "literary" agenda, without any military influence. His idea was to have the Louisiana Seminary of Learning modeled after the rising star in the east: the increasingly popular Virginia Military Institute. Its stated design was to turn out its young men "as thoroughly scientific and accomplished literary gentlemen."[6]

There can be no doubt that Graham's push for a university to contain military training as part of its curriculum was greatly enhanced by sobering national events in 1859. The fanatical attack on the U.S. arsenal at Harpers Ferry by radical abolitionist John Brown shocked the South. Although Brown and his sons were captured and hanged for their treason, Brown's stated goal to arm the slaves and then lead them in an insurrection was a threat to life and property in the South.

Historian Benjamin L. Price wrote that southerners "believed that abolitionists and 'Black Republicans' might attempt by violent means what they had failed to achieve through politics in the election of 1856: the destruction of the southern way of life."[7] And to those who scoffed at this threat to a southern way of life, the harsh reality of the radicalism of John Brown was revealed upon examination of some damning evidence. J. A. Turner, a writer of the day, asked about "maps found in John Brown's possession [with] certain marks designating certain localities all over the Southern Country? What means these marks? Nothing?"[8] The *Louisiana Democrat* wrote that Brown's attack indicated "the necessity of each slave-holding State encouraging and supporting at least one Military school within its own limits."[9]

Graham's influence eventually prevailed, and the arguments by the advocates for a university without any military influence were swept aside. In May 1859, the Board of Supervisors, meeting in Alexandria, voted that the new Seminary would be *"a literary and scientific institution under a military system of government, on a program and plan similar to that of the Virginia Military Institute."*[10] And so LSU was born.

CHAPTER 1

The Ole War Skule

The idea of a young man becoming a "gentleman" in southern society had wide appeal to the local gentry. During this period of the mid-19th century, by all accounts, young males seemed to have been dreadfully lacking in all aspects of manners and gentility.

The *Madison Democrat* pulled no punches in its editorials denigrating youthful exuberance: "in this, our perverted age, almost every youth, scarcely out of his teens, considers himself independent of all moral restraint, and at liberty to do as he pleases."[1] Future Confederate General Braxton Bragg wrote disparagingly to future Union General William Tecumseh Sherman[2] that "The more you see . . . our young men, the more you will be impressed with the importance of change . . . if we expect the next generation to be anything but an aggregation of loafers, charged with the duties of squandering their fathers' legacies and disgracing their names."[3] General Mason Graham wrote that the southern male's worst displays of arrogant independence was "a kind of daredevil masculinity, with an emphasis on prodigious drinking, smoking, [and] skill with a gun."

Graham, who would later be acknowledged as the "Father of LSU," was no stranger to the advantages of discipline, especially military discipline. He had been born in Virginia in 1807 to a famous family. His grandfather had framed the Virginia Declaration of Rights, which became a model for the future U.S. Constitution.

Graham came to Louisiana in 1828 not with lofty ideas for higher education but to check on his father's plantation in Rapides Parish. Later he would purchase his own land, also in Rapides Parish. He had secured a West Point appointment in 1826 through Senator John C. Calhoun but resigned it in 1826 and transferred to the University of Virginia. But he abandoned that in 1831 to pursue his dreams of fortune in Rapides Parish.[4]

By 1839 Graham was a rising star in Louisiana politics and was a factor in the election of President William Henry Harrison. When war broke out with Mexico in 1846, Graham raised a company of volunteers to fight with General Zachary Taylor and became a "first to fight" patriot. He led from the front in battle and was cited for conspicuous gallantry at the Battle of Monterey.[5] In 1852 he was elected brigadier general of the Eleventh Brigade of Rapides and Avoyelles Parishes, and later became Adjutant General of Louisiana.

In discussions about higher education in Louisiana, it was Graham who noted that previous attempts had been a half-million-dollar blunder—money down the drain—with nothing to show for it. Only a military school could provide the discipline to keep brash, youthful exuberance in check.[6] General Mason Graham set his sights on doing just that.

That same year, the Louisiana legislature approved a location for a future institute of higher learning. It was in Pineville, in Rapides Parish, in the woods north of the Red River. It was centrally located and offered easy access by water, and most importantly it was deemed to be a healthful location at a time when yellow fever was on everyone's mind and had been the scourge of cities and urban areas.[7] In 1853, the worst epidemic in the history of Louisiana had killed eleven thousand people in New Orleans alone—one-tenth of the population. Another thirty thousand had been infected. Across the state, others died by the thousands. The pine woods of Rapides Parish seemed to be the perfect place, safe from this deadly pestilence.[8]

But despite having secured a location and authorization to build, Graham's exhortations to create a military school ran amuck with a board that could best be described as the very worst example of a committee. As head of the Board of Trustees, and Vice President of the Board of Supervisors, Graham presided over fourteen members who were at odds over the preference of military or academic predominance. Four wanted a high academic university, four wanted a military school, and six could not make up their minds.[9]

Graham insisted that the college be established with courses similar to that of the Virginia Military Institute (VMI). But a December 1858 initial vote taken by the Board was indecisive and the issue was very much in doubt. A second vote on May 3, 1859, preceded by a fierce struggle and acrimony, squeaked by and finally supported Mason Graham's ideas. The Board passed the final resolution: "Resolved: That the Seminary of Learning of the State of Louisiana shall be a literary and scientific institution *under a military system of government.*" Graham bitterly wrote to Colonel F. H. Smith at VMI of his struggle to gain this hard-won victory. "To get my resolution of organization

Louisiana State Seminary of Learning, Pineville, ca. 1868 (LSU Archives)

George Mason Graham, often called the "Father of LSU"
(LSU Archives)

cost me a great struggle and the *Ancien Regime* folks still badger me greatly, striving now to make a mixtry of it all."[10]

The cornerstone of the building had been laid on March 25, 1855,[11] and construction scheduled for completion in 1858, but shoddy materials and poor construction necessitated tearing down and rebuilding, creating long delays. The completed structure was not turned over to the Board of Supervisors until November 28, 1859.[12] Over its main entrance, carved in marble, were Graham's dedicating words, signifying his appreciation and wish for the future: *By the Liberality of the General Government—The Union—Esto Perpetua.* The structure was indeed impressive: three stories high, with five-story towers. It was 170 feet wide and 117 feet deep, enclosing three sides of a quadrangle. There were seventy-two rooms, and the entire building gleamed in white.

After advertising for faculty, on August 2, 1859, the Board approved William Tecumseh Sherman as the superintendent and professor of engineering, architecture, and drawing. Sherman wrote: "The action of the board was wholly the result of the recommendation of Major Don Carlos Buell, then in Washington, and of Gen. G. Mason Graham, half-brother to my old chief, Gen. [R. B.] Mason, in California."[13] After his confirmation, Sherman soon witnessed the belligerence of southern boys to any form of authority. He wrote to his brother-in-law, "the boys here are *wilful* [*sic*] and govern their parents despotically."[14] General Mason Smith wrote that there was "a reluctance of sons to submit to control or guidance either at home or at school."[15] An article that appeared in the November 1859 issue of the *Louisiana Democrat* sang the praises of a school with military discipline. "A youth's time is so regulated that dissolute and expensive habits cannot be contracted [like] expensive dress, dogs, horses, billiards, etc."[16]

Sherman had finished West Point in 1840, sixth in his class out of forty-three graduates. He wrote:

> At the Academy I was not considered a good soldier, for at no time was I selected for any office but remained a private throughout the whole four years. Then, as now, neatness in dress and form, with a strict conformity to the rules, were the qualifications required for office, and I suppose I was found not to excel in any of these. In studies I always held a respectable reputation with the professors, and generally ranked among the best, especially in drawing, chemistry, mathematics, and natural philosophy. My average demerits, per annum, were about one hundred and fifty, which reduced my final class standing from number four to six.[17]

He then spent thirteen years as a commissioned army officer but saw no combat during the Mexican War.

Desiring to pursue a civilian career, Sherman applied for a government job through his friend Major Don Carlos Buell, who was assigned to an army manpower post. Buell forwarded his name to General Graham when he saw the advertisement for a superintendent for the new Seminary. Mason Graham liked what he saw and wrote in an editorial that Sherman had "standing high in the army as a scholar, soldier, and a gentleman—a man of great firmness and discretion and eminently remarkable for his executive and administrative qualities."[18]

With Sherman at the head, the school opened on January 2, 1860, with five professors and nineteen cadets.[19] The senior professor was Anthony Vallas, a Hungarian and professor of mathematics and natural philosophy. Francis W. Smith, a graduate of VMI and UVA, became professor of chemistry and the Commandant of Cadets. David French Boyd, another Virginian and graduate of UVA, was made professor of ancient languages and English. E. Berte St. Ange, a Frenchman who graduated from Charlemagne in Paris, became professor of modern languages, or, as he called it, "tongues." Rounding out the faculty was Powhatan Clark, assistant professor of modern languages. Sherman wrote, "Indeed on the whole, the professors are above mediocrity."[20]

Late arrivals eventually brought the number of cadets to seventy-three, and Cadet W. S. Bringhurst, one of the original students, described this first class of the Corps' cadets as "Sherman's boys."[21] Another cadet gave his impression of the make-up: "A heterogeneous crowd of matriculates, the sons of wealthy planters from the rivers, and aristocratic Creoles from the south, the nimble, pony riding Cajeans from the prairies, and the diligent, quiet fellows from the pine woods."[22] Despite their poor reputation as spoiled brats, some of them were not. One who greatly impressed Sherman was a cadet named Stokes, who rode his horse 125 miles from Monroe to the location in Pineville "simply to find out whether in fact that such an institution was in existence."[23] When the first session ended in July 1860, of the seventy-three cadets who began, fifty-nine passed the exam. Three months later, on November 1, 1860, the second session began with 130 in the Cadet Corps.[24] And so, what could well be described as a motley crew kicked off what would become a long and distinguished history of military excellence at the Ole War Skule.

As the years went by, it seemed that LSU's military cadets and commissioned officers, teachers, and associates exemplified a famous Marine Corps motto: "First to Fight!" As evidence: of the 562 Confederate officers who held

WILLIAM TECUMSEH SHERMAN, 1860
From a painting by Colonel S. H. Lockett, owned by Louisiana State University

William T. Sherman, first President of LSU
(LSU Archives)

First faculty at LSU
(LSU Archives)

THE FIRST FACULTY
(1) William Tecumseh Sherman; (2) Powhatan Clarke; (3) Anthony Vallas; (4) D. F. Boyd; (5) Francis W. Smith
Dr. Clarke's portrait is of 1910; the others are of 1860. No portrait of Professor St. Ange can be found

any of the four ranks of a general officer during the four years of the Civil War, only eight attained the highest rank of full general. Of those eight, Louisiana provided two: Pierre Gustave Toutant Beauregard and Braxton Bragg.[25] On the battlefields of the Civil War, fifty-six thousand men, out of a military population of eighty thousand Louisianans, served during the years from Manassas to Appomattox.[26] LSU cadets and veterans were in no short supply.

Indeed, few universities can boast of a history that produced two generals who would become Commandants of the Marine Corps. The first was Lieutenant General John Archer Lejeune, the 13th Commandant, serving from 1920 to 1929; the second was General Robert Hilliard Barrow, the 27th Commandant, from 1979 to 1983. (As a young major, Barrow would recruit the author of this work into the Corps). These ranks were indeed lofty heights of military command, but not the loftiest ever produced by the Purple and Gold.

In 1944, during World War II, Congress recreated the ancient rank of "General of the Army," with the accompanying insignia of five stars. During a six-day period from December 16 to December 21, 1944, four general officers were elevated to this most senior rank: Generals George C. Marshall, Douglas MacArthur, Dwight Eisenhower, and Henry "Hap" Arnold. Five years later, during the Korean War, General Omar Bradley would join this prestigious club. Could there ever have been a higher rank?

The answer is yes, and it very much involves LSU. Five previous generals were factually senior to those wearing five stars. The rank of "General of the Army of the United States" was first created and awarded to General Ulysses S. Grant in 1866. The law called for the wearing of four stars for the first time. The uniqueness of this rank was written into law: there could only be one officer, at any one time, that could hold this highest rank. Its exclusivity and singularity placed it one tick above the multiple awarding of the five-star rank in the 1940s and 50s. And the Congresses of the 19th century that originally created the rank of General of the Army, agreed. They said that "the office of General [of the Army of the United States] . . . shall continue until a vacancy shall exist . . . and when such vacancy shall occur . . . by virtue of this act, from thence forward be held to be repealed."[27] Only one could have the rank, and that was General Grant, who wore the four silver stars. When he resigned to run for President, the law creating "General of the Army" automatically repealed the rank. Congress had to pass new legislation in 1869 to "re-create" the rank for Grant's successor, General William T. Sherman, and later again in 1888 for General Philip Sheridan. After Sheridan's death in 1888 the rank was permanently discontinued.

Sherman as General of the Army with his two distinctive insignias (Library of Congress)

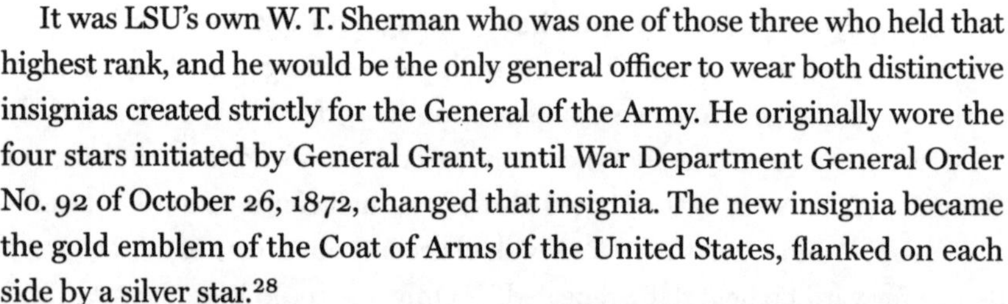

It was LSU's own W. T. Sherman who was one of those three who held that highest rank, and he would be the only general officer to wear both distinctive insignias created strictly for the General of the Army. He originally wore the four stars initiated by General Grant, until War Department General Order No. 92 of October 26, 1872, changed that insignia. The new insignia became the gold emblem of the Coat of Arms of the United States, flanked on each side by a silver star.[28]

However, not to be argued, it was the highest rank of all. In 1919, Congress revived the rank of "General of the Armies" from the time of the Revolution and promoted General John J. "Black Jack" Pershing to it for his superb leadership in World War I. As the insignia for that unique rank, General Pershing chose to wear four gold stars. But Pershing's highest rank created a problem that seemed to give short shrift to Lieutenant General George Washington.

The rank of General of the Armies had been created in 1799 for Washington, but he had not been appointed to it. President Gerald Ford resolved that problem in 1976 by posthumously appointing George Washington to be "General of the Armies of the United States," and specified his seniority over all other officers of the army, past and present.[29] This is military history worth noting, and especially to note that regardless of standards used to rank the top generals of the United States, Sherman's credentials can place him in the top four. It is fitting that LSU should have this recognition of military leadership excellence.

The Great Plain at West Point is adorned with four oversized statues of some of its great military generals looking out over the grassy field: Douglas MacArthur, Dwight Eisenhower, George Patton, and Omar Bradley. If LSU were ever to erect similar monuments honoring its great generals on its Parade Grounds, they would be equally noteworthy: John Archer Lejeune, Claire Lee Chennault, Robert Hilliard Barrow, Graves Erskine, and Troy H. Middleton. All were most heroic and carved their names and exploits into American history.

During World War II, LSU was a leader in supplying military officers for the war effort. It provided over five thousand, more than all the other universities except for Army, Navy, and Texas A&M.[30] Another seven thousand, associated with LSU, served in the war, and more than five hundred were killed. Sixteen of those five thousand who became officers rose to the rank of general.[31] At home and overseas, the names and faces of those distinguished patriots appeared as tributes to their dedication and military excellence. LSU continued this history of military excellence in wars to come. It is fitting to memorialize some of these heroes for future generations.

CHAPTER 2

Awkward and Untutored

A few days after the Seminary's opening, the first attempt at military drill began. For all their pomp, bravado, and male posturing, these southern boys were hopeless at uniform coordination and synchronized marching. Some were humiliatingly bad. "We were an untutored set," wrote a clumsy cadet, "and often provoked the disgust of the officers. Some of us made such slow progress that an '*awkward squad*' was formed, of which I was a prominent member."[1]

The cadet that Sherman finally put in charge to drill this "awkward squad" was John D. Workman, and he recognized from the beginning that he had his work cut out for him. "They are hump-shouldered," he declared of the members of the awkward group, "and their legs bowed two or three different ways, so that it is almost impossible to teach them the position of a soldier."[2] As hopeless as it originally appeared, with sheer determination, and by drilling morning and evening, Cadet Workman finally whipped them into some order of proficiency. Like most of the other Louisiana Seminary cadets in this very first class, Workman's destiny was soon to take him, with his brief training, onto the first battlefield of the Civil War.

Workman would become an elite member of the first combat veterans of the Ole War Skule. He would fight at Manassas. On May 5, 1864, after three years of combat, Workman, as part of the Second Louisiana Brigade, stepped onto a Virginia battlefield known as The Wilderness. He fought in that desperate three-day struggle as the South suffered eleven thousand casualties. Among the fifteen hundred dead was John Workman.[3]

It did not take Sherman long to personally experience the bizarre and undisciplined behavior of the spoiled cadets. He saw that their constant whining and misbehavior was made possible through the pandering of their condescending parents. On one occasion, a boy fastened a pig's tail secretly to the back of another cadet's coat as they marched to their classroom. The snickering and pointing by the other boys led to an outburst of uncontrolled laughter. That uproar, accompanied by the worst display of unruly behavior, caused the beginning of class to be suspended. When order was finally restored, the professor rightfully sent the culprit packing from his classroom. The guilty cadet, now feigning insult, filed a complaint of teacher "*oppression*" and ran to relay his woeful tale to his coddling, sympathetic father. Sherman called the cadet into his office and asked just how he had been oppressed. With a straight face, the boy said that the oppression resulted from his being sent from the classroom. Sherman could only laugh. This schoolgirl whining was too much for him, and he wrote: "Because [he] was ordered to quit the section room, very properly by his professor, he must tell a cock and bull story to his father."[4]

Sherman reserved his harshest criticism for those doting parents who allowed such immature behavior. "Parents while they boast of the hardships they overcame in early life and admire the brave and noble deeds of the past, are willing to listen to, and extend the whims of their boys, who have nobody to wash their faces and comb their hair in the morning."[5]

After the first uproarious month, things settled down. Perhaps it was because Sherman's imposed discipline took effect, but more likely it was the fact that ten to twelve of the worst troublemakers had been singled out and expelled.[6] The very worst incident involved a life-threatening situation. One of the cadets pulled a Bowie knife on another who had called him a liar. In the minds of these boys, being called a liar was considered unforgivable. It was an insult over which duels had traditionally been fought. The potentially deadly fight was broken up before blood could be shed, but Sherman was having none of this and swiftly reacted to this stupidity. He published an order for immediate dismissal, wasting no words in his reasoning:

> Order No. 9. Cadet D. T. H--h, having in an angry controversy with another cadet, drawn a dirk or a Bowie knife, is hereby summarily dismissed.
>
> The superintendent in this connection does not deem it necessary to look to the provocation. Here, no possible provocation can justify such an act.
>
> W. T. Sherman, Superintendent.[7]

Sherman dealt swiftly with rule breakers, and the cadets knew what to expect from him from early on. When two other cadets were found to have contraband tobacco stashed in their rooms, Sherman confiscated the stashes and threw them into the fireplaces. Again, there was a frivolous complaint concerning ill-treatment and a "breach of propriety," charging that Sherman had opened the drawer of the cadet's washstand. Sherman's cold reply to the complaining cadet was short. "I advised him that his concealment and breach of regulations, well known to him, was the breach of honor."[8] Sherman "shipped" them both, and wrote to his wife of the incident. He had no patience for stupidity. "These two last were dull at books and noisy quarrelsome fellows, and a good riddance. We had fifty-three, now fifty-one."[9]

There was also an ill-advised challenge to Sherman's rules and authority by three cadets, who in protest simply walked out of the Seminary without leave. Sherman wrote to Mason Graham, "As these persons have all left the Seminary without leave or authority, and in a spirit of defiance, I shall report them tomorrow as 'deserted.' . . . We have fifty left, one or two more may renew their vain struggle to do as they please."[10]

One by one, the perpetrators fell by the wayside. With those three dismissals, Sherman had cut deeply into the ranks of the cocky troublemakers. Without a military influence and the discipline associated with it, life at the Seminary would have been intolerable. Sherman wrote, "One hundred young men in this building, under a civil government, would tear down the building and make study impossible. [Even] with our frequent roll calls, and the other regulations it is all we can do to keep quiet."[11]

It was obvious that many of the worst troublemakers had been sent to the Seminary by parents who had no disciplinary skills of their own and who were at wit's end in dealing with their delinquent sons. One parent wrote to Mason Graham to complain that his son had been spoken to "in too authoritative a manner."[12] Graham wrote sharply to the complaining parent that a quick, authoritarian tone sometimes "grates harshly on the ears and feelings of boys who have been accustomed to home tones, and who take as long as they pleased to do a thing, or to go to a place that they haven't much fancy." He concluded his letter by directly referring to the problem of parent coddling: "some gentlemen [fathers] have sent chronic cases to this institution as their last hope for a cure, but we do not intend to keep that kind of a hospital."[13]

After the desertions, which Sherman called the "*Emute Horrible*" (horrible disturbance), cadet life settled down. Having put down the cadet "rebellion," he wrote, "I am now rid of five noisy, insubordinate boys. Fifty-one still remain, not a recitation was missed, and I am fully supported. There can be but one master."[14]

However, the parental challenges to his authority had not been completely eliminated. On June 28, 1860, Sherman wrote of the danger of challenges to his authority over the cadets, especially those complaints brought by doting parents who were the source of the disciplinary trouble from the start. He was still subject to the authority of a sometimes fickle board that had the power to overrule him in favor of a whining, permissive parent:

> Last week I dismissed summarily two cadets of good families and large connexions. One has appealed to the Board of Supervisors who may be weak enough to yield to such influences. And if they do, it will severely weaken my power and influence, and may shake my faith in my hold on their confidence. They meet on Saturday. This is Thursday and I will then see whether I am to govern here or be governed by the cast-off boys of rich planters.
>
> . . . like all deliberative bodies they may take some half-way course and recommend me to receive them back on their promising reformation. I will not do so unless they command me, which they have a right to do.[15]

While the Board dithered on the question of Sherman's authority during the first session that ended on July 31, 1860, they clipped his wings while he was on vacation in Ohio. It was indeed the "half-way course" that Sherman feared. The Board wrote: "no cadet shall be expelled, or dismissed, or suspended from the Seminary, or be subjected to other grave punishment, provided in these Rules, without the sanction of the Academic Board, expressed by a majority vote."[16]

General Graham resigned as Vice President in disgust, and Sherman regretted that he had not accepted a recently offered job that would have taken him to London. Sherman had just celebrated his fortieth birthday on February 8. He called this personal milestone "crossing the line," as if he would now be sliding downhill and should accept being called "Old Man" and the "cross old schoolmaster."[17]

The Cadet Corps, minus the troublemakers, was now a much better group. One cadet even wrote to his parents that "these are the most peaceable boys I ever saw, but that is owing in a great degree to the strictness of the rules."[18] Then the uniforms arrived: a dark blue military dress coat with a high collar and embellished with gold gilt buttons bearing the coat of arms of Louisiana. The trousers were a lighter blue with a black stripe welt down the outer seam.

The boys were ecstatic. All the drilling and rules and discipline that they had endured now seemed all worth it for the chance to wear this most stunning uniform. "The boys were 'crazy to show their uniforms to the girls,' so

Sherman gave a dance and invited all the girls in the community."[19] For southern boys who had a proclivity for showing off, this was the chance to prance like peacocks. One cadet beamed, "Our uniforms were showy . . . the hat for dress occasions was a gorgeous affair—high and broad and stiff, with brazen ornaments representing the college building and the coat of arms of the State; and waving, black ostrich plumes. An African prince would have given treasures of ivory and gold dust for such a royal headpiece."[20] Cadet Workman wrote, "Our uniform hats . . . are the fanciest things I ever saw."[21]

Despite his reputation as a stern disciplinarian, Sherman was quite a lover of amusements, music, dancing, and the company of young people. Professor David F. Boyd wrote, "He gave the cadets frequent hops, the planters and their pretty daughters coming in swarms. They soon got to be as fond of Sherman as his cadets were. They were delighted to have him at their homes on the river and bayous, and many an evening did he spend with them."[22]

On the occasion of these social visits, Sherman was usually accompanied by his handsome young Commandant of Cadets, Major Frank Smith. An old adage about war, repeated by all combat veterans, is that no one wants to be the last person killed. In the case of Major Smith, he would claim that dubious fate. He was killed the night before the surrender at Appomattox.[23]

While the Board had indeed clipped Sherman's authoritarian wings, it had not diminished the rigid rules that governed what cadets could and could not do under a military atmosphere. Marriage was forbidden, as was "Reproachful speeches, provoking gestures, unjustifiable tricks, ostracism of a fellow student, profane or indecent language, cards and betting, dueling, and visiting during study hours."[24] The penalty for violations was dismissal. Nor could a cadet have a horse or dog, or smoke, or even have tobacco in his possession. Cooking was forbidden, and having cooked food in his room required permission. Cadet Rene T. Beauregard wrote: "For water to drink or for our ablutions we had to go downhill to the spring, from one to two acres off, then go back to our rooms by way of the interminable stairs, there to resume, tired and out of breath, our studies of sines and cosines."[25]

The remoteness of the Seminary from most civilization translated into a steady diet of poor food. Sometimes it was awful, resulting in open revolt accompanied by the crash and smashing of plates. Waiters were put to flight, pursued by angry cadets who sometimes fired pistols to emphasize their displeasure. When the most violent culprits were eventually caught and dismissed, the food would temporarily slightly improve.[26]

Tobacco and whiskey were another story. Young southern men were notorious consumers of both commodities in large quantities. One cadet simply could not give up tobacco, and after numerous sightings was personally dismissed by Sherman. It was labeled *renvoyé*, a euphemism for being "sent back." Another cadet was completely distraught when Sherman discovered his whiskey jug in his room. Upon confiscation, he threatened the superintendent with his pistol and then ran away, not waiting for his official *renvoyé*.[27]

Still, the cadets gravitated toward Sherman. Despite his stern discipline and unyielding demands for integrity, honor, and gentlemanly decorum, the boys respected him. Professor Boyd wrote: "He was the intimate social companion of the cadets. All loved him. In the 'off hours' from study or drill he encouraged the cadets to look him up and have a talk. And often have I seen his private rooms nearly full of boys, listening to his stories of army or western life, which he loved so well to tell them."[28]

The second session at the Seminary opened on November 1, 1860, with eighty new students applying for admission. The total enrollment was now to be about 130. It could have been many more except for the explosive political climate running rampant in both North and South. Sherman tried to stay above politics even though his brother, in Congress, wrote as an abolitionist. The agitated state of the nation heightened as the nominations for the 1860 presidential elections unfolded.

Sherman had written to his brother in June after learning of the nomination of Abraham Lincoln: "All the reasoning and truth in the world would not convince a southern man that the Republicans are not abolitionists. It is not safe to stop to discuss the question: they believe it, and there is the end of the controversy. . . . But you may rest assured that the tone of feeling is such that Civil War and anarchy are very possible."[29]

After the election of Abraham Lincoln in November, Sherman's outlook for a future without war was most gloomy. He returned to the Seminary without his family because of the national unrest. He told his northern friends that he did not think that Louisiana would secede unless South Carolina started the slide to disunion. Then "the other Southern States will follow and soon general anarchy will prevail."[30]

The prospect of lawlessness was Sherman's foremost concern. He was determined to teach his cadets good discipline. "The law is, or should be our king. We should obey it, not because it meets our approval, but because it is the law. . . . This is the trouble. It is not slavery; it is the democratic spirit which substitutes mere opinion for law."[31] He wrote to his wife, "anywhere from California to Maine any man could do murder, robbery or arson if the people's prejudice lay in that direction."[32]

Rumor began to spread that southern vigilantes were forming in the wake of Lincoln's election. Sherman planned to form his cadets into an armed military force and lead them to suppress any and all such illegal bodies. He was not sure he had the authority to compel his cadets to form such a unit, but that question would not deter him. He saw it as his duty to try.[33]

Sherman despised the dithering of the Buchanan administration that emboldened the South Carolinians. A strong federal stance would give South Carolina pause in its rush into the pitfall of emotionalism. "Oh, for just one hour of Andy Jackson,"[34] he lamented, referring to Jackson's 1832 thwarting of a South Carolina attempt to nullify federal power.

Most of the young cadets were filled with the spirit of adventure and glory that an armed conflict offered. No young man ever equates that allure with the devastation of death and destruction. But there was one cadet who submitted a very thoughtful paper on the subject and was complimented by Professor Boyd as "very good in thought and language."[35] Cadet W. J. Brown wrote: "If the Union is dissolved it will ruin both the North and South. A dissolution can hardly be accomplished without bloodshed, and if blood is once shed it will be the precursor of a long and bloody war. . . . If the North and South would stand by the Constitution, and wait until it was violated before they hurried into extreme measures, I think the Union would stand a great deal longer."[36]

Sherman had no intention of leaving the Seminary, or Louisiana, unless the state seceded. And even then, he would stay if there was no conflict. But if there was armed conflict, he was most adamant of his actions: "If Louisiana assumes a position of hostility toward the government, then this Seminary becomes an arsenal, and a fort, and I quit . . . I will do no act, breathe no word, think no thought hostile to the government of the United States."[37]

On Christmas Eve, 1860, while the cadets were off on holiday leave, Sherman sat in his room, sorely missing his family, and was joined by his friend and associate Professor David Boyd. They enjoyed this moment of camaraderie when a rare mail delivery arrived with letters and a five-day-old newspaper. Sherman unfolded the paper and saw the headline. It blared the unthinkable—South Carolina had seceded!

He was thunderstruck even though it was not a complete surprise. But it shattered his last hope that there could be a peaceful solution to the long-simmering emotional national angst. He rose sullenly from his chair and began pacing the floor, back and forth, the sound of his heels echoing in the deadly silent room. His movements were those of a Shakespearean actor, pacing an empty stage, delivering a tragic soliloquy to an unseen audience. Boyd remembered it well as Sherman poured out his thoughts:

> You people of the South. . . . You don't know what you are doing. . . . This country will be drenched in blood, and God only knows how it will end. It is all folly, madness, a crime against civilization!
>
> You people speak so lightly of war; you don't know what you're talking about. War is a terrible thing! You mistake, too, the people of the North. They are a peaceable people but an earnest people, and they will fight, too. They are not going to let this country be destroyed without a mighty effort to save it. . . .
>
> Besides, where are your men and appliances of war to contend against them? The North can make a steam engine, locomotive, or railway car; hardly a yard of cloth or pair of shoes can you make. You are rushing into war with one of the most powerful, ingeniously mechanical, and determined people on Earth-right at your doors.
>
> You are bound to fail. Only in your spirit and determination are you prepared for war. In all else you are totally unprepared, with a bad cause to start with. At first you will make headway, but as your limited resources begin to fail, shut out from the markets of Europe as you will be, your cause will begin to wane. If your people will but stop and think, they must see in the end that you will surely fail.[38]

Later, the Buchanan government sat, once again stupefied, while Mississippi and Alabama seized more fortresses and arsenals. On January 8, 1861, Sherman wrote to his wife of his revulsion with federal inaction. "This disgusts me and I would not serve such a pusillanimous government. It merits dissolution."[39] On January 10, Braxton Bragg's forces seized the Baton Rouge arsenal. And on January 13, Sherman wrote, "The revolution has begun and the national government has shown weakness in all its attempts. . . . It will be a triumph to South Carolina to beat Uncle Sam."[40]

On January 18, 1861, he wrote his letter of resignation to Louisiana's Governor Moore. On the 26th, Louisiana seceded. But it was not until late February that all of Sherman's administrative details were completed. On February 20, William Tecumseh Sherman departed the Seminary. It was a most tearful separation: "The morning he left us he had his battalion formed. Stepping out in front of them, he made them a short talk, and then, passing along the line, right to left, bade each and every officer and man—not a dry eye among them—Then, approaching our sad group of professors, he silently shook our hands, attempting to speak, broke down, and, with the tears trickling down his cheeks . . . he could only lay his hand on his heart and say: 'You are all here,' then turning quickly on his heel, he left us, to be ever in our hearts."[41]

Sherman proceeded to Alexandria for his final official act: a parting visit to Governor Thomas Moore. In a final moment of jocularity as they shook hands, Sherman told him, "Well, good-bye, Governor, I hope that if I should go into the army, I'll not catch you, for I should certainly hang you."[42] With that, he was gone.

CHAPTER 3

Reunion at Manassas

When Superintendent Sherman finally departed Alexandria, he proceeded to New Orleans, arriving on February 22, 1861, and completed his banking business with Dr. S. A. Smith of the Bank of New Orleans. That cut his final ties to his Louisiana adventures, and after he left the bank he melancholically wrote: "I stood free and discharged of any and every obligation, honorary or business, that was due by me to the State of Louisiana."[1]

During his several-day stay in New Orleans, he roomed at the luxuriously domed, St. Louis Hotel on the corner of Royal and St. Louis streets. And he dined with Colonel Braxton Bragg and his wife. Bragg wore his uniform, and he told Sherman that he had garrisoned Forts Jackson and St. Philip as well as the arsenal at Baton Rouge. Sherman also wanted to bid a final farewell to Beauregard, who had two sons now enrolled at the Seminary, and to whom Sherman had afforded special care at their father's request.

"I went to his usual office in the Customhouse Building and found him in the act of starting for Montgomery, Alabama."[2] President Jefferson Davis had sent for Beauregard, presumably for an important command in the Confederate Army. Sherman then took a final, nostalgic walk around the city. "I walked the streets of New Orleans and found business going along as usual. Ships were strung for miles along the lower levee, and steamboats above, all discharging or receiving cargo. The Pelican flag of Louisiana was flying over the Custom House, Mint, City Hall, and everywhere. At the levee ships carried every flag on earth except that of the United States. . . . I therefore bade adieu to all my friends, and about the 25th of February took my departure by railroad, for Lancaster, via Cairo and Cincinnati."[3]

When Sherman departed, it had been a month since the Louisiana delegation had voted to adopt the Ordinance of Secession on January 26, 1861.

While secession had been the hot topic among all Louisianans for most of the year, the actual splitting apart was a matter of relief and celebration. The vote was 113–17, and it triggered rejoicing and parades. Despite lots of bravado, few thought war was in the offing. On February 12 there had been a public inauguration of the National Flag of Louisiana. Two brigades of artillery were formed in Lafayette Square, and as the large flag was smartly raised on the roof of City Hall, "the city bells rang out a peal, the Battalion of Washington Artillery fired a royal salute of twenty-one guns, and the entire troops presented arms, and the populace cheered and re-cheered."[4] A week later there were more celebrations—this time, Washington's Birthday. Two military parades enthralled the onlookers and then made a grand entrance at the racetrack where twenty thousand gathered to witness the splendor. One of the military units, the Orleans Cadets, was led by the dashing Captain Charles D. Dreux.[5]

Proof that most of the South thought that its legal right of secession would simply lead to a status of peaceful coexistence with the North is seen in an event that concluded on April 8, just four days before General Beauregard would open fire on Fort Sumter. Lincoln had been inaugurated on March 4, and shortly afterward the new Confederate government sent three emissaries to Washington to negotiate peaceful solutions. John Forsyth of Alabama, Martin J. Crawford of Georgia, and Andre Bienvenu Roman of Louisiana presented their credentials but were put off. This ignoring by President Lincoln continued until April 8, when his staff finally informed the three Confederate diplomats that he would not recognize them.[6]

Sherman meanwhile had secured a low-paying job as the President of the St. Louis Railroad Company: a horse-drawn street railway company paying only $60 a month, while his bare-minimum expenses totaled $70. His salary was half of what he had made at the Louisiana Seminary.

He wrote to his friend Professor Boyd at the Seminary and voiced his complete disgust with the entire affair unfolding in Washington. "Lincoln was organizing his administration on pure party principles; [I] concluded it was no place for me who profess to love and venerate my whole country and not a mere fraction."[7] In early April, because of his extreme dissatisfaction with the partisan political maneuvering in Washington, Sherman turned down a lucrative offer to come to Washington and become the chief clerk of the War Department, with the promise to become the Assistant Secretary of War. But Sherman was still miffed and his terse answer to authorities was simply, "I cannot accept."[8]

In a letter to Boyd, written on May 13, Sherman proclaimed: "I believe

[Lincoln] sincere in his repeated declarations that no dismemberment shall be even thought of. The inevitable result is war, and an invasive war. . . . No one now talks of the negro. The integrity of the Union and the relative power of state and general government are the issues in this war."[9] The following day, Sherman's life changed in the blink of an eye. On May 14, he received a telegram "telling me to come on at once, that I had been appointed colonel of the Thirteenth Regular Infantry, and I was wanted in Washington immediately."[10]

On the day of Sherman's telegram, as if the two events were linked by destiny, three of the five original professors at the Louisiana Seminary in Pineville resigned. They had all been charter members of the staff with Sherman, and now Professors Francis W. Smith, David F. Boyd, and Powhatan Clarke threw their lot in with the Confederate Army. The remaining two were foreigners. Boyd enlisted as a private in Company B of the 9th Louisiana Regiment of Volunteers, Confederate States Army, under General Leroy Stafford, and almost immediately went to Virginia with the regiment.[11] "[Anthony] Vallas had a wife and seven children to support, and [E. Berte] St. Ange, even though he was a foreigner, had once been an officer of marines in the French navy, and tried in vain to get a commission in the Confederate artillery."[12]

On March 29, 1861, the first troops from Louisiana departed, heading east toward Virginia for an anticipated showdown with the Federal army. The deployment contained four companies of Zouaves, and they were followed on April 11 by the 1st Louisiana regulars commanded by Colonel A. H. Gladden and the Orleans Cadets with the gallant Captain Dreux commanding. On April 19–20, the Caddo Grays followed, along with the Crescent Rifles and the Louisiana Guards.[13]

During that fateful summer of 1861, the few remaining students at the Seminary joined Professor Boyd to enlist. They went to Camp Moore, across Lake Pontchartrain, and enlisted in the Louisiana regiments.[14]

The Louisiana Zouaves soon developed a reputation of their own in the Confederate Army. They were especially fearsome in appearance, and became known as Wheat's Tigers, after their heroic leader, Major Chatham Roberdeau Wheat. They were especially noted for their blood-curdling, screaming charges that terrified both friend and foe alike. To the Yankees, "they were a vision straight out of hell."[15]

Their outlandish Turkish uniforms were bright red shirts under blue jackets, embroidered with red trim, and "billowing knee-length blue-and-white-striped pantaloons and white gaiters, all capped with tasseled red fezzes." Some wore flashing red pantaloons, but regardless of uniform, they were highly visible to all. They killed and looted with practiced aplomb, and their ferocious charges were made under the banners of the traditional Louisiana and Confederate battle flags. But they also sported a special flag, a flag that bore the image of an innocent lamb, cleverly placed just after the words *Gentle as a. . . .*[16] Some painted slogans and taunts on their fezzes: "Lincoln's Life or a Tiger's Death" and "Tiger in Search of a Black Republican" and "Tiger Bound for the Happy Land."[17]

Rob Wheat was the very essence of a perpetual warrior always seeking a fight. At six feet, four inches and weighing 250 pounds, he was a mountain of a man. "Wheat embodied the image of a man of breeding, yet [a] student of poetry and literature . . . and a Southern gentleman."[18] Wheat's menacing troops were known as a battalion of tigers for their ferocity; "men of desperate courage but questionable morals . . . whose fierce passions were kept in abeyance only by the superior discipline of their commander."[19] His was hardly the persona expected of a young man who had begun the study of law, but when the trumpet of war called to fight against the Mexican army, Wheat was first in line.

"Gen. John A. Quitman, under whom Wheat served in Mexico, called him 'the best natural soldier' he ever knew."[20] After the war, Wheat completed his law studies and was admitted to the Louisiana bar in 1848—a profession that bored him stiff. He heard any and all calls to arms, and from 1849 to 1857 fought in three Cuban invasions and a Mexican expedition as a soldier of fortune. When he ran out of foreign conflicts to join, he returned to New Orleans for an elected stint in the Louisiana House of Representatives, as a twenty-six-year-old. Then there was another Cuban adventure, and then one in Mexico, where he was a general in a conquest of Mexico City. At twenty-nine he was off to Nicaragua with forty volunteers on an expedition that came to naught. In 1857 he was fresh out of wars and returned home.[21] For two years he tried his hand as an inventor, developing a bigger and better cannon that would exceed the power and range of anything in existence. In 1859, at the age of thirty-three, war found him again, this time in Mexico, and then on to Sicily.

The Civil War was perfect for him, and he set up his customary recruiting station at 64 St. Charles Avenue, where he had recruited for his far-flung expeditions. The pay for the infantry in the Confederate Army was $11 a month,[22] but that did not deter the Rob Wheat disciples. They flocked to his station and scratched their marks on the enlistment papers. More than five hundred soon fleshed out Wheat's Special Battalion. There were Irish and German immigrants, and most of the officers were unskilled in the art of gentility. They were best described as the dregs of humanity: "The bulk consisted of street thugs and wharf rats," wrote one historian.[23]

One particularly vicious company was under the command of Captain Alex White. It called itself Tiger Rifles, and Captain White was hardly a saint; he was described as "a man who had served a term in the penitentiary."[24] Five companies would make up this battalion of Louisiana Tigers. They completed their organization on June 6, 1861, and hustled off to Virginia to be available for combat. The five companies, totaling 416 men, each bore distinctive names: Company A was Walker Guards; Company B was Tiger Rifles; Company C was Delta Rangers; Company D (1st) was Catahoula Guerillas; and Company D (2nd) was Old Dominion Guards. The only company to wear those fearful Zouave uniforms was Company B, Tiger Rifles,[25] but in short order all of Wheat's battalion were called "Tigers."[26]

On May 21, four batteries of the famous Washington Artillery left Louisiana to join Wheat's battalion in Virginia. The buzz among the gathering forces was that a decisive battle was to shortly begin that could determine the outcome of the war. But before that battle could be joined, Louisiana lost a favorite son. On July 4, 1861, in a minor skirmish around Young's Mill, Virginia, Charles Dreux, now a lieutenant colonel commanding the 1st Louisiana Battalion, became the first Louisianan killed in the Civil War. While trying to ambush some Federal officers marauding in the area, Dreux himself was fatally shot dead on the battlefield. Weeks later, thirty thousand mourners and dignitaries attended his funeral in New Orleans.[27] His massive procession would wind its way from Gallier Hall to St. Louis Cemetery on Esplanade Avenue.

There was no shortage of military enthusiasm in Louisiana in 1861. Young men beat a path into the ranks of the Confederate Army in a never-ending stream. During the year, 60,726 Louisiana men, far in excess of the quota of 10,000, shouldered arms to fight against the northern invaders.[28] At the Louisiana Seminary, the call to arms created a stampede. More than half of the more than 115 cadets that had begun the semester in April were gone by June.[29] The rest were shortly to follow, all entering the Confederate Army—all except Cadet Taliferro, who enlisted in the Union Navy.[30]

On April 15, the day after the bombardment of Fort Sumter, President Lincoln called for seventy-five thousand volunteers for a ninety-day enlistment. On May 3, the President called for an additional enlistment of three-year volunteers. William Tecumseh Sherman scoffed at the tiny commitment and said, "Why, you might as well attempt to put out the flames of a burning house with a squirt-gun."[31]

Lincoln soon realized that the first seventy-five thousand enlistments were to expire at the end of July. With that date threatening the integrity of his army, he pushed his generals for early offensive action. The Confederates sat just across the Potomac River in Virginia, almost within a cannon shot, and the public outcry was "On to Richmond." On July 16, Lincoln ordered his army to march south. They were green, but the President rationalized that, despite their inexperience, they could certainly be no worse off than the green Confederate troops defending on the other side.[32]

The first attrition to Mr. Lincoln's army came not from the Confederates but from the dreaded expiration of the ninety-day enlistments. Three days after the twenty-five-thousand-man Federal army under General Irwin McDowell began its march south, the 4th Pennsylvania and 8th New York Militia suddenly about-faced and headed home; and no amount of pleading to remain for a few days more could deter their retreat back to Washington.[33]

McDowell lamented that with every passing day of expiring enlistments his army became weaker, and the enemy's grew stronger.[34] Within the ranks of his five attacking divisions was the brigade of Colonel W. T. Sherman: the 3rd Brigade of the 1st Division. It was composed of militia who had signed on as three-month volunteers: the 13th, 69th, and 79th Regiments of New Yorkers, and a battery of artillery. The two-day, twenty-five-mile march from Washington to Centerville, located six miles north of Manassas Junction, had been most frustrating for Sherman and best described as an exercise in herding cats.

Sherman's first days in combat were not auspicious. The July sun was blistering hot and turned the roads into choking dust. Discipline quickly broke down. Unit formations disintegrated and the unruly, scattered troops mingled with each other, yelling at their officers, who helplessly rode back and forth shouting unobeyed orders. Occasionally the officers fired shots into the air, all to no avail. Sherman continually sent couriers back to his undisciplined troops to reinforce his own orders: "stay in ranks; close the formation up; stop chasing pigs and chickens." Those couriers were greeted with disdain and catcalls, and told in no uncertain terms, especially by the thick-brogued Irishmen of the 69th Regiment, to "Tell Colonel Sherman, we'll be having all the water, pigs, and chickens that we want. And who the hell are ye anyway!"[35]

Discipline among the rest of the McDowell's brigades was no better. Thirsty soldiers broke ranks to climb farm fences and gather around wells, just as if they were stopping at a local bar. Or they splashed and dipped in the streams until their thirsts were quenched, and only then reluctantly returned to the march. They knocked apples from the trees, chased chickens, shot pigs, and even killed a cow or two. With razor-sharp knives, the soldiers cut their own steaks from the flanks of the slaughtered animals as they marched past them.[36] Officers and sergeants could only watch, and Sherman said in disgust, “No curse could be greater than invasion by a volunteer army.”[37]

On July 18, after two days of this rowdy trek, the Federal army pulled into Centerville, on the northern edge of what was to become the battlefield. Six miles to the south was the railhead at Manassas Junction, where the Confederate Army of the Potomac was digging in and frantically being reinforced by the arriving soldiers from General Joseph E. Johnston’s Army of the Shenandoah.

Days earlier, at his command post at Winchester across the Blue Ridge Mountains, sixty miles from Manassas Junction, General Johnston had received a frantic message from Richmond: “General Beauregard is attacked. To strike the enemy a decisive blow, a junction of all your effective force will be needed.”[38]

In a brilliant tactical move, Johnston left the barest of forces to convince his Union adversary to the north that he was still in the valley. But he secretly disengaged his nine-thousand-man army and slipped out. Then it was a race through the Blue Ridge to the Piedmont rail station on the Manassas Gap Railroad line. There his men entrained, jamming every inch of space in every variety of rail car available. They balanced on the roofs, hung from the sides, and stood in the aisles for the thirty-five-mile ride to Bull Run.

Historian Clifford Dowdy wrote of this scene at Piedmont Station: “While they were being loaded into freight cars, cattle cars, and flatcars, ladies came to the station with home-cooked food and fresh lemonade.”[39] A former cadet from the Louisiana Seminary, B. C. Cushman, now of the 1st Louisiana Regiment, wrote to his old professor, David F. Boyd, of the exciting experiences while moving east to the battlefield. His greatest fear had been that the war would end before he and his friends could get into the fight, and now it was apparent that he would be in the very first battle. “At every little town and village,” Cushman wrote, “the inhabitants greeted us with cheers, and welcomed us to their soil, opened their doors to us all, and treated us to the best fare they had without charging any one a single cent.”[40] Cushman also wrote that in Virginia the Louisianans were treated “more in the manner of the Prince of Wales than as common soldiers, and we would find [the stations] thronged with ladies waving their handkerchiefs, tossing us flowers, and bidding us to be of good cheer, and fight like brave fellows.”[41]

The train movement of Johnston's Army had been a remarkable feat of arms. His soldiers had been in the valley at Winchester on the morning of July 18 facing Union General Robert Patterson's force, and on the evening of the next day were in Manassas Junction facing McDowell's entire Federal army.

When Johnston's *coup de main* maneuver to reinforce Manassas Junction was finally discovered by the out-maneuvered Federals, a furious General-in-Chief, Winfield Scott, telegraphed his extreme displeasure to General Patterson, who was supposed to be confronting Johnston and engaging him, and keeping an eye on him. "I have certainly been expecting you to beat the enemy," Scott seethed. "If not, to hear that you had felt him strongly, or, at least, had occupied him by threats and demonstrations. . . . Has he not stolen a march and sent reinforcements toward Manassas Junction?"[42] Having incurred the General-in-Chief's caustic wrath, Patterson's military future was doomed. Before the end of July the seventy-year-old general had been mustered out.

The battle lines were now set, and set on a textbook battlefield that could have been a classic model to teach military tactics. The attackers were advancing from the north, at Centerville; the defenders were several miles to the south at Manassas Junction. There were many hidden intersecting roads on this battlefield to allow both clandestine and vast, sweeping maneuvering, and there were open fields intermingled with dense areas of concealing foliage. The entrenched Confederates were behind a formidable barricade, the now sluggish but sometimes rain-swollen stream called Bull Run. It ran for eight meandering miles diagonally across the battlefield, from northwest to southeast.

There were eight crossing points along Bull Run: seven fords and a stone bridge. On the extreme right of the Confederate line was Union Mills Ford. From there, stepping to the extreme Confederate left were McLean's Ford, Blackburn's Ford, Mitchell's Ford, Island Ford, Ball's Ford, Lewis Ford, the massive two-span Stone Bridge where the Warrenton Turnpike crossed Bull Run, and finally, Sudley Springs on the far left. At each of these potential crossing sites Beauregard positioned defending units. Especially strong were the Confederates deployed at Blackburn's Ford under the command of Brigadier General James Longstreet, and at the Stone Bridge under the command of Colonel Nathan G. "Shanks" Evans. At Evans's position were the 4th South Carolina Volunteers and the notorious Louisiana Tigers with their charismatic leader, Major Rob Wheat.

There was actually another remote defensive position that was not along Bull Run; it was an unarmed observation post. Signal Hill was three-quarters of a mile behind Union Mills Ford on the right. On top of Signal Hill the Confederates had built a thirty-foot wooden signal tower that allowed an observer a bird's-eye view of most of the battlefield. Captain Edward P. Alexander manned that tower. He had recently perfected what became known as the "Wigwag" signal. It involved the use of a single flag to communicate in Morse code, as opposed to the traditional two-flag semaphore system.[43]

General Beauregard personally described his own view of the battlefield from the south side of Bull Run. "The banks for the most part are rocky and steep, but abound in long-used fords. The country on either side, much broken and thickly wooded, becomes gently rolling and open as it recedes from the stream. On the northern side, the ground is much the higher, and commands the other bank completely."[44]

The great battle would indeed be waged by green troops on both sides, and the hoped-for prize was to win the war in one climactic battle. *On to Richmond* was the north's cry—*Drive them into the Potomac* was the South's.[45]

The Confederate armies under the commands of Generals P. G. T. Beauregard (Army of the Potomac)[46] and Joseph E. Johnston (Army of the Shenandoah) had previously formed a mutually supporting defensive line, separated by the Blue Ridge Mountains, to block any Federal advances along either of those fronts. McDowell's immediate threat had served to join these two armies at the critical rail hub at Manassas Junction, where the Manassas Gap line intersected with the Orange & Alexandria line. This critical railhead allowed the Confederate Army to quickly move in and out of the valley and up and down the corridor between Washington and Richmond. Manassas Junction in Federal hands would be a devastating blow to the South and could indeed open the road to Richmond and end the war quickly.

But there was much more at stake in this upcoming battle than the loss of a strategic position. General Beauregard pondered it all on the eve of battle:

> There was much in this decisive conflict about to open . . . which pervaded the two armies and the people behind them and colored the responsibility of the respective commanders.
>
> The political hostilities of a generation were now face to face with weapons instead of words. Defeat to either side would be a deep mortification, but defeat to the South must turn its claim of independence into an empty vaunt.[47]

⚜ ⚜ ⚜

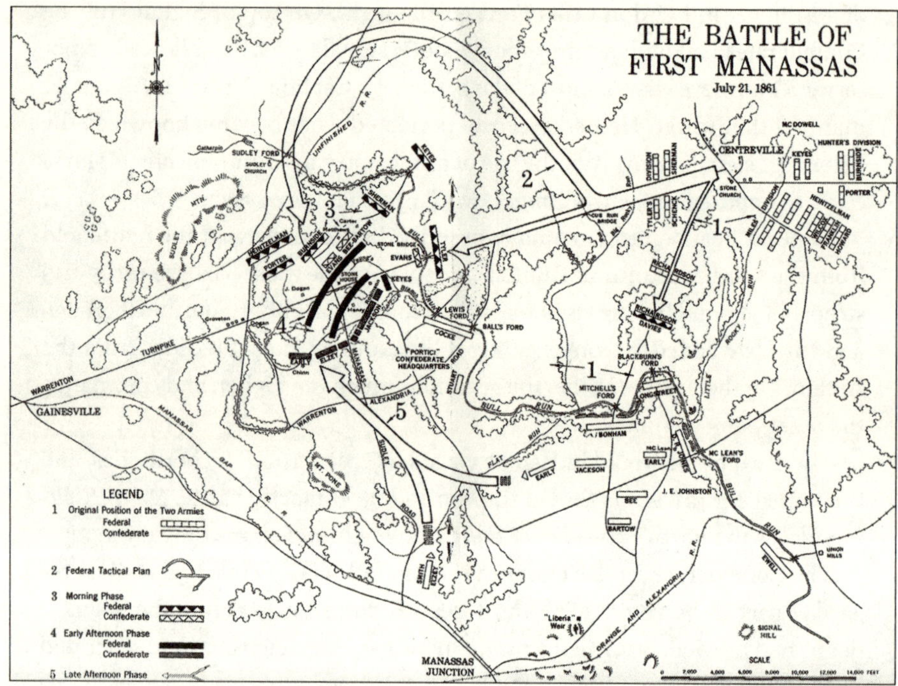

Manassas battlefield map (National Park Service)

On July 18, General McDowell ordered a reconnaissance to feel out where the Confederates' strength was. He chose to reconnoiter Blackburn's Ford and fired a few artillery shots to see if he could get the Confederates to return fire. The ford was defended by troops under the command of General James Longstreet, recently reinforced by the 7th Louisiana Regiment from Jubal Early's Sixth Brigade. The Louisianans had been moved forward when the Union Army was seen advancing upon Blackburn's Ford. The disciplined rebels did not take the bait nor reveal their positions, so the Federal pickets moved closer. That brought on a sporadic artillery duel, but still no revealing fire. They pressed even closer toward the northern bank of the ford and were finally greeted with a hail of steel and whining musket balls that stopped the skirmishers dead in their tracks. In that number of Confederates hurling back the Union advance at Blackburn's Ford was First Lieutenant W. P Harper and his Company H of the 7th Louisiana Infantry with some cadets from the Seminary. Company H suffered two men killed and two wounded.[48]

The threat of a follow-up Confederate counterattack forced Sherman's Brigade to move forward, just in case. The riddled Union skirmishers fell back in disarray. At first it was a trot toward the rear, and then an all-out run. Wounded men stumbled back past Sherman as enemy cannon shot ploughed the ground and tore through the trees, splintering and shattering the limbs. Sherman would forever remember the scene: It was a "sickening confusion as one approaches a fight from the rear." For thirty minutes his brigade covered the disorganized stampede to the rear and endured the long-range Confederate fire that took down a few of his own troops. Sherman also witnessed the ghastly sight of "a cannon ball demolish a man with a splashing thud."[49]

The first confrontation of Sherman, the former superintendent, with cadets from the Seminary was decisive. With the repulse at Blackburn's Ford on the 18th, McDowell changed his attack plan. Realizing the enemy's right was too strong, he focused on the Confederate left, where it seemed they were weakly arrayed, especially to the west of the Stone Bridge and Sudley Ford. He decided that he would hurl the weight of three of his divisions in a giant turning movement and roll up Beauregard's flank. Two divisions would cross at the remote Sudley Springs Ford, while the third division forced passage at the Stone Bridge. As those divisions pressed the enemy's left, Sherman's Brigade would stand ready to attack, where needed, along Bull Run.

The attack began in the dark at 2:30 a.m. when McDowell's flanking divisions began their twenty-mile trek around the Confederate left. It was slow, much slower than anticipated, and some of the concealing roads were thickly overgrown. Meanwhile, Sherman's Brigade was arrayed in the open fields north and west of the Stone Bridge, awaiting orders to attack. But Sherman was especially troubled and wrote to his wife concerning his misgivings. It had the tone of a final farewell.

Historian Lloyd Lewis wrote: "His chief danger was being shot by his own awkward men . . . he would do the best he could . . . the volunteers were apt to stampede . . . his best love to all at home . . . his faith in Ellen was perfect and whatever happened on the morrow, the children were in a fair way to grow up in goodness and usefulness . . . goodbye for the present."[50]

Daybreak on the 21st found Sherman's force occasionally firing a cannon to remind the Confederates of its presence. At 9:30 he heard the roar of cannons to his distant right. He concluded that McDowell's envelopment had begun, but by noon he still had not received orders to attack. He gazed forward, his glass to his eye, trying to determine where would be the best place to cross Bull Run when the time came. Unseen by him, concealed by the heavy foliage surrounding the stream, Confederate forces were not dug in, but frantically moving units to their left. Suddenly a solitary Confederate figure on horseback emerged from the thick woods on the north bank of Bull Run. He wore a dark blue, double-breasted coat with blue trousers, looking very much like a senior Federal officer. He also sported a general's sash and a red French kepi laced with gold braid.[51]

This splendidly uniformed figure reined in and gazed upon the Federal line. He was in plain view and began waving his clenched fist and hurling every manner of insult in Sherman's direction. As some of his men brought their arms up into a firing position, Sherman raised a restraining hand. The officer taunting Sherman's Brigade was none other than Major Chatham Roberdeau Wheat, commanding the Louisiana 1st Special Battalion, Wheat's Tigers. Wheat's typical bombast was full-throated for several minutes. After this ostentatious display, Wheat turned his back to Sherman and reentered the stream crossing. Sherman watched and made note of where Wheat recrossed, concluding that it would be his best fording point.[52]

On the other side, the scene was one of marching and side-stepping to the left. The Confederates had received a very late warning of a gathering storm pressing in on their left flank. From that high tower on Signal Hill, Captain Edward Porter Alexander frantically wigwagged his signal over and over. "Look out on your left. You are turned!"[53] he could see the advancing dust clouds sent up by the marching of thousands of Federal boots.

General Shanks Evans, at the Stone Bridge, picked up Anderson's signal and, without waiting for orders, seized the initiative and began sidestepping part of his force to the left so as to hurl them into the path of the advancing Federal juggernaut. Bull Run had been crossed, not in the anticipated frontal assault but at Sudley Ford on the extreme left in a gigantic turning movement! From 8:30, Evans had also watched that distant dust cloud, wispy at first but constantly developing into a dust storm. It inexorably moved toward Sudley Ford. When Anderson's frantic signal confirmed his fear that the main attack would be on his flank and not at the Stone Bridge, he was ready to move.[54] Evans left only four companies to cover the bridge and marched his remaining eleven hundred men upstream at the double quick. His six companies of the 4th South Carolina and the five companies of Wheat's Louisiana Zouaves raced three-quarters of a mile, dragging with them a pair of six-pound howitzers.

General Beauregard described the Union force bearing down on his left flank: "The Federal turning column was about 18,000 strong, with 24 pieces of artillery."[55] That annihilating column was already approaching along the Sudley Springs Road, where it would intersect with the macadam-surfaced Warrenton Turnpike. Evans formed his hasty line at right angles to Bull Run to form a defensive hook. In front of him were open fields and distant woods that could hide the advance of the massive Federal force. Already he could hear the sound of soldiers crashing through the foliage. Evans set his two guns on either flank of his makeshift line. His gunners rammed both guns with grapeshot.[56]

Wheat deployed most of his men behind a split rail fence and then personally led his gray-clad Catahoula Guerillas to an advanced, exposed position and placed them as pickets. His entire force bristled with the very accurate "Model 1841 'Mississippi' rifles that had been seized from the U.S. arsenal at Baton Rouge."[57] Stuck in their belts were large Bowie-style knives.

At 9:45, the flanking Union infantry burst out of the concealing woods, reinforced with six rifled artillery pieces. Wheat's Tigers volleyed them with their Mississippi rifles and kept up a steady fusillade to stop the attack and hold it at bay for almost an hour. Evans's two guns poured shredding grapeshot into the attacking Federals and shattered them.[58] The Confederate brigade of General Bernard Bee now rushed up to join Evans to extend the left line, bending it farther back. But the weight of the Union attack hammered Evans and Bee, and their lines wavered. Some men bolted to the rear. Sherman's Brigade, having patiently waited for the order to attack in the fields north of Bull Run, now crossed the stream to the west of the Stone Bridge, moving directly to join the attack.[59]

The combined shock by the ever-increasing Union mass was too much. The Confederate line staggered back, bending until it resembled a fishhook. Major Rob Wheat fell, shot through both lungs, and was carried off the field. Without their inspiring leader, the Louisiana Tigers wavered.[60] They had suffered eight killed and thirty-eight wounded. Two were missing.[61]

Outnumbered and outgunned, southern soldiers retreated up the western slope of Henry Hill. Their shouting officers tried to rally them. On top of the rise, Colonel Thomas Jonathan Jackson's First Brigade of Virginians arrived after its own leftward march. As Evans's and Bee's retreating forces stumbled up the exposed face of Henry Hill, General Bee saw Jackson's Brigade at the top, solid in their hastily formed defensive line. He shouted back to his men, "Look! There stands Jackson, like a stone wall! Rally behind the Virginians!"[62]

Meanwhile, General Beauregard personally directed reinforcements to his beleaguered left flank. Colonel F. S. Bartow rallied a part of his previously shattered 7th Georgia and raced to the left of Jackson's Virginians. Then three companies of the 49th Virginia dug in to the left of the Georgians. Then the 2nd Mississippi fell in on the left of the 49th Virginia.[63]

Beauregard was a blur of a figure, riding back and forth along the newly formed defensive line shouting encouragement. "I sought to infuse . . . the determined spirit of resistance to this wicked invasion of the homes of a free people," he said. ". . . we fought for our homes, our firesides, and for the independence of our country."[64]

Despite his exhortations, many terrified southerners bolted, convinced that all was lost. But enough stayed to man the now-growing line. Through the smoke on Henry Hill, the Union Army advanced to deliver the hammer stroke. Federal artillerists from two powerful batteries unlimbered eleven guns. They would smash the Confederates into submission.[65] But the supporting infantry was slow to follow the guns, and in that delay a Confederate cavalry charge struck the lagging infantry and scattered it.

Now in view of Jackson's line, but without its supporting infantry, the first guns fired. The 33rd Virginia turned its attention to this deadly threat. Seeing the exposed Union artillerists aiming, from point-blank range, the Virginians unleashed a deafening volley and charged forward.

Further to the right, other soldiers saw what was unfolding and focused their fire upon the Federal cannons. The wall of steel ripped into the gunners and cut them down to the man. The eleven guns were silenced. Union colonel Samuel Heintzelman, who had led his 3rd Division up the slopes, witnessed the devastation:

> We soon came in sight of the line of the enemy. . . . Soon after the firing commenced the regiment broke and ran. I considered it useless to attempt to rally them. The want of discipline in these regiments was so great, that most of the men would run from fifty to several hundred yards to the rear.
>
> During this time Ricketts' battery . . . was finally; lost, most of the horses having been killed; Captain Ricketts being wounded, and First Lieut. D. Ramsay killed. Lieutenant Kirby behaved with great gallantry and succeeded in carrying off one caisson.[66]

At this point, General Beauregard fed the newly arrived 6th North Carolina Regiment from the Shenandoah Valley into his bending left flank. This Confederate line then surged forward as one and swept Henry Hill clear of Union forces. But the success was short-lived. A Federal counterattack reclaimed the lost ground, and the Battle of Bull Run raged on. A second Confederate charge, attempting to hurl back the northern invaders, was also defeated and General Bee was killed.[67]

It had been five hours since General Shanks Evans had seen Captain Anderson's wigwagged warning, and still the issue was in doubt. Now the 28th Virginia stumbled onto the Henry Hill battlefield to further extended the Confederate left. Victory or defeat for either side hung by a thread. Which side could throw more reserves into the line the fastest? Three regiments from General Johnston's Army, having just detrained from the Shenandoah Valley, rushed to the sound of the guns.

The Confederate left flank was now grotesquely bent backward. It would take a reserve force—from either side—to break the stalemate, and Beauregard knew that the Federals had more reserves than he. "I received a dispatch from Captain Alexander . . . warning me to look out to the left, a large column was approaching," said General Beauregard.[68] Alexander added that it could very well be the Federal forces from General Patterson's command that Johnston had confounded in the Shenandoah Valley. "At that moment, I must confess, my heart failed me," Beauregard said. "I came, reluctantly, to the conclusion that after all our efforts we should at last be compelled to yield to the enemy, the hard-fought and bloody field."[69]

Beauregard stared through his glass, but no amount of staring caused the flag of the advancing column to unfurl and reveal whether the force was friend or foe. Standing next to him was the heroic General Shanks Evans. He, too, focused through his strong glass, but the approaching banner remained obscured. Beauregard informed him to prepare for the worst—to fall back to a reserve defensive line. As Evans departed to inform General Johnston of the new defensive plan, he halted for a moment as Beauregard took one last look at the approaching column. "It had now come within full view," said Beauregard. "A sudden gust of wind shook out its folds, and I recognized the stars and bars of the Confederate banner. It was the flag borne by . . . Colonel Hays of the gallant Seventh Louisiana."[70]

Colonel Hays's regiment was the vanguard of the arriving brigade of Colonel Jubal Early. Early had marched his men to the sound of the guns toward Chinn Ridge, and now led the column of reinforcing Confederates from General Joseph Johnston's Army of the Shenandoah. Johnston had plucked it from its defensive position on the extreme right flank along Bull Run and sent it to the left. In those ranks of Louisianans were former cadets from the Ole War Skule.

The Confederate battle line now outnumbered the attacking Federals. The entire southern line rose as one, screaming and shouting, and surged forward, first step by step, and then on the dead run into the Federal line. In the face of this charge, the Union line collapsed. The Confederates poured a continuous fire into the scattering enemy, and Confederate guns blasted them as they ran. The collapse turned into a rout; and that rout would continue all the way back to Washington.[71]

"Such a rout I never witnessed before," lamented the Union's Colonel Heintzelman, whose 3rd Division had been shattered. "No efforts could induce a single regiment to form after the retreat was commenced."[72] General Irvin McDowell officially reported, "The retreat soon became a rout, and this soon degenerated still further into a panic."[73] The 7th Louisiana had fired the first shots of the battle on July 18, and now delivered the last shattering blow in the late afternoon of July 21 to decide the outcome of the battle.

CHAPTER 4

The Flush of Victory and the Seminary

Late in the afternoon of July 21, 1861, the Manassas battlefield was littered with the wreckage of war. The Union Army was for the most part in full flight back to Washington. Covering the retreat in a somewhat orderly fashion was a brigade whose soldiers had been formed into a hollow square to defend against Confederate cavalry. Its commander was in the center to direct maneuver. This was William Tecumseh Sherman's Brigade, and its ranks had been substantially thinned on the slope of Henry Hill.

At 5 o'clock on that late summer afternoon, the question was whether there could be a Confederate pursuit to try to finish off McDowell's army. There were still three hours of daylight left, and although both sides were exhausted, there were some units—especially among the Confederate forces—that were fresh. They had remained in defensive positions along Bull Run and were capable of making a concerted attack toward Centerville.

That town was now the center of confusion as retreating Federals jammed and blocked the roads. A Confederate battery (the Alexandria Artillery) under the command of Captain Delaware Kemper now rattled across Bull Run near the Stone Bridge to join with infantry following the Federal retreat. Part of that battery was a detachment of the Crescent Blues, Company B, which mustered in New Orleans and arrived in Virginia on June 17. It was composed of men from the Pelican Hook and Ladder Fire Company of New Orleans[1] and had started its day at Bull Run at the Stone Bridge before racing to the left in the afternoon. At that critical moment, the Crescent Blues had arrived with Jubal Early's Division to take up positions on Henry Hill to blast the final assault of the Union attack.[2] It now pressed forward in a general pursuit toward Centerville. When Kemper's Battery came to the top of a rise on the Warrenton Turnpike, the artillerists looked down upon the suspension bridge crossing Cub Run, south of Centerville. Tantalizing targets were strewn in disarray. "Cannon, caissons, ambulances, wagons, and other vehicles, most of them abandoned by their drivers, were congregated in a confused mass around the turnoff to Sudley Ford,"[3] wrote historian V. C. Jones.

Kemper ordered two guns unlimbered and rolled into place and aimed at the mass jamming the crossroads. He signaled sixty-seven-year-old Edmund Ruffin to step forward. The white-haired South Carolinian was a legend. He had volunteered to fight the British in the War of 1812 but never got into action. He had been in the line that day along Bull Run, and in April had fired one of the first guns at Fort Sumter. The old soldier had hitched a ride with Kemper's Battery from Stone Bridge to not miss out on the kill. There was no room for him on the caisson, so he rode astride the barrel of one of the guns, as if on a horse.[4]

Without hesitation, Ruffin stepped forward and pulled the friction lanyard that touched off the black powder. The cannon roared, and its first shot crashed dead center on the bridge. It overturned a wagon, killing the horse. The bridge was now completely blocked, and the previous panic turned into uncontrollable terror. Soldiers and teamsters freed the horses from their harnesses and hitches and rode them bareback.[5] Two more rounds crashed into the terrified pack.

The Federals left behind fourteen pieces of artillery, complete with ammunition, thirty wagons, and almost fifty horses. Also left behind were panicked congressmen, now sorry that they had ever left the calm and safety of their Washington offices. In that mass confusion were many "spectators." Civilians and members of the Federal government had followed the Union Army out of Washington and, completely confident of the destruction of the Rebel forces, turned the operation into an opportunity for a misguided festival. The anticipated celebration of victory had turned into a stampede. Included in its numbers were gentlemen in top hats and ladies wearing their finest and sporting fashionable colorful parasols to ward off the hot rays of the summer sun. Now these ladies and gentlemen were desperately holding on to the side rails of their speeding carriages and buggies while their horses galloped to the rear.

Pursuit was certainly on the minds of the southern senior commanders. Four companies of Confederate cavalry had already crossed Ball's Ford. Colonel Jeb Stuart, fresh from his fighting on the slopes of Henry Hill, was hustling his own cavalry toward Sudley Ford, where the Union forces had secretly crossed, and were now recrossing. Other southern regiments were following the retreating enemy to the east, up the Warrenton Turnpike, as they recrossed the Stone Bridge and raced for Centerville.[6]

Beauregard galloped to meet with Johnston, and Johnston favored the continued chase. But the Federals retreated faster than the Confederates could chase them, and Stuart's cavalry was soon bogged down with an enormous collection of prisoners. He broke off his pursuit, and eventually all pursuit was called off. General Joseph E. Johnston rationally explained: "Our army was more disorganized in victory than that of the United States in defeat; that there were strong fortifications, well manned, to cover the approaches to Washington, and prevent the establishment of our guns on the south bank of the [Potomac]. We had no means of cannonading the capital."[7]

But not all commanders dismissed the possibility of pursuit so readily. General James Longstreet, commanding at Blackburn's Ford on the right, had been ordered to pursue the fleeing enemy shortly after the rout began. He led his brigade across Bull Run and advanced on Centerville, joined by General Bonham's brigade from Mitchell's Ford. Longstreet observed, "Through the abandoned camps of the Federals we found their pots and kettles over the fire, with food cooking; quarters of beef hanging on the trees, and wagons by the roadside loaded, some with bread and general provisions, others with ammunition."[8]

In the bright light of the late summer sun, he could observe the Federal columns passing through Centerville, and as soon as the Confederates got into artillery range, the infantry stepped to the side of the road to allow the batteries to deploy. But just as the batteries were to open fire, a messenger arrived informing Longstreet that the enemy was not in retreat but marching around to attack the Confederate right. Supposedly orders were on the way to return to the defensive positions behind Bull Run.

Longstreet scoffed. "I denounced the report as absurd . . . and ordered that the batteries open fire."[9] To return to Bull Run would be absurd. But another rider arrived from General Johnston's staff and said, "In the name of General Johnston, I order that the batteries not open."[10] When Longstreet asked if Johnston had sent him to give that order, the rider said he had not; and Longstreet renewed his previous order for the batteries to engage. But before the guns could fire, General Bonham rode to Longstreet's side and asked to delay the order to fire. Since Bonham was the ranking officer, the guns remained silent. While the other brigade withdrew back to Bull Run, Longstreet did not and assumed that his order to advance would be reissued. He remained at his advanced position in front of Centerville until 10 p.m. Only then did he reluctantly withdraw. Longstreet lamented the opportunity lost because of this bogus report. "And upon this report one of the staff-officers sent orders, in the names of the Confederate chiefs, revoking the orders for pursuit."[11]

Later, Colonel Richard Taylor of the 9th Louisiana Regiment would write on the argument of pursuing McDowell's army: "Napoleon held that, no matter how great the confusion and exhaustion of a victorious army might be, a defeated one must be a hundredfold worse, and action should be based on this. Assuredly, if there be justification in disregarding an axiom of Napoleon, the wild confusion of the Confederates after Manassas afforded it."[12]

First Manassas, or the Battle of Bull Run, was over. The Union and Confederate armies had met on near-equal terms: Union forces were approximately 35,000 to the Confederates' 32,000. Union casualties were 2,896, of whom 460 were killed, while the Confederates lost 1,982, of whom 387 were killed.[13]

General Sherman now covered the Union retreat. He had been in on the final attack to crack the Confederate left flank, having maneuvered to Henry Hill. And like the other Federal forces, he had been repulsed. He wrote in his final report:

> The firing was very severe, and the roar of cannon, muskets, and rifles incessant. It was manifest the enemy was here in great force, far superior to us at that point. The Sixty-ninth held the ground for some time, but finally fell back in disorder. . . . Up to that time all had kept their places, and seemed perfectly cool and used to the shells and shot that fell comparatively harmless all around us; but the short exposure to an intense fire of small-arms at close range killed many, wounded more, and produced disorder in all the battalions. About 9 o'clock at night, I received . . . the order to continue the retreat to the Potomac. This retreat was by night, and disorderly in the extreme.[14]

In Washington, the veterans who had been involved in the battle were easily distinguished from those who jammed the capital waiting for assignments or encamped in the many military enclosures. The steady disorderly stream of panicked soldiers and unfortunate observers recrossing the Long Bridge from Alexandria was a sight not soon forgotten.

Private Warren Lee Goss of the 7th New York had just arrived in Washington the week before the Manassas battle and saw the results of the full-blown retreat. As he walked the streets, he ran into Private Jim Tinkham, whom he had met earlier, prior to his enlistment. Tinkham had been a store clerk, but was now part of the retreating Federal column. He was sunburned, and his uniform was reddish with the stain of battle wear. At their earlier meeting, Goss and Tinkham had argued as to how long the war would last. Tinkham was confident and predicted, "Sixty days—the Government would soon blockade all the rebel ports and starve them out." He even proposed a wager of a fine supper "if the rebels did not surrender before snow came that year."[15] Now the rebels had routed him, and he confessed his military underestimation to Private Goss—and his personal desire to run and keep running. He said, "After getting the order to retreat at that battle, he should not have stopped short of Boston if he had not been halted by a soldier with a musket after crossing the Long Bridge."[16]

⚜ ⚜ ⚜

When news of the victory at Manassas reached the Louisiana Seminary at Pineville, there was unbridled jubilation, and it was the talk for days among the few students remaining, and how they wished to have been among their cadet comrades who had experienced the battlefield. Professor St. Ange, a French citizen, took keen interest in the war and the events unfolding in Virginia. He even hoped to join up and command an artillery battery. A cadet who was very close to the professor remembered: "just after the battle of Manassas, so many in the South were disappointed because the Confederates did not advance upon Washington. Major St. Ange expressed great dissatisfaction with the situation in Virginia. 'Ah!' said he, 'if I had ten thousand Frenchmen, I would jump over zee trees and go to Washington.'"[17]

On June 30, 1861, the Seminary ended its second session with no fanfare—no examinations, no parties, no formalities. The Board announced that all activities would be suspended for an indefinite period, and designated Professor Vallas to manage the property. At their meeting, one of the Board members moved that the marble slab placed over the main entrance when the building was constructed be removed. The words on that slab were *By the Liberality of the General Government, the Union, Esto Perpetua.* Dr. Vallas reported that "the inscription over the chief entrance of the Seminary was obliterated by a resolution of the Board of Supervisors."[18] There was no perpetual union. The slab was removed and broken into pieces.

During the summer, Professor Boyd along with several cadets went to Camp Moore, on the north shore of Lake Pontchartrain, and enlisted as privates in the Stafford Guards, to be later attached to the 9th Louisiana Regiment, commanded by Colonel Richard Taylor. Taylor was the only son of former President Zachary Taylor, and Boyd had enlisted as a private so to be free to return to his beloved Seminary at the end of his enlistment.[19]

Others joined Confederate commands already in the field, leaving fewer than ten cadets at the Seminary. Faced with those numbers, and with Vallas and St. Ange as the only professors left to instruct, exercises at the Seminary were suspended indefinitely.[20] However, that indefinite suspension only lasted until April 1862, when the Seminary opened a new session. Rev. W. E. M. Linfield was made "Superintendent *pro temp,* for the session of 1862."[21] He was an enthusiastic supporter of the war, and on the day of the Louisiana secession, had taken to the floor of the convention and offered a prayer for success.

The new session opened with very young cadets. The age limit had been dropped to fourteen for new arrivals, and admission was allowed for any young man who was exempted from the draft. On November 1, 1862, another academic session began. One hundred and twelve students enrolled, but only three were over eighteen.[22] The youthfulness of the boys led to uncontrollable mischief, and Linfield was no Tecumseh Sherman when it came to discipline. All the boys were eager to go war, and on April 1, 1863, three months before the battles at Vicksburg and Gettysburg, "the cadets, early one morning, decided to break up the school. They took the dishes, knives, and forks from the mess hall, and threw them into the well; and then destroyed the kitchen furniture."[23] That was the end of Superintendent Linfield.

His successor was Professor William A. Seay, but his term lasted only three weeks before the approach of the Union Army under General Nathanial Banks brought all activity to a halt. The cadets "were dismissed and told to go fight the enemy. Their bedding was offered to General Kirby Smith [and] the students dispersed to their homes or joined the Confederate forces."[24] The Seminary remained closed for the rest of the war.

✵ ✵ ✵

The Louisiana forces that fought at Manassas had distinguished themselves in a manner like that of Colonel Jackson and his Stonewall Brigade. It was part of the 7th Louisiana Regiment that had defended at Blackburn's Ford and forced McDowell to abandon his original plan. And it had been Major Wheat's battalion of Tigers that had first slowed McDowell's advancing juggernaut so that Beauregard had time to react. And then it was Colonel Hays's 7th Louisiana Regiment, along with the Crescent Blues, that marched with Jubal Early to deliver the hammer stroke that broke the Union line.[25]

The terribly wounded Major Rob Wheat, having survived his gunshot wounds despite the dire predictions of doom from the field surgeons, became a living legend. Wheat had asked the surgeons about his chances of survival, saying that he wasn't ready to die. But the doctors pulled no punches and bluntly told him, "you can't live 'til day. . . . There is no instance on record of recovery from such a wound."[26] Wheat was undaunted by this proclamation and shot back, "Well, then I will put my case on record."[27] His heroism and remarkable recovery made him a celebrity with both dignitaries and the press, and Wheat's Tigers turned some of their former notoriety into fame.

Countless other Louisianans were terribly disappointed to have missed out on the fight at Manassas Junction. This was especially galling for the men of Colonel Richard Taylor's 9th Louisiana Regiment, who heard of the impending battle just as the regiment arrived in Richmond on the afternoon of July 20. Colonel Taylor would write of their frustration: "The town was filled with rumor of battle away north at Manassas. . . . Officers and men were delighted with the prospect of active service, and largely supplied want of experience by zeal. Ammunition was served out, three days' rations were ordered for haversacks, and all camp equipage not absolutely essential was stored."[28]

At 5 p.m., Taylor walked over to the War Office and reported to General Pope Walker of Alabama that he and the 9th Louisiana were ready to go, combat ready with full ammunition and rations. Walker was delighted since he had been anxious to send more troops forward to Manassas Junction but had been handicapped by ammunition shortages. The general thought a train could be ready to move Taylor's regiment at 9 p.m. Taylor later wrote:

> Accordingly, the regiment was marched to the station, where we remained several weary hours. At length, long after midnight, our train made its appearance. As the usual time to Manassas was some six hours, we confidently expected to arrive in the early forenoon; but this expectation our engine brought to grief.
>
> It proved a machine of the most wheezy and helpless character, creeping snail-like on levels, and requiring the men to leave the carriages to help it up grades. As the morning wore on, the sound of guns, re-echoed from the Blue Ridge Mountains on our left, became loud and constant. At every halt of the wretched engine the noise of battle grew more and more intense, as did our impatience. [Finally] At dusk we gained Manassas Junction.[29]

CHAPTER 5

Tigers in the Valley

After the Manassas battle, the Louisiana forces were reorganized and placed in General Joseph E. Johnston's newly formed Army of Northern Virginia. On October 21, 1861, the Confederate government created the Department of Virginia, commanded by General Johnston, and subdivided it further into three districts: the Valley District, commanded by Stonewall Jackson; the Potomac District, commanded by Beauregard; and the Aquia District, commanded by Major General Theophilus H. Holmes.[1]

A year after the Manassas victory, the Confederate Army faced an entirely different set of circumstances. In March 1862, General Johnston had withdrawn from Manassas to confront General George McClellan's Federal army that had landed on the Virginia Peninsula and threatened the Confederate capital at Richmond. The original Union army from Bull Run was now a collection of Union armies—not a single army advancing on a single axis from Washington, but four armies facing off against General Johnston, each vastly superior to his in numbers and equipment. The plan was to actively pursue him on multiple fronts and eventually converge on Richmond and end the war.

General Jackson was in the Shenandoah Valley with a force of eighty-four hundred infantry and one thousand cavalry (Turner Ashby's) and was facing his own horde of superior forces.[2] Johnston's options were few and all fraught with difficulty and uncertainty. But the bottom line was simple: "Richmond must be held; the battles must be waged as far from the city as possible. And the first opportunity to strike the enemy 'must' be taken."[3] Fittingly, this list of "musts" was written in the passive voice, because that is exactly what it was: a passive plan with no bold commitment to the offensive.

In Jackson's Valley District, forty-five thousand Union troops prowled about in different parts of the Shenandoah Valley, all seeking to defeat him and then link up with McClellan's massive one-hundred-thousand-man force gathering in front of Richmond. Jackson took it upon himself to act independently and prevent any Union forces from leaving the Shenandoah to reinforce McClellan. He wrote to General Johnston on March 24, 1862, that he had indeed acted. But this first independent engagement was not victorious. In fact, he had been repulsed while frustrating a Union attempted to exit the Shenandoah:

> As the enemy had been sending off troops from the district, and from what I should learn were still doing it, and knowing your great desire to prevent it, and having a prospect of success, I engaged him yesterday about 3 P.M. near Winchester, and fought until dusk, but his forces were so superior to mine that he repulsed me with the loss of valuable officers and men killed and wounded; but the obstinacy with which our troops fought from their advantageous position I am of the opinion that his loss was greater than mine.[4]

This glimmer of light shining on an otherwise gloomy military outlook encouraged Richmond to move to reinforce him. The War Department dispatched General Ewell's Division to Jackson, increasing his force to thirteen thousand men.

Confederate General John D. Imboden wrote: "[Richmond's] only salvation depended upon Jackson's ability to hold back Fremont, Banks, and McDowell long enough to let Johnston try doubtful conclusions with McClellan."[5] *Doubtful conclusions!* Hardly confident words of survival.

So, having missed the Manassas battle, Richard Taylor, now a brigadier general, and his newly formed First Louisiana Brigade were placed in General Richard Ewell's Division. The brigade consisted of the 6th, 7th, 8th, and 9th Louisiana Regiments and Rob Wheat's Tiger Battalion.[6] And in those numbers were cadets of the Ole War Skule. Ewell's entire division moved to reinforce Stonewall Jackson. The Louisianans began their difficult march into the Shenandoah Valley. The Seminary had, as a matter of fact, sent a larger proportion of its students and faculty into the army than any other American institution, and that included West Point.[7]

The geography of the Shenandoah Valley seemed to have been designed to especially favor the Confederate Army. The parallel lines of the three mountain ranges that formed the valley ran southwesterly along the Washington-

Richmond north-south axis. Each southerly step of a Federal army that entered the valley from Washington took it away from Richmond. Conversely, the northerly march of any Confederate force entering the valley brought it closer to the environs of Washington.

Jackson made it his business to become an expert on the complicated geography in the valley. All of its roads, intersections, and "gaps" that allowed passage across the three major mountain chains were etched in his mind. The first was the Blue Ridge, and then the Massanuttens, with the Luray Valley separating them. West of the Massanuttens was the final chain, the Allegheny Mountains, with the Shenandoah Valley in between. This entire mountain and valley belt was fifty miles wide, and Jackson, who had been uncertain concerning all the twists and turns, commanded Major Jedidiah Hotchkiss, an expert topographical engineer, "to make me a map of the Valley from Harper's Ferry to Lexington, showing all the points of offense and defense between these points."[8] In a letter to his closest friend he had said, "If this valley is lost, Virginia is lost."[9]

On April 30, 1862, Ewell's Division of eight thousand, including the Louisianans, crossed the Blue Ridge and camped at Conrad's Store, where a bridge spanned a swift-running stream. On that same day, many miles to the west, Jackson's force, having created the deception that it was quitting the valley to reinforce Richmond, actually raced in the opposite direction. On May 8, he fell upon and routed the unsuspecting forces of General Robert Milroy at McDowell, sending him fleeing to Franklin. Jackson's surprise attack had been reinforced by two hundred cadets from VMI.[10]

For the Louisianans, crossing into the valley at Swift Run Gap was like stepping into an idyllic paradise. Away from the fields of war-torn Manassas, away from the mud and slop of the trampled Virginia roads, the pristine valley was a picture of perfect peace. General Richard Taylor described his first view. "The great Valley of Virginia was before us in all its beauty. Fields of wheat spread far and wide, interspersed with woodlands . . . quaint old mills, with turning wheels, were busily grinding the previous year's harvest; and grove and eminence showed comfortable homesteads."[11]

The Blue Ridge in the east and the Alleghenies to the west bound this enormous valley and its population in an isolated world. "These were thrifty, substantial farmers, and, like their kinsmen of Pennsylvania, expressed their opulence in huge barns and fat cattle. The devotion of all to the Southern cause was wonderful," wrote General Taylor. "Jackson, a Valley man by reason of his residence at Lexington . . . was their hero and idol. The women sent husbands, sons, lovers, to battle as cheerfully as to marriage feasts."[12]

In this mix of early "colonial" inhabitants, General Taylor's Brigade stood out in stark contrast. Colonel Hays's 7th Louisiana Regiment was well known for its gallant attack to secure the victory at Manassas, and its Blue Pelican Flag had gained equal fame. Wheat's Tigers' reputation preceded them wherever they went; and mostly struck fear into the hearts of friend and foe alike. "So villainous was the reputation of this battalion that every commander desired to be rid of it," lamented Taylor, "and General Johnston assigned it to me, despite my efforts to decline the honor."[13] The 6th Louisiana Regiment, commanded by Colonel Isaac Seymour, was composed mostly of Irishmen, "stout, hardy fellows, turbulent in camp and requiring a strong hand, but responding to kindness and justice, and ready to follow their officers to the death."[14] The 8th Louisiana Regiment was commanded by Colonel Henry Kelly, and the soldiers were Acadians—Cajuns as they were popularly called: "A home loving, simple people; few spoke English," said Taylor. Fewer still had ever been more than ten miles from their native homes, and wherever they went, their regimental band accompanied them. Even after long marches "they would waltz and polka, in couples, with as much zest as if their arms encircled the supple waists of the [girls] of their native Téche."[15] The 9th Louisiana Regiment was commanded by Colonel Leroy Stafford, from North Louisiana, and its soldiers were completely opposite to those of the other regiments. They were the spoiled socialites. They were "planters, or sons of planters, many of them men of fortune; soldiering was a hard task to which they only became reconciled by reflecting that it was 'niddering' [cowardly] in gentlemen to assume, voluntarily, the discharge of duties, and then shirk."[16]

After only two days in their new camp at Conrad's Store, Taylor's Brigade was ordered to link up with Stonewall Jackson's forces at New Market, on the western side of the Massanutten Mountain. It was a march of over twenty miles, often in choking dust raised by the shuffling of animals and men. Finally the Louisianans approached their camp, only to be motioned to continue up the road. Even after this twenty-six-mile trek, there was not one straggler. Their marching ranks drew the stares of impressed onlookers as the setting sun danced its rays off of the polished steel of three thousand bayonets on shouldered arms.

This was so impressive that thousands of Jackson's men "gathered on either side of the road to see us pass,"[17] said Taylor. And there was spontaneous clapping and cheering as the Blue Pelican Flag and the green Irish flag passed, as if in a parade review. The soldiers gathered to watch the brigade's arrival were treated to the strains of music from Wheat's band playing "The Girl I Left Behind."[18]

Jackson himself sat some distance away, on the top rail of a nearby fence. His disheveled, slouched cap was pulled low over his eyes, and as he sipped water he sucked on a lemon. Known as a man of few words, he used the fewest to address General Taylor: "You seem to have no stragglers," he said in a low, soft voice. Taylor replied that he didn't allow stragglers. Jackson replied, "You must teach my people; they straggle badly." And then hearing the Creoles start up their music and dance, he turned his attention to them and said, "Thoughtless fellows for serious work."[19]

But Jackson was surely most impressed with the Louisianans, because that very evening he visited Taylor in his tent and confided to him that his brigade would lead the way on the army's next march. Where and when? He didn't say. His reputation was to reveal nothing to no one. Taylor was shocked to learn that even General Ewell, Jackson's second in command, knew nothing of his plans. When he confronted Ewell, the division commander replied, "If General Jackson were shot down, I wouldn't know anything of his plans."[20] And later he added "that he never saw one of Jackson's couriers approach without expecting an order to assault the North Pole!"[21]

Using his cavalry to screen all movements, Jackson left New Market on May 21, and Taylor's Brigade was in the lead. Jackson rode with them. The army headed north, but soon turned to the east, through a gap through the Massanutten Mountain. It entered Luray Valley, and Taylor then turned the column to the north and followed the South Fork of the Shenandoah River. The objective was the town of Front Royal.

General Robert E. Lee, the senior military advisor to President Jefferson Davis, had been clear in a May 16 dispatch to Jackson. "Whatever movement you make against Banks, do it speedily, and if successful drive him back toward the Potomac, and create the impression, as far as practicable, that you design threatening that line."[22] That night, the Louisianans crossed the river by bridge and camped on the east side. Taylor examined his map and saw that he was only a short distance from Conrad's Store, from where he had begun days earlier. Perhaps that was proof of Jackson's devotion to security. For Taylor to have moved the short distance directly from Conrad's Store to his present encampment would have signaled exactly where he was going.

The next day, May 23, the Confederates advanced upon Front Royal, sur-

Richard Taylor
(LSU Archives)

prising Federal pickets and easily drove them in. On top of a rise looking down into the small town, the column paused. In the town could be seen a flurry of activity of running soldiers and galloping horses as the Federals reacted to the complete surprise of Jackson's arrival. Front Royal had, in 1861, been designated as a hospital for wounded Confederate soldiers, but it was now occupied by the Federal army.

Jackson remained mounted, while Generals Taylor, Ewell, and Jackson's Assistant Inspector General, Captain Henry Kyd Douglas, flanked him.[23] It was Captain Douglas who first spotted a young girl in a white bonnet slipping out of the main part of town and running toward them. "She glided swiftly out of town on our right and, after making a little circuit, ran rapidly up a ravine in our direction and then disappeared from sight," wrote Doug-

A Louisiana "Pelican"
(Public domain)

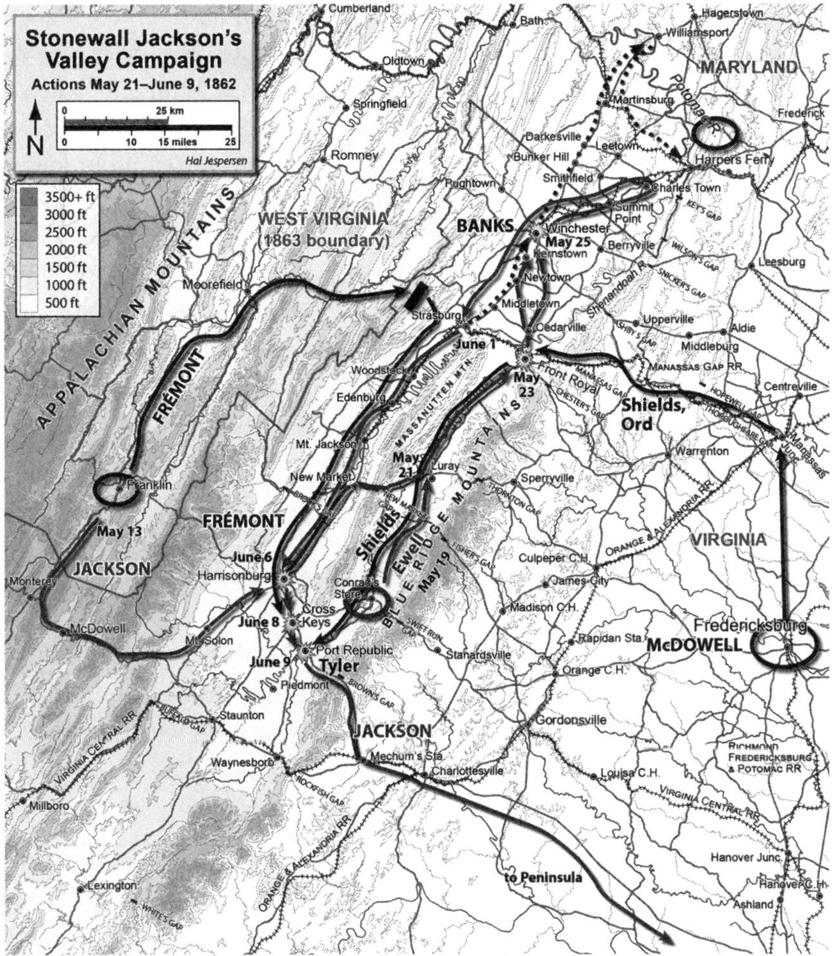

Jackson's Valley Campaign, 1862 (National Park Service)

las. The girl seemed not to be slowed by any obstacles of thick weeds or any intersecting fencing but raced forward at her best speed, waving her bonnet to them. Douglas pointed in her direction, but just as the others looked, she had disappeared behind a slight dip in the rolling terrain. "At General Ewell's suggestion, [Jackson] sent me to meet her," said Captain Douglas. "That was just to my taste, and it took only a few minutes for my horse to carry me to meet the romantic maiden whose tall, supple, and graceful figure struck me as soon as I came in sight of her."[24]

The girl was Belle Boyd, whom Douglas had known from childhood. She had just had her eighteenth birthday. Struggling to catch her breath, she finally gasped out in short sentences: "I knew it must be Stonewall when I heard the first gun. Go back and tell him that the Yankee force is very small—one regiment of Maryland infantry and several companies of cavalry. I know, for I went through the camps and got it out of an officer. Tell him to charge right down and he will catch them all. . . . My love to all the dear boys—and remember if you meet me in town, you haven't seen me today."

She also informed them of details that could only come from a practiced spy, and General Taylor recorded the information written on a small, folded sheet: "The Federals camp was on the west side of the river, where they had guns in position to cover the wagon bridge but none bearing on the railway bridge below the former; that they believed Jackson to be west of Massanutten, near Harrisonburg; that General Banks, the Federal commander, was at Winchester, twenty miles northwest of Front Royal, where he was slowly concentrating his widely scattered forces to meet Jackson's advance, which was expected some days later."[25]

Belle Boyd's exciting escapade with Jackson's army that day at Front Royal had been the result of a remarkable sequence of terrible events. The teenager had been spying since she was seventeen, and her father was in Jackson's army. In July 1861, the Union Army had occupied her town of Martinsburg to the west of Winchester, a gateway to the Shenandoah Valley. The retreating southern army had evacuated and left two of its sick behind. Miss Boyd was caring for them when out-of-control Union soldiers broke in and surrounded the men and hurled insults at them and threatened to bayonet them. Belle interceded on their behalf, and a Union officer restrained the perpetrators.[26]

"The Yankees were in undisputed possession of Martinsburg . . . ," she said, "and whatever might have been the intentions of the officers, they had not the inclination, or they lacked the authority, to control the turbulence of their men. The doors of our houses were dashed in; our rooms were forcibly entered by soldiers who might literally be termed 'mad drunk.'"[27] Homes were ransacked and left as "mere wrecks, utterly despoiled and mutilated. Shots were fired through the windows; chairs and tables were hurled into the street."[28]

It seemed that the cruelty of these rampaging soldiers of the occupying Federal army knew no bounds. In several extreme instances, trembling old ladies, sifting through the wreckage of their destroyed homes and possessions, requested of an occupying soldier to be allowed to retrieve a special memento, only to be denied "with a volley of blasphemous curses and horrid imprecations," recalled Boyd. "Words from which the mind recoils with horror, which no man with one spark of feeling would utter in the presence even of the most abandoned woman, were shouted in the ears of innocent, shrinking girls."[29]

These despicable incidents enraged young Belle but were nothing compared to a terrible event that was to forever be etched in her mind. "A party of soldiers, conspicuous, even on that day, for violence, broke into our house and commenced their depredations; this occupation, however, they presently discontinued, for the purpose of hunting for 'rebel flags,' with which they had been informed my room was decorated. Fortunately for us, although without my orders, my negro maid promptly rushed upstairs, tore down the emblem, and, before our enemies could get possession of it, burned it."

The soldiers had brought with them a large Federal flag that they now gleefully intended to raise over the Boyd home, gloating that it signified the rebels' submission to their authority. But Belle's mother stepped forward and defied them, and said that she would not allow this in her house. "Upon this, one of the soldiers, thrusting himself forward, addressed my mother and myself in language as offensive as it is possible to conceive. I could stand it no longer; my indignation was roused beyond control; my blood was literally boiling in my veins; I drew out my pistol and shot him."[30]

The other perpetrators carried the dying man away, and then returned to burn down the Boyd home. Incendiary material was stacked along the walls of the house, but at the last minute an officer stopped it and arrested the would-be arsonists. When the commanding officer later came to investigate, he interviewed all the witnesses and concluded that seventeen-year-old Belle Boyd had "done the right thing" in self-defense, and in defense of her mother's life.[31]

This was the Belle Boyd who after a year of harsh Federal occupation—and countless acts of criminal misconduct—now ran as fast as she could across the grassy fields toward Stonewall Jackson and his mounted officers. She carried vital information, scribbled on a folded paper, that she had heard in a secret military meeting among the top Union officers. She had moved from Martinsburg and occasionally stayed at Front Royal, in her aunt's house. As she recalled, "out of thc frying-pan into the fire."[32] Numerous arrests (and releases) as a spy had become an unpleasant routine of her daily life, and Belle just happened to be in Front Royal the night before General James Shields departed, announcing to all that he "was about to 'whip' Jackson."

"A council of war was held in what had formerly been my aunt's drawing-room," she said. "Immediately above this was a bedchamber, containing a closet, through the floor of which I observed a hole had been bored."[33] As the war council began, Belle had crept softly up the stairs, and was on the floor in that dark closet with her ear pressed to that hole. She remained in that motionless position for the several hours of the meeting, which lasted until 1 a.m. She wrote it all down, in cypher, as she remembered the important details.

On the afternoon of May 23, Belle was sitting at her upstairs window reading to her grandmother and cousin when shrieking and shouting was everywhere in the street. Grabbing her opera glasses, she rushed to the balcony and could observe "the advance guard of the Confederates at the distance of about three-quarters of a mile, marching rapidly upon the town."[34] "I was in possession of much important information, which if I could only contrive to convey to General Jackson, I knew our victory would be secure. . . . I put on a white sunbonnet, and started at a run down the street, which was thronged with Federal officers and men. I soon cleared the town and gained the open fields, which I traversed with unabated speed, hoping to escape observation until such time as I could make good my way to the Confederate line, which was still rapidly advancing."[35]

The vanguard of Jackson's force and the Federals defending at Front Royal broke out in sporadic fire, and soon Federal cannon fire erupted, crashing into the closing Confederate column. The Union artillery was up higher than the level of the approaching road, and the infantry had taken up positions in the large hospital and delivered constant musket and rifle fire. Soon Jackson's artillery was in position and able to pour solid shot into those very windows.

Through this crossfire, Belle Boyd had made her run; but she also attracted the attention of the Federal pickets. One by one, they began to fire at her fleeing figure. Seeing this, the troops in the hospital also began to shoot at her. "I was exposed to a crossfire from the Federal and Confederate artillery, whose shot and shell flew whistling and hissing over my head," she said. "A Federal shell struck the ground within twenty yards of my feet; and the explosion . . . sent the fragments flying, in every direction around me. I shall never run again as I ran on that memorable day."[36]

Armed with Belle Boyd's information, Jackson ordered Taylor to advance, and his First Louisiana Brigade moved forward on both sides of the road. Already he could see Federal infantry retreating and crossing over the wagon bridge, just above the railroad bridge, to take up positions on the west bank. Just as the girl had told him, Taylor noticed that the artillery was indeed aimed at the wagon bridge, and he suggested to Jackson that the Louisianans might cross the railroad bridge instead by stepping on the crossties.[37]

Jackson concurred, and the 8th Louisiana, with Colonel Henry Kelly in the lead, were soon stepping across, one crosstie to the next, under heavy musket fire. Several men were hit, and without a bridge flooring, fell into the swirling waters beneath. Colonel Kelly's men soon gained the opposite bank, and the Federals, seeing they were outflanked, ignited the combustibles placed in the center of the wagon bridge. Jackson signaled Taylor, who ordered Wheat's Tigers to advance with him. At the double quick they raced across the burning bridge.[38] "Concealed by the cloud of smoke, the suddenness of the movement saved us from much loss," said General Taylor, "My horse and clothing were scorched, and many men burned their hands severely while throwing brands into the river."[39]

With Wheat's Battalion racing for the bridge, a witnessing soldier wrote: "[Wheat] was riding at full gallop, yelling at the top of his voice, his big sergeant-major running at top speed just after him, calling to the men to come on . . . all running . . . all yelling . . . their peculiar Zouave dress & wild excitement made up a glorious picture."[40]

The Confederate victory was complete; the Union forces were routed and in full retreat toward Winchester. Jackson pursued them for four miles and finally encamped at Cedarville. Taylor's casualties were only among Wheat's Battalion: eight wounded, and the two men killed who had been shot and fell into the river. That night, General Jackson again met with Taylor around his campfire. His brigade of Louisianans would again lead the way tomorrow.[41]

At 6 a.m. the Confederate Army was on the move, pursuing General Bank's army that had been surprised at Front Royal. Banks was retreating along a parallel road to stave off Jackson's enveloping move. If he could not pass through Middleton and Winchester before Jackson cut the road, he would be enveloped. Jackson's army would be between him and his escape route to the Potomac River.

Once more, Jackson and Taylor were in the front with a small force of the Tigers. They raced ahead in hopes of reaching Banks's retreat route to cut his passing army in half. Jackson ordered the Louisiana Tigers to keep up with the artillery, so "they trotted along with the horse and artillery at Jackson's heels . . . and we speedily came upon a moving spectacle."[42] Jackson's lightning-quick advance intersected Banks's column at Middleton, twelve miles south of Winchester. The sight was exactly what Jackson had hoped for. A long train of wagons protected by a small detachment of cavalry moving at escape speed toward Winchester. He ordered the Tigers to attack, and, as General Taylor described, "The *gentle* Tigers were looting right merrily, diving in and out of wagons with the activity of rabbits in a warren; but this occupation was abandoned on my approach, and in a moment they were in line, looking as solemn and virtuous as deacons at a funeral."[43]

Several hundred Union supply wagons and a hundred prisoners were taken, but because of the distraction of looting and grabbing the spoils of war, further pursuit of the enemy fleeing toward Winchester had to be called off. But Jackson was not ready to call off the advance of his army. He pressed on, even into the night. For some miles the road was illuminated by the light of burning Union wagons, set aflame by the retreating Federals, then darkness surrounded them.

"Whenever the column stopped for a minute," said Captain Douglas, "the sleepy officers and men would throw themselves on the ground. And thus, skirmishing and halting, marching while sleeping, the night wore away, and within hours of daylight, we halted within two hours of Winchester."[44] The rest lasted only an hour before Jackson ordered the men aroused to continue to Winchester. It was 4 a.m., Sunday, May 25, when the column approached the outskirts of the town. The high ground to the south was "occupied by a heavy mass of Federals with guns in position. A Virginian battery . . . was fighting at a great disadvantage, and already much cut up."[45] The Federal line was in a strong position facing the advancing Confederate force. Jackson moved to envelop it. He raced to General Taylor, "pointed to the ridge and said, 'You must carry it.'"[46]

At 6 a.m., Taylor maneuvered to attack the Federals and their murderous artillery. He wrote: "The column faced to the front and began the ascent. . . . As we mounted we came in full view of both armies, whose efforts in other quarters had been slackened to await the result of our movement."[47] As soon as Taylor's assault line came into view, the Federal line unleashed a torrent of grapeshot, solid shot, and musketry into it. In front of the line, General Taylor rode with drawn sword. He occasionally turned in his saddle to encourage his men forward. But even as the grapeshot ripped holes in their line, the Tigers held their fire.

> About half way up, the enemy's horse [cavalry] from his right charged; and to meet it, I directed Lieutenant-Colonel Nicholls, whose regiment, the *8th*, was on the left, to withhold slightly his two flank companies. By one volley, which emptied some saddles, Nicholls drove off the horse.
>
> Closing the many gaps made by the fierce fire, the brigade, with cadenced step and eyes on the foe, swept grandly over copse and ledge and fence, to

> crown the heights from which the enemy had melted away. Loud cheers went up from our army.[48]

Witnessing this attack from her house near the Federal defensive line was the wife of a Confederate colonel—a Mrs. McDonald. She could see from her front door a long line of Union soldiers lying down just below the crest of the hill. On the crest of the ridgeline, above their heads, artillery fired rapidly at the advancing Confederates. "Suddenly I saw a long even line of gray caps," she said. "Then appeared the gray forms which wore them, with the battle flag floating over their heads! The cannons ceased suddenly, and as the crouching forms . . . rose to their feet, they were greeted by a volley from their assailants."[49]

Captain Douglas said, "I have rarely seen such a beautiful charge. This full brigade, with a line of glistening bayonets bright in that morning sun, its formation straight and compact, its tread quick and easy as it pushed on through the clover up the hill, was a sight to delight a veteran."[50] The Confederates swept into a jubilant Winchester, and Banks's army was driven to the Potomac. Jackson would write to his wife: "My precious darling, an ever-kind Providence blessed us with success at Front Royal on Friday, between Strasburg and Winchester on Saturday, and here with a successful engagement yesterday (Sunday). Our entrance into Winchester was one of the most stirring scenes of my life."[51]

⚜ ⚜ ⚜

Jackson's Valley Campaign was a brilliant stroke of generalship. Playing a huge part in the entire campaign were the Louisianans: former cadets and volunteers under the command of General Taylor and Major Wheat. After confounding the three Union armies in the Shenandoah, Jackson recrossed the Blue Ridge Mountains to participate in the defense of Richmond.

On June 27, Taylor's Brigade was again in the attack at a place called Gaines' Mill, but without General Taylor in command. He had been evacuated with terrible head and back pains. Colonel Isaac Seymour of the 6th Louisiana Regiment took over command.

Major Wheat also seemed not to be himself. He talked of his impending death and, as if giving away a precious object that he would not need in the hereafter, passed his brandy flask to David F. Boyd, formerly of the Seminary and now a major on Ewell's staff corps in the Army of Northern Virginia. He then read a prayer from a small book. Boyd recalled: "Wheat would cry like a child, take another drink, and read from his little book again."[52] He made everyone promise to bury him where he fell. While riding on the march to battle, he talked about his mother and was sure this was his last day and that he would die before sunset. One of the prayers he repeatedly said was heard by many: "Lord, I commend myself to Thee. Prepare me for living, prepare me for dying."[53]

At 2 p.m. the sound of battle was to the front, and Wheat seemed to snap out of his depression. The Tigers formed for the attack, and coming up from the rear was none other than Stonewall Jackson, who had an affection for the Louisianans. When Major Wheat saw him, he rode over and cautioned Jackson that this forward, exposed position was no place for the general. Jackson shook his hand in appreciation, and Wheat, his combat juices once again flowing, rode next to Major Boyd and proudly pointed to his formed line of Tigers. "Major, just look at my Louisiana planters! I'd like to see any 5,000 button makers stand before them this day."[54]

The brigade swept forward "and became the target of searching artillery shells that shattered treetops that rained branches and shrapnel down upon the advancing line."[55] A fusillade of minié balls then ripped into them. Colonel Seymour fell dead. Some of the dazed Tigers fell back, and Major Wheat, seeing them falter, urged his horse forward to gallop in front of his men and toward the Union line. Suddenly the Tigers saw the unthinkable. Both Major Wheat and his horse fell just yards short of the smoking enemy guns. A ball had torn through Wheat's eye and out the back of his head. He was dead before he hit the ground.

Just as at the Battle of Manassas, when Wheat had charged to the front and been shot, his men faltered. Now, when he fell, the attacking line again stopped. Major David Boyd was in the attacking line behind him. A man suddenly bolted to the rear, and Boyd grabbed him and asked why was he running away? With tears running down his face, the youthful Tiger sobbed, "They have killed the old Major, and I am going home. I wouldn't fight for Jesus Christ now."[56] The leaderless Tigers fell back to their starting point. The fighting continued for another month, and when the Peninsula Campaign in front of Richmond was finally over, a thousand Tigers had become casualties.

General Taylor was promoted to major general and transferred to the Trans Mississippi Department, and Wheat's Tiger Battalion was disbanded. Because of its heavy losses, it was reduced to a level where it could no longer function

Major David French Boyd
(LSU Archives)

as an independent command.[57] Major David French Boyd soldiered on with Jackson, whom he affectionately called Parson Stonewall, and was with him through all his battles, including his final one at Chancellorsville. Afterward he transferred with General Taylor to the Trans Mississippi Department west of the Mississippi. He served as his captain of engineers.

Boyd served along the Red River until 1864, when he was unfortunately captured by a lawless, renegade faction of the Union Army known as the "Jay hawkers." They took him to Shreveport and sold him to the Federal authorities for $100.[58] These outlaws were well described by historian Tony O'Brian. "By the time the war ended . . . the term 'Jayhawkers' became synonymous with Union troops led by abolitionists from Kansas, and 'jayhawking' became the generic term for armies plundering and looting from civilian populations nationwide."[59]

It was as a prisoner at Shreveport that Major Boyd would experience a welcome reunion with—and redemption by—his old friend and mentor from the

Copyright by Review of Reviews Co.

THE LAST TO LAY DOWN ARMS

Recovered from oblivion only after a long and patient search, this is believed to be the last Confederate war photograph taken. On May 26, 1865, General E. Kirby Smith surrendered the troops in the Trans-Mississippi Department. Paroled by that capitulation these officers gathered in Shreveport, Louisiana, early in June to commemorate by means of the camera their long connection with the war. The oldest of them was but 40. The clothes in which they fought were worn to tatters, but each has donned the dress coat of an unused uniform carefully saved in some chest in the belief that it was to identify him with a victorious cause and not as here with a lost one. The names of those standing, from left to right, are: David French Boyd, Major of Engineers; D. C. Proctor, First Louisiana Engineers; unidentified; and William Freret. The names of those seated are: Richard M. Venable; H. T. Douglas, Colonel of Engineers; and Octave Hopkins, First Louisiana Engineers.

"The Last to Lay Down Arms" (LSU Archives)

Ole War Skule: General William Tecumseh Sherman. An eyewitness providentially wrote of this meeting. "Commodore Porter was sitting in conversation with Sherman, who was leaning on the mantel and talking very earnestly, and when Boyd was ushered in, Sherman grabbed him by both hands and exclaimed, 'How do you do, Professor Boyd! I am very glad to see you, indeed sir! Very glad.'"[60] Turning to Commodore Porter, as if introducing him to a fellow Union general, Sherman said, "This is Professor Boyd, one of my assistants in the State Seminary and Military Academy of Louisiana. He is Professor Boyd. He calls himself Major. He is no such thing. He is Professor Boyd, sir."[61] Then Sherman rambled on about his dilemma of how to deal with his prisoner. "Boyd, what in hell am I going to do with you? They tell me I have got to send [you] up yonder to Fort Douglas. . . . Sit down here and stay with me, I will guard you . . . I don't want to send you up yonder. Tell you what I'm going to do. I am going down to New Orleans and I will take you along with me down there and turn you over. Just sit down and make yourself comfortable."[62] Professor Boyd was exchanged through the good graces of General Sherman for three Union staff officers and eventually was paroled in June 1865 at Shreveport with the last Confederate troops.[63] He posed for a rare, final photograph, and one month later was the "acting superintendent" of the Seminary and preparing to reopen the Ole War Skule in the fall.[64]

CHAPTER 6

Raphael Semmes

SUMTER AND *ALABAMA*

On February 14, 1861, Captain Raphael Semmes, a commanding officer in the Union Navy and future professor of moral theology at the Ole War Skule in Louisiana, relaxed at his home in Washington when a telegram arrived from Montgomery. It was from the chairman of the newly formed Confederate Committee on Naval Affairs, and its message was short: "I beg leave to request that you will repair to this place at your earliest convenience. Your obedient servant, C. M. CONRAD, Chairman."[1]

Semmes replied immediately that he would hasten to Alabama and that he stood ready to serve in the new Confederacy but had previously cautioned: "My fate is cast with the South; but I should be unwilling, unless invited, to appear to thrust myself upon the new Government until my own State has moved."[2] Born in 1809 in Maryland, Semmes's adopted state was Alabama. He had relocated there following his service in the Mexican War. He practiced law and wrote a well-received historical work: *Service Afloat and Ashore During the Mexican War.* He had first entered the U.S. Navy in 1826, and now at the age of thirty-five, having resigned his Federal commission, he hastened to answer the call at Montgomery.

On February 20, 1861, just prior to Abraham Lincoln's inauguration, Semmes met with Provisional President Jefferson Davis, who broached a most disquieting subject that the new Confederacy would face if the Lincoln government attacked or invaded. Davis had no doubt that, should hostilities commence, the southern population would heed a call to arms and rise to defend their land. But the means to obtain the necessary materiel of war to build a navy was remote. If there was to be a common defense, that included defense of the great Mississippi River. "He then explained to me," Semmes said, "his plan of sending me back to the city of Washington, and thence into the Northern States, to gather together, with as much haste as possible, such persons

Raphael Semmes in 1861
(LSU Archives)

and materials of war as might be of most pressing necessity. . . . We had not even percussion caps enough to enable us to fight a battle, or the machines with which to make them,"[3]

Semmes traveled to Washington and, on March 4, as Lincoln delivered his inauguration address "with," as he described, "triple rows of bayonets between him, and the people to whom he was speaking."[4] He then departed for New York on a quest to acquire the necessary materiel of war. For the next three weeks, Semmes visited all the armament shops in New York, Connecticut, and Massachusetts. "I purchased large quantities of percussion caps in the city of New York, and sent them by express without any disguise, to Montgomery. I made contracts for batteries of light artillery, powder, and other munitions, and succeeded in getting large quantities of the powder shipped. It was agreed between the contractors and myself, that when I should have occasion to use the telegraph, certain other words were to be substituted . . . to avoid suspicion."[5]

Unfortunately, proposed contracts for heavy equipment used in the rifling of cannons, and other machinery, were not available because of the early outbreak of hostilities. "Men were becoming more shy of making engagements with me," said Semmes, "and the Federal Government was becoming more watchful. The New York, and Savannah steamers were still running, curiously enough carrying the Federal flag at the peak, and the Confederate flag at the fore."[6]

Captain Semmes returned to Montgomery on April 4, just eight days before the firing upon Fort Sumter, and the same date that the State of Virginia voted 90–45 to not secede. But on April 12, after the Federal occupying forces at Fort Sumter refused to leave the newly declared Confederate land, the surrounding Confederate batteries fired on the fort, forcing its surrender. "On April 15, Abraham Lincoln called for 75,000 troops to suppress the rebellion. The Virginia convention reversed itself and voted 88–55 for secession . . . voters ratified the ordinance on May 23."[7]

As war began, recruiting for the army in the South was enthusiastically fast. Emotion trumped the realities of military weaknesses, and the most glaring weakness was the nonexistence of a navy. Ever the realist, Raphael Semmes perfectly described the South's Achilles' Heel:

> The Atlantic and the ports of America were ruled at that time absolutely by President Lincoln. The South had not a voice upon the sea.
>
> The merchants of New York and Boston looked upon the war as something which concerned them very little. Not a dream of any damage possibly to be inflicted on them, disturbed the serenity of their votes for the invasion of the South.[8]

This absolute overconfidence of complete naval supremacy was born out by the fact that the Federal forces had the largest merchant fleet, second only to the fleet of Great Britain. And its supremacy over the South in surface ships was 300 to zero. "The South had not a single vessel. Here and there a packet-steamer might be caught up and armed, but what would they avail against such fleet and powerful ships as the *Brooklyn,* the *Powhattan,* and dozens of others? . . . they were to strike the blow, and no one was to give them the value in return."[9]

The only way to combat this overwhelming superiority would be to wage war against the Federal navy in an irregular way; and that was to attack its merchant shipping. "Wealth is necessary to the conduct of all modern wars," wrote now Commodore Semmes, "and I naturally turned my eyes . . . to the enemy's chief source of wealth. It became an object of the first necessity with the Confederate States, to strike at his commerce."[10]

For that purpose, a naval board met in New Orleans to examine what might be available in light, fast steamers that could be so outfitted. Unfortunately, there were none without major defects. Semmes reviewed one of the written reports examining a 500-ton, small propeller vessel. On the plus side, it could make 9 or 10 knots and could be modified to carry four or five guns. But it had no accommodations for a crew and had space to carry only five days of coal.[11]

But Semmes's expert eye saw past these negatives. He told the board's secretary, "Give me that ship; I think I can make her answer the purpose."[12] So thus the Confederate States' steamer *Sumter* came into being, and Semmes was appointed by the Secretary of the Navy to command it. He would say in a biblical comparison, "I had accepted a stone which had been rejected of the builders."[13]

Raphael Semmes went to New Orleans to supervise the ship's conversion and was shocked to see the city as he had never seen it. The great scene of bustling commerce was now gone, and the river was mostly vacant. "The levee in front of the city was no longer a great mart of commerce, piled with cotton bales, and supplies . . . it was not infrequent to see [the river] without so much as a skiff in motion."[14]

While the South still held out hopes for a peaceful arrangement with the North, Abraham Lincoln committed the North to war. On May 24, 1861, news reached New Orleans that the Federal army had crossed the Potomac and invaded Virginia, and on the 27th the USS *Brooklyn* made her appearance at the mouth of the Mississippi to begin the blockade. The following day, the frigates *Niagara* and *Minnesota* joined her; and on June 1 the USS *Powhatan* blocked the last exit from the river into the Gulf of Mexico.[15]

It took two months to convert *Sumter* into a warship, and the work did not have the luxury of being done in a shipyard. "Everything had to be improvised,"[16] wrote Semmes. That included producing plans, making drawings, manufacturing gun carriages, strengthening the decks for the additional weight of the heavy guns, and altering the ship's rigging for appropriate sails. *Sumter*'s battery was impressive: one 8-inch shell gun to be pivoted amid ship and four 32-pound howitzers, each weighing 1,300 pounds, arranged in broadside.[17]

Although this irregular warfare against enemy shipping had been a mainstay for the United States during both wars with Great Britain and was recognized by the world powers as a means of war between combatants, it seemed that the very threat of it now seriously worried Abraham Lincoln. Confederate President Jefferson Davis had issued a proclamation, following Virginia's secession, for the purpose of raising an irregular naval force. Interested parties were invited to apply for *letters of marque* and reprisal, and to be fitted out as privateers to wage war against the enemy's commerce. Several had already been outfitted and even taken a few small prizes off the mouth of the Mississippi. "Even this small demonstration seemed to surprise, as well as alarm the Northern government," wrote Commodore Semmes, "for President Lincoln now issued a proclamation declaring the molestation of Federal vessels, on the high seas, by Confederate cruisers, *piracy*."[18]

Lincoln's piracy claim, while an excellent example of propaganda, was absurd in the world of law, especially international law. There was nothing to support such a proclamation. The United States Supreme Court, in the *Prize Cases,* had left no doubt as to the distinction between *rebellion* and *belligerent;* and by all definitions the Confederates were *belligerents.* Historian James Randall wrote: "The word *rebellion* denotes an attempt to overthrow the government itself. *War* is a conflict conducted between recognized belligerents. Justice Robert Grier defined it as 'that state in which a nation prosecutes its rights by force.'" And in a later case, Supreme Court Justice John Harlan reaffirmed the status of the Confederates: "The Confederate States were belligerents in the sense attached to that word in the law of nations."[19]

Lincoln's ill-advised use of the word "pirate," however, would expose him to domestic and international ridicule, especially when one of his own ships was caught in a real act of piracy on the high seas. In November 1861, two Confederate diplomats, James Mason and John Slidell, escaped the Charleston blockade in an effort to travel to Europe to present their case for British and French support and diplomatic recognition. Their ship landed in Cuba, and they further embarked on a British Royal Mail ship, RMS *Trent.* But the USS *San Jacinto,* under the command of Captain Charles Wilkes, discovered that the two Confederate envoys were aboard *Trent,* and tracked the British vessel after it departed from Havana. His intention was to stop it, board it, and remove the two diplomats just as if they were contraband. On November 8, 1861, that is exactly what he did, ignoring the vigorous protests of the British captain, Mason, and Slidell.

While northern enthusiasts cheered the outrageous action, it did not take long for cooler heads to compare Wilkes's actions with the very actions that had led President James Madison to declare war in 1812. With complete disregard for the law, Britain had, at that time, routinely boarded ships and removed American seamen, and impressed them into British service. There was no justifying law to do so; it was, in Britain's mind, its God-given right. Now it was Great Britain's turn to be outraged. In London, the outrage was so vehement that some even called for war against Lincoln. The *London Standard* saw the *Trent* Affair as "one of a series of premeditated blows aimed at this country . . . to involve it in a war with the Northern States."[20]

Even though war against Britain would have been a disaster for Abraham Lincoln, he still wanted to keep the captured diplomats in jail. The British demanded their release, and as the clamor of public opinion turned against Lincoln, he reassessed his position. After a long delay, he reluctantly conceded, and his administration heaped the blame on Captain Wilkes and disavowed his actions; but there was never an apology.

If there was a plus in all of this for the Confederates, the *Trent* Affair solidified its status as a belligerent in the eyes of the world, even though it was forever denied by the Lincoln administration. "Mr. Lincoln, and his Secretary of State, [tried] to convince the European governments that the job which they had on their hands was a small affair; a mere family quarrel, of no great significance," said Commodore Semmes. "But the truth would not be suppressed, and when it became necessary to declare the Confederate ports in a state of blockade, and to send ships of war thither, to enforce the declaration, the sly little game . . . was all up."[21]

And in fact, it was President Lincoln's very actions that exposed the folly of his claims. Although he steadfastly refused to acknowledge the Confederacy as a belligerent throughout the war, his actions were the very essence of that very acknowledgment. He knew that internationally the Confederacy was a recognized belligerent. He had been forced to declare a blockade of southern ports, an action justified not against a *rebellion,* but only against a *belligerent;* he treated captured enemy soldiers as prisoners of war, not as insurgents, and captured seamen were not hanged as pirates. At war's end, southern combatants were afforded treatment as enemy soldiers and not as rebels attempting to overthrow the government—no one was declared a traitor as propaganda demanded, and Lincoln's own vetoing of the draconian Ironclad Oath revealed his inner thinking.

On June 18, after her two-month conversion, *Sumter,* with eight officers and ninety-two men, moved downriver to Forts Jackson and St. Philip and anchored there for three days while taking on ammunition and supplies. It would then wait for an opportunity to run the blockade. Semmes wondered

Raphael Semmes and officers of CSS *Sumter*, 1862. *Seated, left to right:* First Lieutenant William E. Evans; Commander Raphael Semmes, Commanding Officer; and First Assistant Engineer Miles J. Freeman. *Standing, left to right:* Surgeon Francis L. Galt; Lieutenant John M. Stribling; First Lieutenant John M. Kell, Executive Officer; Lieutenant Robert T. Chapman; and First Lieutenant Becket K. Howell (Marine Corps).
(Courtesy *Naval History* #42383)

why the powerful *Brooklyn* did not simply enter the river and, by her presence alone, effectively blockade instead of remaining outside to cover all the passes.[22]

For almost two weeks *Sumter* waited for that opportunity. Finally, on June 30, *Brooklyn* left her post at Pass à l'Outre, the easternmost pass to the Gulf, to chase after another vessel. Commodore Semmes fired his boilers and surged forward for his blockade run. But as the Confederate ship approached within six miles of the bar, *Brooklyn* spotted her maneuver and reversed course from its previous chase to be able to engage her. The race was on, each ship was equidistant from the tiny opening in the bar that led to the deep water of the Gulf.

Sumter rounded the mud flat at high speed and steamed to the east. *Brooklyn* fired her pivot gun but the shot splashed woefully short.[23] Both vessels were now under full sail and steam, and thick, black smoke billowed from their funnels. *Sumter*'s steam engine drove its brass propeller at sixty-five revolutions per minute, but the boilers suddenly "foamed," robbing the engine of its power, and sent a sheet of "spit" from the stack. The high revolutions had caused too great a demand for steam, and impurities in the water made it foam.

Semmes quickly analyzed the situation and realized he would have to press every tiny advantage of his smaller ship to outrun *Brooklyn*. He had one advantage: he could sail more efficiently into the wind. "I knew I could lay nearer the wind than she," he said. "I was able to brace my yards sharper, and had . . . larger fore-and-aft sails, stay-sails, tri-sails, and a very large spanker. I resolved at once to hold my wind, so closely, as to compel her to furl her sails, though this would carry me . . . a little nearer to her, for the next half hour."[24]

During this complicated maneuver a sudden rain squall developed and blinded both ships. When this rainy veil of obscurity finally lifted, *Brooklyn* was menacingly on top of *Sumter*. Semmes stared in awe of the enemy ship that closed upon him like a shark on its hapless prey. "I could not but admire the majesty of her appearance," he said, "with her broad flaring bows, and clean, and beautiful run, and her masts, and yards, as taunt and square, as those of an old-time sailing frigate. The Stars and Stripes of a large ensign flew . . . and we could see an apparently anxious crowd of officers on her quarter-deck, many of them with telescopes directed toward us."[25]

Following this hypnotizing moment, Semmes quickly recovered and ordered the public chest and papers readied to be thrown overboard. When all seemed lost and *Brooklyn* steadily closed the gap, *Sumter*'s engineer suddenly stumbled up from below and shouted that the foaming had stopped and was ready for full steam. With funnels streaming black, *Sumter* slowly inched forward.

"I began to perceive that I was 'eating' the *Brooklyn* 'out of the wind.' . . . I knew that as soon as she fell into my wake, she would be compelled to furl her sails."[26] And that is exactly what happened. "I have witnessed many beautiful sights at sea, but the most beautiful of them all was when the *Brooklyn* let fly all her sheets, and halliards, at once . . . and furled all her sails."[27] *Sumter* now surged ahead, and after four hours the chase was over. *Brooklyn* gave up. "We fired no gun of triumph in the face of the enemy," said Semmes, "my powder was too precious for that—but I sent the crew aloft, to man the rigging, and three such cheers were given."[28]

Semmes set his course for the south end of Cuba to begin his mission as instructed from the Secretary of the Navy: "to do the greatest injury to the enemy's commerce, in the shortest time."[29] His plan was to raid any enemy ships on the south side of Cuba, then put in at a convenient location for coal, sail over to Barbados, refuel once again, and then attack toward the Brazilian coast.[30]

On July 3, 1861, as he ran along the Cuban coast, the first cry of "Sail ho," was heard. The lookout reported two ships directly to the front, which *Sumter* overtook. The first proved to be a Spanish ship, and she was quickly released. But the second one sailed along without showing her colors, and Semmes had no doubt that it was American. "The American character impressed upon every plank and spar of the ship."[31]

A blank shot from *Sumter*'s gun soon had the other vessel heave to and raise the Stars and Stripes. *Sumter* approached, flying the British flag, but now struck those colors and unfurled the Confederate banner. A small boat was lowered, and a Confederate boarding party soon retrieved the American captain and his ship's papers. The prize was *The Golden Rocket* out of Maine, only three years old and worth $40,000,[32] seeking a sugar cargo from Cuba. The astonished captain said, "A clap of thunder in a cloudless sky could not have surprised me more, than the appearance of the Confederate flag in these seas."[33]

Semmes announced that the American crew would be taken off and well-treated aboard *Sumter,* as demanded by the rules of war, but the ship was doomed and would be destroyed as a victim of war. "At about ten o'clock at night, the order was given to apply the torch to her,"[34] wrote the Confederate commodore.

Sumter's crew hugged the rails of their vessel, hoping to see the men of their scuttling boat board *The Golden Rocket,* but the blackest of nights hid the other vessel, and only when the sound of the oars of the boarding boat ceased did they know it had reached its objective. Hidden from their view, the boarding officer applied the flame to the cabin, the hold, and the forecastle. Within minutes, fingers of fire groped up the rigging and masts as if feeling its target. A fiery explosion shattered the silence and *The Golden Rocket* detonated in spontaneous combustion. The boarding crew was still in the process of shoving off from the ship when the ignition took place, and Semmes recorded the spectacular destruction:

> The decks of this Maine-built ship were of pine, calked with old-fashioned oakum, and paid with pitch; the wood-work of the cabin was like so much tinder, having been seasoned by many voyages to the tropics, and the forecastle was stowed with paints, and oils.

> The indraught into the burning ship's holds, and cabins, added every moment new fury to the flames, and now they could be heard roaring like the fires of a hundred furnaces, in full blast.[35]

The flames that had raced up the masts and out along the yards ignited the tinder of the sails, cordage, and ropes. These hundreds of burning fragments floated skyward, propelled along by a gentle breeze. They drifted like fiery kites in spectacular patterns until gravity returned them to the black, glassy surface of the sea with an extinguishing hiss. Not so quiet was the collapse of the three giant, flaming masts that went over the side "like that of the sturdy oak of the forests when it falls by the stroke of the axeman."[36] *Sumter* retrieved its boarding boat and proceeded on its war patrol.

On board, the captain of *The Golden Rocket* "was invited to mess in the wardroom, and . . . the officers generously made him up a purse to supply his immediate necessities."[37] The crew was put into a separate mess with their own crew and had their own cook, and shared equally in the rations of *Sumter*'s crew.

Semmes entered into his log a refutation of the propaganda used by the North: "The '*pirate*' whom the enemy denounced with a pen dipped in gall . . . set the example . . . of treating prisoners of war, according to the laws of war."[38] *Sumter* proceeded on. "The next day was the 'glorious Fourth,'" wrote Semmes. "I had turned into my cot, late on the previous night, and was still sleeping soundly, when, at daylight, an officer came below to inform me, that there were two sails in sight from the mast-head."[39]

The adventures of the warship *Sumter* over the next two days were nothing short of astounding in the annals of war patrols at sea. She became a scourge to the U.S. merchant fleet, as well as to other ships transporting northern cargoes. In a continuing attack, *Sumter* collected a small armada.

On July 6, leading the way into the Cuban port of Cienfuegos, was Lieutenant Robert T. Chapman. He was to arrange for supplies and coal, and to present his respects to the governor along with a letter from Commodore Semmes saying: "Sir, I have the honour to inform your Excellency of my arrival at the Port of Cienfuegos with seven prizes of war. These vessels are the brigantines *Cuba, Machias, Ben Dunning, Albert Adams* and *Naiad;* and barques *West Wind* and *Louisa Kilham,* property of citizens of the United States, which States, as your Excellency is aware, are waging an unjust and aggressive war upon the Confederate States."[40]

He went on to say that he had arrived in this Spanish port with his prizes expecting that Confederate States' cruisers would receive the same friendly reception that had been extended to the northern states. He asked if Spain could resolve a dilemma by taking custody of his prize ships until their status could be "determined . . . in the Prize Courts of the Confederate States."[41] "I have the right to destroy the vessels, but not the cargoes, in case the latter should prove to be Spanish property—but how am I to destroy the former, and not the latter? It will be for your Excellency to consider and act upon these grave questions, touching alike the interests of both our Governments."[42]

Semmes realized that his well-worded letter was meaningless to this small port's governor, and even to the Captain-General of Havana. His real hope was that it would find its way to the Spanish queen, detailing the violation of international law practiced by the exclusion of Confederate prizes from neutral ports.

Months later, the Spanish crown gave its answer and excluded all prizes from her ports. As such, since the port of Cienfuegos had accepted the Confederate prizes in good faith as custodian, it seemed logical that they would be returned to Commodore Semmes with the order to remove them. But that was wishful thinking. As soon as Havana received the queen's order concerning prize ships, the Captain-General "deliberately handed back all my prizes to their original owners,"[43] wrote Semmes.

This unlawful action of a neutral power would now affect how Raphael Semmes would wage war. The tight Union blockade of Confederate ports was almost airtight. "*Every little nook and inlet was either in possession of the enemy, or had one or more ships watching it,*" lamented Semmes.[44] And now the supposed neutral foreign ports were also closed. "I had not so much as a single port open to me, into which I could send a prize to avoid the necessity of burning my prizes."

On July 7, *Sumter* left Cienfuegos to continue her southerly patrol. Semmes had been able to keep up with current events through captured northern journals and newspapers. The southern privateer *Savannah* had been captured, and her crew were hurried off to New York, confined in cells like convicted felons and afterward brought to trial and convicted of piracy under Mr. Lincoln's proclamation.

"My lieutenant brought back with him, from the shore . . . an American newspaper . . . stating that the *Savannah* prisoners had been released from close confinement and were to be treated as *prisoners of war.* I was gratified to learn that my Government had taken a proper stand on this question," wrote Semmes.[45]

Indeed, President Jefferson Davis had personally confronted Lincoln's irresponsible proclamations that had declared Confederate sailors to be pirates. "President Davis, as soon as he heard of the treatment to which the *Savannah* prisoners had been subjected, wrote a letter . . . to President Lincoln, threatening retaliation, if he dared execute his threat of treating them as pirates." Davis's letter to Lincoln read in part: "This Government will deal out to the prisoners held by it, the same treatment, and the same fate, as shall be experienced by those captured on the *Savannah*."[46] For the remainder of the war, while Lincoln would continue to deny the legitimacy of the Confederacy, he abandoned his proclamations concerning piracy, and no other Confederate sailor was ever again charged.

Sumter continued south, sailing along the coast of Venezuela, and crossed the wide mouth of the Amazon River. It stopped at Port of Spain in British Trinidad, and then French and Dutch Guiana before heading east to Spain.

Three more prizes fell to *Sumpter: Arcade, Vigilante,* and *Ebenezer Dodge.* All carried contraband goods and were condemned and torched. A severe storm damaged the Confederate warship, and on January 4, 1862, entered the port of Cadiz for repairs. But Semmes had no funds to purchase fuel, and *Sumter* was ordered out of port.[47]

On January 17 he sailed for Gibraltar, powered by wind and a scant supply of leftover coal. When he finally anchored at the famous rock, two Federal warships quickly blockaded him. But the blockade was the least of his problems; his boilers were shot. "*Sumter*'s boilers were very much out of condition, and we had hoped . . . that we should find here, machine and boiler shops sufficiently extensive to enable us to have a new set of boilers made."[48] But that was not the case, and Semmes was out of business. On April 11, 1862, Commodore Semmes was aboard the British mail steamer *Southampton,* bound for London. Proceeding west, he hoped to meet with Confederate officials but could not run the Union blockade to enter Confederate territory. Capture would have been a certainty. So, on June 13, 1862, he disembarked and landed at Nassau, a dismal but neutral site 180 miles off the tip of Florida.

"The island of New Providence, of which Nassau is the only town, is a barren limestone rock, producing only some coarse grass, a few stunted trees, a few pine-apples and oranges, and a great many sand-crabs,"[49] he wrote. "Commerce it had none, except such as might grow out of the sponge trade, and the shipment of green turtle and conch-shells."[50] But this bleak, rocky place was transformed by the war, and it became the perfect spot for a neutral off-shore meeting ground for anyone participating in any aspect of the war—including those seeking to profit from it. Confederates, Union officers, and merchants and businessmen from all over the world rubbed elbows as if they were traders on the floor of a mercantile exchange. It was a strange setting where friends and foes hammered out deals and trades. Yankee merchants, seeking to import valuable cargoes into the South, did a brisk business aiding Confederate blockade runners, and giving advice on slipping past the Union blockade. (Eighty years later, in 1942, the Hollywood industry captured a similar scene in the movie *Casablanca.*)

Nassau's harbor was jammed with ships, and its warehouses and wharves crammed with cargoes that the war demanded. The epicenter of this swirling international intrigue was the Victoria Hotel, and it was there that Raphael Semmes would sit, face-to-face, with Captain G. T. Sinclair of the Confederate Navy. Sinclair produced an envelope from his breast pocket containing Semmes's orders. "I was directed by this letter, to return to Europe, and assume command of the new ship, which was being built on the Mersey, to be called the *Alabama.*"[51]

Alabama was a new state-of-the art vessel, built by Laird and Sons of Birkenhead. She was a 1,040-ton steam and sail sloop, not an ironclad, and built for speed. Two 300-horsepower engines gave her a racehorse's advantage. On the main deck, she brandished twelve guns, with two large pivot guns located amidships. To man this powerful ship required 120 men.[52]

On July 31, 1862, in the dead of night, *Alabama* weighed anchor and slipped out of the confining British waters and on to the high seas. The word was leaked that she was on her way to Nassau, but that was to deceive Federal cruisers that lurked about to pounce.[53] Instead, *Alabama* sailed into the North Atlantic. On August 10, sometimes at a speed of 13 knots,[54] she sailed to the protected port on the island of Terceira in the Azores and was outfitted with her guns, which converted her into a ship of war.

On Sunday, August 24, 1862, there was a formal dedication at sea on *Alabama*'s exposed deck. Captain Semmes read aloud his commission as Commander of the Confederate States Steam Sloop *Alabama.* He addressed the crew; their main mission was "to cripple the commerce of the enemy."[55] But he hastened to add, "He was not to fight a fifty-gun ship, but when the opportunity offered to engage, on anything like equal terms, *Alabama* would . . . accept the combat."[56] There was a rousing cheer, and the boilers' fires were reduced to a minimum. With much fanfare, the massive sails snapped in the wind and billowed out to accelerate *Alabama* to cruising speed. Her destination? To the northeast, back toward Portugal and the shipping lanes—especially the whaling shipping lanes, where she expected to become a predator prowling among unsuspecting prey.

On September 5, she made her first kill: an American whaler, *Ocmulgee*, registered out of Massachusetts. It was at anchor, stripping a large sperm whale that was lashed to the ship's side and partially hoisted by the ship's tackle. In short order, *Alabama*'s boarding crew took the thirty-six crewmen as prisoners of war and confiscated all vital supplies and food. The boarding crew postponed torching the ship. That could wait until morning, lest the blazing fire alert other ships to *Alabama*'s presence.[57] The next day's torching provided a spectacular scene of an exploding, roaring inferno fueled by barrels of whale oil.

For the next eleven days *Alabama* roamed about freely and became the scourge of the U.S. whaling fleet. Seven other whaling ships fell victim. *Alabama* could run down anything on the high seas. These hapless vessels were overhauled by the Confederate raider as easily as a cheetah runs down a young antelope. The ensuing warning shots fired across their bows forced their masters to submit them as prizes. The prizes were *Starlight, Ocean Rover, Alert, Weather Gauge, Altamaha, Virginia,* and *Elijah Dunbar.* Their value alone was almost a quarter of a million dollars, and that excluded their cargos and materiel.[58]

In each case, Semmes removed the crews to his own ship, took what he needed, and placed the ships under the control of his own prize crews. When *Alabama* became overcrowded with prisoners, he would return to Flores Island and disembark them. He always removed the small boats from the captured ships and allowed his captives to take whatever they wanted from their vessels. When it was time to send them to Flores, they went ashore in their own boats with all that they could carry. He wrote: "I had two motives in thus landing my prisoners in their own boats. . . . It saved me the trouble of landing them myself; and, as the boats were valuable, and I permitted the prisoners to put in them as many provisions as they desired, and as much other plunder as they could pick up about the decks of their ships . . . the sale of their boats and cargoes to the islanders, gave them the means of subsistence, until they could communicate with their consul on the neighboring island of Fayal."[59] Semmes would later say of his prisoners released on Flores Island: "I had thrown ashore there, nearly as many Yankee sailors as there were original inhabitants."[60]

The whaling season in the Azores came to an end, and Raphael Semmes changed his hunting ground. He now moved to the northern waters off Newfoundland to throw himself astride the shipping lanes between there and the east coast of the United States. He entered the waters in search of what the seafaring men of his crew called the "Great American 'junk' Fleet."[61] This referred to a familiar scene that annually unfolded in the waters around China. During the autumn, when the great harvest had been gathered, thousands of "junks," ladened to their marks with grain, swarmed the seas and rivers to deliver to eager markets. In America it was now October 1862, and *Alabama* prowled the Atlantic Ocean searching for the American "junk" fleet.[62] "It was this grain, laden in Yankee ships, which it was my object now to strike at,"[63] declared Captain Semmes. That fleet had previously carried valuable southern cotton to eager European markets, and now carried the rich harvests of grain from "Western New York, Ohio, Indiana, Illinois, Michigan, Minnesota, and Iowa"[64] as a substitute for the cash of King Cotton.

Alabama and its eager crew did not have long to wait for the vanguard of the American junk fleet to arrive in Newfoundland waters. On October 8, 1862, two "fine, large ships, with a profusion of tapering spars and white canvas . . . came on, with all their sails set, from truck to rail."[65] *Alabama* had no need to take up its usual chase since these two merchantmen sailed directly toward her, without the slightest suspicion that the only Confederate sloop of war anywhere on the high seas was actually in front of them, waiting like the spider for the fly. "We did not hoist any colors," said Semmes, "until the vessels were nearly abreast of us, and only a few hundred yards distant, when, suddenly wheeling, we fired a gun, and hoisted the Confederate flag."[66]

There was mass confusion on the decks of the two American ships, with men running about trying to grasp the situation and officers panicked as to what to do. The answer was eventually obvious, and the captains submitted to the boarding. The two ships were both from New York with similar grain cargoes: *Brilliant* was bound for London, and *Emily Farnum* for Liverpool.

The cargo of *Emily Farnum* was immediately documented to be neutral property, and the ship was released. But *Brilliant*'s grain cargo was not neutral. It was U.S. cargo being sent to a supposed neutral nation in ships not allowed into neutral ports. It was subject to the rules of war, and Semmes condemned it and burned it. This northern trade between the United States and England violated all English neutrality laws and was proof of the old saying that all is fair in love and war. But much to the chagrin of the Lincoln administration, and especially to Secretary of State William Seward, *Alabama* exposed the chicanery of the American arrangement, including Seward's hypocrisy in insisting that those very neutrality laws that he brazenly violated should be rigidly enforced against the South.

The loss of *Brilliant,* worth $98,000, and her cargo of equal value seemed more than Seward could bear. He pushed for indemnity against the British, as if the loss of U.S. ships and cargoes was England's fault. Captain Semmes wrote in mocking humor, "The owners of the ship have since put in a claim . . . which Mr. Seward has pressed with so little effect hitherto against the British Government, for indemnity for the 'depredations of the *Alabama.*'"[67] This U.S. effort to hold Great Britain responsible for the loss of shipping and cargoes stemmed from the Lincoln administration's steadfast refusal to recognize that the war with the Confederate States was not a rebellion but a war between two belligerents. The rest of the world knew that was so.

The feigned indignation by the American administration for the depredations of *Alabama* all came to a laughable conclusion following the introduction of a lengthy American resolution in the British House of Commons. It accused Great Britain of dereliction, and detailed what the northern states claimed were "piratical acts"[68] in its building of vessels like *Alabama;* and that Britain had not "put a stop to the aggressions of the pirate."[69] In continuing to supply *Alabama* "with coal and ammunition, *Alabama* is enabled to pursue her piratical courses against American commerce."[70]

The American speaker continued to lament that this war on American commerce is "not rebuked by the British press generally; is not discouraged by the public sentiment," and not "stopped by the British government."[71] Finally, the speaker revealed the true nature of his complaint: it was all about losing the edge in shipping. The result of *Alabama*'s twenty-seven prizes—not counting *Sumter*'s eighteen prizes—against Federal merchant shipping had meant that the "carrying trade" was no longer done in American ships. That trade was now being carried in British vessels; and, as the American speaker protested, "this great peril, threatening to drive American commerce from the ocean, is of British origin."[72] It was all Britain's fault. There was no sympathy or cheers from the British assembled ministers to any of the American's remarks. Next, John Laird, the builder of *Alabama,* took to the speaker's podium for his say. It was a terse rebuttal. He had indeed built the Confederate ship, but its arming had occurred elsewhere. He mocked the American denunciation:

> If a ship without guns and without arms is a dangerous article, surely rifled guns and ammunition of all sorts are equally—(cheers)—and even more dangerous. (Cheers.) . . . I have obtained from the official custom-house re-

turns, some details of the sundries exported from the United Kingdom to the Northern States of America, from the 1st of May, 1861, to the 31st of December, 1862. There were—Muskets, 41,500—(hear! Hear!)—rifles, 341,000—(cheers)—gun-flints, 26,600—percussion-caps, 49,982,000—(cheers and laughter)—and swords, 2260 . . . the Northern States have been well supplied from this country (Hear! Hear!)[73]

After two months in the northern waters, *Alabama* again returned to the Caribbean seas in search of further prey. While boarding the vessel *Mina*, which proved of no consequence since it carried only English cargo and was released, Captain Semmes and his crew received a most satisfying report in the form of a recent northern newspaper. A New York letter had been published in the *Baltimore Sun* and revealed *Alabama*'s prowess: "The shipments of grain from this port during the past week have been almost entirely in foreign bottoms, the American flag being for the moment in disfavour in consequence of the raid of the rebel steamer *Alabama!*"[74]

⚜ ⚜ ⚜

On January 5, 1863, *Alabama* abandoned her usual mission of waging war against the Federal merchant fleet and set her eyes on offensive action against the armed enemy. Upon initially taking command of the ship, Captain Semmes had promised his crew that should the opportunity arise to engage the enemy in offensive action, they would not be denied.

Alabama had been in the sheltered waters of the island of Barranquilla in the southern Caribbean, just off the northern coast of Venezuela, taking on coal, when captured northern newspapers revealed that a great seaborne northern invasion was being planned. It was to be under the command of General Nathaniel Banks, who had suffered major humiliating defeats at the hands of General Stonewall Jackson in the Shenandoah Valley. This military outfitting was taking place in Boston and New York and involved an invasion force of no fewer than thirty thousand men. The expedition was to rendezvous off Galveston, Texas, currently in northern hands.[75] Captain Semmes wrote: "Much disorder and confusion would necessarily attend the landing of so many troops, encumbered by horses, artillery, baggage-wagons and stores. My design was to surprise this fleet by a night-attack, and if possible, destroy it, or at least greatly cripple it. The Northern press, in accordance with its usual habit of blabbing everything, had informed me of the probable time of the sailing of the expedition, and I designed so to time my own movements as to arrive simultaneously . . . or. At least a day or two afterward."[76] Those same northern newspapers reported to their readers that the invasion would not be troubled since there were no Confederate cruisers in the Gulf and "*Alabama was well on her way to the coast of Brazil and the East Indies.*"[77] January 10, 1863, was the expected date for Banks's expedition to arrive at Galveston.

On January 11, Semmes and *Alabama* were thirty miles off the Galveston coast. His lookouts kept a sharp eye for the anticipated immense northern fleet anchored near the Galveston lighthouse. But the lookouts could only report the presence of five Federal warships—no transports. Suddenly one of the warships fired a shell that arched and burst over the city. Semmes and his fellow officers were puzzled but quickly figured it out. "The enemy would not be firing into his own people, so we must have recaptured Galveston, since our last advices," said Semmes, and the officers agreed. Major General John B. Magruder with the able assistance of Captain Leon Smith and his river steamers, bristling with sharpshooters, had indeed overwhelmed and driven to sea the enemy's entire invasion fleet and recaptured Galveston.[78]

Semmes was disappointed with the loss of the chance for ship-to-ship combat, but before he had time to reflect on this setback one of the Federal warships set on a course to engage *Alabama.* Semmes immediately ran before the enemy ship, hoping to lure it away from the rest of the fleet and then turn to attack. The pursuit went on even as night fell. When the two adversaries were twenty miles from the other ships, it was time for action. Semmes wheeled to meet his enemy, and as *Alabama* approached within a hundred yards of the other, both stopped their engines. In the silence of the night, the enemy captain hailed, "What ship is that?"[79]

Semmes replied, "This is her Britannic Majesty's steamer *Petrel,*"[80] and immediately demanded to know who the other ship was, but the answer was unheard, just that it was a United States ship. That was all that Semmes wanted to hear. The other captain said that he would send a boat over to verify identity.

Alabama's crew heard the order to lower a boat, and then heard the creaking of the tackles as the boat was lowered into the water. Semmes turned to Lieutenant McIntosh Kell, his executive officer standing at his side, and asked, "I suppose you are all ready for action?" The lieutenant answered that the men were eager and awaited Semmes's order.

"Tell the enemy who we are" ordered Semmes, "for we must not strike him in disguise; and when you have done so, give him the broadside!"[81]

Kell shouted out, across the water in his booming voice, "This is the Confederate States steamer *Alabama!*" And quickly turned to his gunners and shouted, "Fire!"[82]

The broadside was deafening, and the balls slammed into the Federal ship, which immediately returned fire, as its gunners too had expected a fight. The two ships maneuvered, and both turned to a course moving off in the same direction. *Alabama* fired her starboard guns, and the enemy fired its port guns.

The high-speed, running gunfight was fought at close quarters—sometimes at twenty-five yards, never much more than that. Each ship fired fifty broadside shots, and since the distance was so close, sailors with rifles and pistols fired to rake the opposite decks. The naval duel was furious but lasted only thirteen minutes before the enemy's ship, discovered to be the USS *Hatteras,* signaled surrender.[83] An officer on board *Alabama* recorded his observations of the fight:

> A tremendous volley from our whole broadside [was] given to him, every shell striking his side; the shot striking being distinctly heard on board our vessel, and thus [we] found that she was iron.
>
> Pistols and rifles were continually pouring [fire] from our quarter-deck. . . . The distance during the hottest of the fight not being more than forty yards! It was a grand, though fearful sight, to see the guns belching forth in the darkness of the night, sheets of living flame, the deadly missiles striking the enemy with a force that we could *feel.*
>
> Then, when the shells struck her sides, especially the percussion ones, her whole side was lit up, and showing rents of five or six feet in length. One shot had just struck our smoke-stack . . . when the enemy ceased his firing. The order was given to "Cease firing." This was at 6:52. A tremendous cheering commenced.[84]

Alabama's gunners had delivered two hammer-stroke shots that decisively ended the fray. One smashed through the side, midships, and lodged in a hold, setting it on fire, and the second exploded in an adjoining compartment after first passing through *Hatteras*'s sick bay. A second fire soon roared out of control. Both shells had entered the doomed ship at the waterline, tearing away great sheets of iron, allowing the sea to pour in.[85] *Alabama* dispatched its boats to rescue the *Hatteras*'s crew and had just retrieved the last of the men when, two minutes later, she went down by the bow. *Alabama* scurried off before the arrival of the other Federal warships, and by the time they steamed onto the scene the sea was empty.

Ten days later, on January 20, *Alabama* arrived at Port Royal, Jamaica, off-loaded her prisoners, and made a few repairs before again setting out to sea to wreck the Union supply line. On January 26, she took her twenty-eighth prize, the USS *Golden Rule* out of New York, and Semmes burned it and sent it to the bottom. That sinking began her South Atlantic Expeditionary Raid, which would last for six months and become her most successful patrol. *Alabama* overwhelmed a staggering twenty-nine ships while prowling along the coast of Brazil, taking the last one on July 6.

The entire world became *Alabama*'s battleground. Her South African Expedition and Indian Ocean Expedition claimed three more prizes between August and November of 1863. Finally, she cruised to the South Pacific to run down and destroy seven more prizes. By December she had turned around to steam to France for much needed repairs.

But now the sea seemed empty, stark proof that American shipping had been vanquished. On March 25, 1864, *Alabama* rounded the Cape of Good Hope, bound for France. It had been three years since Captain Semmes had left his family to become the essence of the entire Confederate Navy on the high seas. On April 23 and April 27, *Alabama* overhauled, captured, and burned her final two prizes: the merchant ships *Rockingham* and *Tycoon*.[86] In those three years patrolling the oceans of the world, Raphael Semmes had been the scourge of the northern merchant fleet: sixty-three prizes with *Alabama* and eighteen with *Sumter*.[87]

On June 11, 1864, *Alabama* dropped anchor in the French port of Cherbourg. Captain Semmes wrote, "In my two ships, the *Sumter* and *Alabama*, I had had . . . five hundred men under my command. The ships were small and crowded. As many as two thousand prisoners were confined, for longer or shorter periods, and yet, out of the total of twenty-five hundred, I had not lost a single man to disease."[88]

Three days later, USS *Kearsarge* steamed into Cherbourg Harbor, dropped off a landing boat, and steamed out again to take up a threatening position off the breakwater. "It was generally understood among my crew that I intended to engage her," said Semmes. "I addressed a note to our agent, requesting him to inform Captain Winslow [of *Kearsarge*] . . . that if he would wait until we had received some coal aboard . . . I would come out and give him battle."[89]

"The *Kearsarge* mounted seven guns: two eleven-inch Dahlgrens, four

32-pounders, and a rifled 28-pounder. The *Alabama* mounted eight guns: one 8-inch; one rifled 100-pounder; and six 32-pounders."[90] Although *Alabama* had one more gun than her adversary, *Kearsarge* had the advantage of a heavier weight of metal in a broadside. But *Kearsage* had another huge advantage unbeknown to Captain Semmes: she was ironclad. Captain Winslow "had hung all his spare anchor cable over the midship section of the *Kearsarge*, on either side" and covered it over to camouflage it. *Kearsage* was "as effectually protected as if she had been armored with the best of iron plates."[91]

In the olden days of knights and chivalry, this deception would have been an egregious violation of honor. It would have been, as Captain Semmes described, "as if two men were to go out to fight a duel, and one of them, unknown to the other, were to put a shirt of mail under his outer garment."[92] There was no chivalry to be practiced here, but this imminent showdown had all the trappings of an honor duel. *Kearsarge* had entered Cherbourg Harbor, violating French rules of neutrality, showed herself briefly to Semmes, and quickly departed to await *Alabama* at the breakwater. The challenging gauntlet had been dropped, and *Alabama* picked it up. She would shortly be out for the showdown at sea.

The French townspeople were giddy with anticipation of a glorious naval duel for all to witness. They greatly admired the daring exploits and adventures of *Alabama* and her legendary captain and crew. On Sunday morning, June 19, "all Cherbourg was on the heights above the town and along the bastions and the mole. Never did a knightly tournament boast a more eager multitude of spectators."[93]

Semmes steered *Alabama* out of the harbor's western exit, accompanied by the French ironclad *La Couronne* and a British yacht, *Deerfield*. She made straight for *Kearsarge* waiting in the distance. It took forty-five minutes to close to battle range, and Semmes summoned his crew to gather round him as he mounted a gun carriage: "You have, at length, another opportunity of meeting the enemy—the first that has been presented to you since you sank the *Hatteras* . . . you have destroyed, and driven for protection under neutral flags, one-half of the enemy's commerce, which, at the beginning of the war, covered every sea. . . . The name of your ship has become a household word wherever civilization extends. . . . Remember that you are in the English Channel, the theatre of so much naval glory . . . and that the eyes of all Europe are at this moment upon you. . . . Go to your quarters."[94]

The final fight was almost anticlimactic. *Kearsarge* engaged *Alabama* at long range, and Semmes's ship was not the greyhound she had been. Attempts to close the range were easily parried by *Kearsarge*'s superior maneuverability. Still, *Alabama*'s gunners were every bit as proficient as they had ever been, and in the first thirty minutes pounded the Union vessel with shot and shell. *Alabama*'s ordinance struck the sides of *Kearsarge* twenty-eight times.[95] But the side armor now showed its full advantage. Shot and shell had bounced harmlessly off and into the sea. But one shell did find a most vulnerable spot on the Union ship. A rifled percussion shell smashed into *Kearsarge*'s stern post where there was no armor.[96] It lodged into the hull but failed to detonate. Had it exploded, *Kearsarge* would have been doomed. The explosion would have ripped off most of the ship's stern.

Kearsarge continued to pound *Alabama* from long range for an hour and ten minutes, with devastating effect. *Alabama* shuddered under the blows. "We fought her until she would no longer swim," said Semmes, "and then we gave her to the waves."[97]

> The ship settled by the stern, and as the taffarel [the upper rail on the stern] was about to be submerged, Kell and myself threw ourselves into the sea, and swam out far enough from the sinking ship to avoid being drawn down into the vortex of waters.
>
> We then turned to get a last look at her and see her go down. Just before she disappeared, her main-topmast, which had been wounded, went by the board; and, like a living thing in agony, she threw her bow high out of the water, and then descended rapidly, stern foremost, to her last resting-place.[98]

The British yacht *Deerhound* had circled the waters as a close-up observer of this most-anticipated naval battle. Now it steamed into action to help rescue some of the sailors who struggled for their lives. *Deerhound*'s vigilance allowed the rescue of Captain Semmes along with thirteen of his officers and twenty-seven men who would have been lost.[99]

⚜ ⚜ ⚜

Nine months later the Civil War was over. Two years after that, Superintendent David Boyd announced that he had hired Admiral Raphael Semmes as part of the new Seminary faculty to teach Moral Philosophy. He became one of the legendary contributors to the lore of the Ole War Skule.

In January 1910, as the school celebrated the fiftieth anniversary of its founding, there was a three-day festival of celebrations. Included in those celebrations was the presentation of two portraits. One was of former professor Admiral Raphael Semmes, the charismatic captain of CSS *Alabama,* and the second was of a former cadet, Confederate John David Workman. As a cadet in 1860, Workman had famously been in charge of the "awkward squad," and he was later killed in the Wilderness in 1864. The entire event "was a celebration of the sacrifices of LSU's founders."[100]

CHAPTER 7

John Archer Lejeune

A CADET AT THE NEW WAR SKULE

Although Superintendent Boyd had resumed the running of the Seminary at war's end, it continued almost as if nothing had happened; but it was four years before the cadets would be allowed to resume wearing uniforms or drill while bearing arms. That prohibition was the result of the orders of General Philip Sheridan, the Commander of the 5th Military District.[1]

Still, Boyd insisted on the regimen of strict military discipline in his effort to turn out gentlemen. And it paid off. One young man who entered the campus on an errand was extremely impressed with the students' civility: "The Seminary is very different from any other school. Why, at other schools, when a stranger goes there the boys generally see how many pranks and tricks they can play on him, but over at the Seminary everybody is as polite as they can be."[2]

The Seminary's remote location in the pine woods of Rapides Parish offered little opportunity for mischief or excitement. "Three hours a day were set apart for exercises in the open air. The grounds were extensive, and the students were not permitted to leave them except on Saturdays when those who were in good standing were allowed to hunt and fish in the nearby forests and streams."[3] And then there was the Hop! It was most popular and was touted in advertisements to attend the Seminary. "To give the cadet the benefit of the society of ladies there is a Cadet Hop at the Seminary on the last Friday evening of each month. The 'Hop' has worked like a charm during the past year in polishing the manners and refining the feelings of the cadets and in maintaining good order and discipline."[4]

But all of this rebuilding and new hope for the future of the Seminary came to a flaming end on the night of October 15, 1869, when the old school burned to the ground. The administration was forced to move to a new location in Baton Rouge, supposedly as a temporary solution, but it proved to be anything but temporary. After eighteen years, Baton Rouge became the institution's

LSU cadet, 1885
(Howard Lytle Collection, LSU)

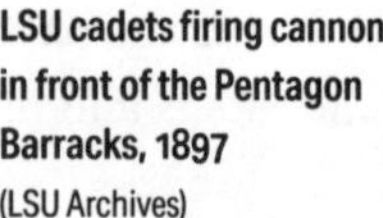

LSU cadets firing cannon in front of the Pentagon Barracks, 1897
(LSU Archives)

permanent home. "On March 16, 1870, the name of the school was changed to Louisiana State University."[5]

LSU had barely survived during the difficult time of Reconstruction (1865–1877). Historian Alcée Fortier labeled it a "hideous nightmare."[6] Finally, with the election of Rutherford B. Hayes as President of the United States, those long years of Reconstruction ended on April 24, 1877. And with that end came the end of the federal military occupation of Louisiana.

Within the framework of this new enthusiasm, LSU published its academic criteria; and by any standard, it was daunting. For the degree of Bachelor of Arts, the following courses were required:

> English, French, German, Latin, and Greek; chemistry, physics, zoology, botany, mineralogy, and geology; mathematics and astronomy; mental and moral philosophy, history, and English literature; military science and tactics.
>
> For the degree of Bachelor of Science: English, French and German; physics, general chemistry, agricultural chemistry or analytical and practical mechanics, drawing, geology, botany, zoology, anatomy, physiology, and mineralogy; mathematics and astronomy; mental and moral philosophy, history, and English literature; military science and tactics.[7]

In 1881, a most unlikely fourteen-year-old student, John Archer Lejeune, ventured into this crucible of academic challenge. He had been born in 1867, in Raccourci, Pointe Coupée Parish, Louisiana, and into a world best described as grinding poverty—a world of need and want following the Civil War. His recollections of the first five years of his life were grim. "Clothing was limited to actual necessities," he said, "luxuries did not exist so far as we were concerned, and money was unknown."[8] When he was seven, his mother began to teach him since there were no schools in his little corner of the world. In the spring, while most of the country looked forward to throwing off the cold and gray of winter, in Raccourci the people lived behind protective levees that separated them from the avalanche of the Mississippi flood waters racing by on the other side. In their isolation, they were all on their own. There was no government to help them should the rampaging river roar over the protecting mud wall. In fact, there was no government to serve the people in any way.

"No Louisianian whose memory extends to all or of a part of the ten-year period preceding 1877, can fail to have vivid recollection of the corruption, and the failure to function, of the state and local governments during that time," wrote Lejeune. "There was no such thing as justice, and no serious effort was made to protect the people or their property from the criminal element of the population. Robbery and stealing were prevalent. Cattle, hogs, and sheep were killed by night marauders."[9]

Lejeune's home schooling ended when his mother sent him to a boarding school run by an uncle. It was 1880 and young John had just turned thirteen. "I suffered the extreme pangs of homesickness," he said, "which were accentuated by my intense bashfulness. To be with strangers was torture."[10] John progressed steadily in his studies, but at the end of his second year his family could see that he was completely isolated. He had no contact with other boys his own age, and in fact had little contact with anyone other than his family. His parents saw this was not good for his social development and decided to send him to LSU in Baton Rouge. They investigated the two-year preparatory curriculum, and John's grades qualified him to enter the advanced class.

"Owing to the lack of schools in Louisiana during the Civil War and the twelve-year period following, it was necessary for nearly all the boys entering the University to receive a part of their preparatory training there. . . . When I entered the University, Colonel William P. Johnston, the son of [the late] General Albert Sidney Johnston, was the President."[11] Lieutenant General Albert Sidney Johnston had been killed at Shiloh in 1862.

John Lejeune was an instant fit at the Ole War Skule. "We were instructed daily in close order drill up to and including the School of the Battalion, and in the ceremonies of Dress Parade and Guard Mounting, following . . . the rules and regulations prescribed in Upton's Military Tactics. The company officers and non-commissioned officers were required to be experts in drill and devoted considerable time to the study of Upton's incomparable manual."[12] *Upton's incomparable manual!* For a young student to be so moved by the study of these military writings was most impressive. That LSU was teaching them to the likes of John Lejeune and his fourteen-year-old cadet classmates was a tribute to the excellence and dedication of the school.

Indeed, Colonel Emory Upton had been to the Union Army every bit the revolutionary tactician as had been the great military genius Carl von Clausewitz to the art of war. Although Upton had only been twenty-one years old on the battlefield at Manassas, his critical eye had recognized the flawed tactics of the Union Army. The following year, he had watched General Burnside's Army get cut to ribbons in its relentless uphill assaults against Lee's entrenched army at Fredericksburg.

By May 1864 Upton was a colonel leading an eight-thousand-man brigade in General John Sedgwick's Sixth Corps. After newly appointed General

Ulysses S. Grant's Army had crossed the Rapidan River on May 4, Upton's brigade approached Spotsylvania Court House, deep in the Wilderness. Lee's Confederates were heavily entrenched. By this time the young colonel had developed an enormous disdain for much of the Union Army's officer corps, especially those senior generals who seemed oblivious to the folly of squandering soldiers' lives. He would bitterly write to his sister: "My dear Sister: . . . I am disgusted with the generalship displayed. Our men have, in many instances, been foolishly and wantonly sacrificed. Assault after assault has been ordered upon the enemy's entrenchments, when they [the generals] knew nothing about the strength or position of the enemy."[13]

Upton had seen it all and had witnessed Lee's Army succumb to these same life-squandering tactics at Gettysburg, where his long, gray line was shattered after a devastating frontal assault. Upton was astounded that such antiquated tactical thinking should still prevail over the mentality of present-day Union Army generals. He further wrote, "Thousands of lives might have been spared by the exercise of a little skill. . . . So long as I see such incompetency, there is no grade in the army to which I do not aspire."[14] The day following his first letter to his sister, he hurriedly wrote again, as if he had forgotten to mention all that had irked him about incompetent generals. "Some of our corps commanders are not fit to be corporals. Lazy and indolent, they will not even ride along their lines; yet, without hesitancy, they will order us to attack the enemy, no matter what their position or numbers. Twenty thousand of our killed and wounded should to-day be in our ranks."[15]

Nor had Upton been silent among his superiors about his observations, and they were well aware of his feelings. He was confident that he could do better. As General Grant's new offensive unfolded, several futile attempts had already been made to probe Lee's defensive line, but to no avail. Union attacking forces had been repulsed and thrown back in confusion. An air of defeat prevailed.

But Colonel Upton had formulated his own tactical plan. He was confident that, if called upon, he could hurl his brigade against those same enemy positions and break the line. On May 10 his opportunity came. His eight regiments were beefed up to twelve, and he would command this entire force. "On the afternoon of the 10th, an assault was determined upon," he wrote, "and a column of twelve regiments was organized. . . . The point of attack [and] . . . the entrenchments were of a formidable character; with abatis [felled trees] in front and surmounted by heavy logs; underneath the logs were loopholes for musketry. About a hundred yards to the rear was another line of works, partly completed, and occupied by a second line of battle."[16]

From his brigade's jumping off position, there was a little road that ran up into the tangle of pine woods that finally gave way to an open field. It offered a view of this elevated enemy position. From the inner edge of the woods, his attack would be uphill, and would have to carry for two hundred yards—through the abatis—up to the enemy trench line. At the trench line, the heavily armed Confederates shouldered weapons that bristled menacingly through the loopholes. Upton took his regimental commanders forward to observe.

His plan was not to attack in waves of sweeping, shoulder to shoulder offensive lines, as had been the army's standard practice. He planned to attack on a narrow front, in a column of twelve regiments, like a massive spear to smash through the defensive line. Three regiments, massed abreast, tightly packed to form the first attacking line, and followed by three more identically configured.

This massive attacking column would move forward, silently, up the small road through the woods, with one regiment from each line on the left side, and two on the right. When the first line reached the edge of the open field, the men were to quietly lie down.[17] When the order to attack was given, they would rise up, again quietly, and then race forward with loud yells. Nor were the attacking men to stop and fire their weapons along the way That would only slow down the impetus of the attack. "No commands were given in getting into position," said Upton. "The pieces [rifles] of the first line were loaded and capped; those of the others were loaded but not capped; bayonets were fixed."[18]

If the charge up to the enemy position proved successful, the first assault line was to branch off left and right. The second line of regiments, following the first, was to halt at the trenches and only then insert their firing caps. The third line was to lie down behind the second and await further orders. The fourth and final line was to only move to the edge of the woods and be prepared to react to the results of the previous charges.[19] "All the officers were directed to repeat the command '*Forward*' constantly from the commencement of the charge till the works were carried,"[20] wrote Upton.

At 6:10 p.m. the columns rose up and the attack began "with a wild cheer." The Union artillery had been peppering the Confederate works for the previous ten minutes, and now lifted for the assault. Upton wrote: "Through a terrible front and flank fire the column advanced, quickly gaining the parapet. Here occurred a deadly hand-to-hand conflict. The enemy, sitting in their pits, with pieces upright, loaded, and with bayonets fixed, ready to impale the first who should leap over, absolutely refused to yield the ground. The first of our men who tried to surmount the works, fell, pierced through the head with musket-balls."[21]

As Upton's men clamored up upon the trench line, some held their weapons at arm's length in front of their bodies and fired down into the Confederate trench, while others hurled their bayonetted rifles, as if they were spears. "The struggle lasted but a few seconds," said Upton, "Numbers prevailed, and, like a resistless wave, the column poured over the works. . . . Pressing forward, and expanding to the right and left; the second line of entrenchments, its line of battle, and the battery, fell into our hands. The column of assault had accomplished its task: the enemy's lines were completely broken."[22]

However, total victory eluded Upton's grasp when a follow-up attack by another division did not materialize, and the Confederates quickly mended their lines and drove the Union forces from their hard-won gains. "Our officers and men accomplished all that could be expected of brave men," he said. "They went forward with perfect confidence, fought with unflinching courage, and retired only upon the receipt of a written order after having expended the ammunition of their dead and wounded comrades."[23]

For his actions, Major General George Meade promoted Emory Upton to the rank of brigadier general.[24] "In 1867 Upton's tactics were officially adopted by the U.S. Army and then printed in a book titled 'A New System of Infantry Tactics.'"[25]

⚜ ⚜ ⚜

This new system of tactics was heady stuff for the LSU cadets, especially John Lejeune, who was enthralled with anything and everything bearing Colonel Emory Upton's name. In the fall of 1882, at the end of his year-long preparatory curriculum, he shifted into LSU's regular B.A. program and spent the next two years learning all that he could.

"For a number of years, my father had frequently discussed the possibility of my going to the United States Military Academy at West Point," he said, "So it came about that when I acquired the age, [he] took steps to secure me an appointment."[26] But there were no appointments to be had at the Military Academy, and the young cadet was extremely disappointed. "My chief reason for my willingness to go to West Point was to lift the heavy financial burden of my education from the shoulders of my parents."[27]

Before the disappointment had time to set in, Lejeune's congressman notified him that while West Point was not available, an appointment to the Naval

Academy at Annapolis was. "It was early in 1884 [and] I promptly resigned from the University . . . near the end of my sophomore year, and went home, where I spent a month in self-preparation for the entrance examination."[28]

The new navy plebes slept and lived on an old sailing ship, USS *Santee.* It was fast in the mud next to the dock and served as a dormitory. All the 4th Class Midshipmen reported in wearing whatever they had. Lejeune had no money and reported in wearing his gray LSU cadet's uniform—after first removing his cadet lieutenant's shoulder straps. The gray in a sea of blue made him most conspicuous to the upperclassmen, who spared no opportunity to haze the new arrivals.

The curriculum was brutal, and only by a strict regimen of study and discipline could one hope to advance to the next year. "There was no way for one to camouflage his ignorance," said Lejeune. "Each of us had to satisfactorily pass the monthly and semi-annual examinations or be 'bilged.'"[29] That was the slang vernacular for thrown out. "The slaughter was great, and only thirty-five of the original ninety-five of my class (1888) received diplomas at the end of the four years' course."[30]

Although he had his diploma in hand, a postgraduate course was required for commissioning and to determine a final class ranking, which would greatly affect his future career. This postgraduate course, still as a naval cadet, was a two-year cruise. Lejeune spent it in the waters of the vast Pacific Ocean, around Hawaii and Samoa. His voyage included the unexpected terror of surviving a hurricane on his capsized ship and watching many of his crew drown in the boiling waters of the storm.[31]

In the spring of 1890, the naval cadets who had begun in 1884 finally returned to Annapolis for final examinations and class rankings. Lejeune's marks were very high, and he improved his 1888 graduate ranking in the class from twelfth place to sixth, raising his confidence about being assigned to his top choice of service. He had listed his top two choices, first, the Marine Corps, and second, Navy Line. He had eliminated the third option: the Engineering Corps. It held no allure for him—"I had no bent for mechanical engineering."[32] But it was precisely to the Engineering Corps that he was assigned. Lejeune was furious. He called it "the sting of injustice."[33] He had achieved his high class ranking over six years of hard work, and felt that he had earned the right to be assigned to the Corps. "The wishes of my juniors in class standing had . . . been granted."[34]

He began a campaign of knocking on doors up the chain of command. But at each level the answer was the same: nothing could be done. He threatened to resign, but that threat impressed no one. It seems that his unwanted assignment was the result of the Commodore, who was the Engineer-in-Chief of the Navy, and who was tired of routinely receiving cadets from the bottom of the class. He decided to hand-pick one from the top, and that turned out to be John A. Lejeune.[35]

As a last resort, Lejeune appealed to the superintendent. The explanation he received for his bad luck was far from satisfactory. He was the only one in the top who had not applied for a commission in Navy Line; the Engineering Corps needed graduates of ability; and the Board considered Lejeune to be too high in the class to be assigned to the Marine Corps. End of story.

When all seemed hopeless, fate suddenly intervened in the form of another cadet's poor eyesight. That flaw made him physically unfit for commissioning into the Navy Line, but the engineers did not have such a restriction, and he took Lejeune's spot in the Engineering Corps.

On June 25, 1890, a jubilant Second Lieutenant John Archer Lejeune was commissioned in the United States Marine Corps and joined four other commissioned Marines from the starting class of 1884. It was the beginning of a long career. He served in typical assignments ashore and at sea that included orders to Panama, the Philippines, Cuba, and Mexico. In 1910 he graduated from the Army War College. In 1914 he was promoted to the rank of colonel and served in the occupation of Veracruz. But throughout his varied career he was greatly disappointed that, although there were fireworks wherever he went, he seemed to always just miss the combat. He was always in the area where conflicts happened, but never quite made the battle line.

But 1914 brought to the civilized world a cataclysm the likes of which it had never seen. It began on June 28 with the assassination of Austria's Archduke Franz Ferdinand and his wife, and because of entangling national alliances signed among the European powers, those powers suddenly found themselves at war with each other in what was to be a fight to the death. British writer H. G. Wells famously labeled the war's savagery as "The War that will end all Wars."[36] General Billy Mitchell, the great American aviation hero, would later describe it as a "slaughterhouse."[37]

For the first two and a half years of the conflict, the United States adopted a policy of strict neutrality and stayed off the battlefields, even while it provided material support to its allied combatants. Its citizens kept up with the war through highly censored newspapers that attempted to put on a good face for national confidence. Mostly the public went merrily about their business, thankful to have emigrated away from those countries that always were at war.

General John A. Lejeune, USMC
(LSU Archives)

But the newspapers purposely hid the ghastliness of the battlefields, especially at places like the Somme, Passchendaele, the First Battle of the Marne, and Verdun. Those slaughterhouses inflicted upon the youth of Europe more than three and a half million casualties.[38]

Although the weapons of war had drastically changed, and now featured powerful artillery and quick-fire machine guns, the tactics of the infantry attack had not. Those tactics still called for frontal assaults across a shattered No-Man's land and into the teeth of these powerful weapons, pitting deadly steel against flesh and bone. It was never a contest. On the first day alone at the Battle of the Somme, July 1, 1916, the British Army sustained 57,000 casualties. By the end of that ineffective five-month offensive the number had swollen to 420,000.[39] On that dreadful first day, the ground attack began after a seven-day artillery bombardment in which British artillery poured 1.5 million shells into the opposing German line. There was confidence among the British that nothing could survive that bombardment; and it would simply be a rush over to the German trenches behind a creeping barrage to occupy their position. That would prove to be a deadly overestimation. British lieutenant Alfred Bundy, who made that "simple" rush to the German trenches, described the horror of the attack. "Went over top . . . after an interminable period of terrible apprehension. Our artillery seemed to increase in intensity and the German guns opened up on No Man's Land. The din was deafening, the fumes choking and visibility limited owing to the dust and clouds caused by exploding shells. It was a veritable inferno. I was momentarily expecting to be blown to pieces. My platoon continued to advance in good order without many casualties and had reached halfway to the Boche [German] line."[40]

But the creeping artillery barrage that was supposed to keep the Germans pinned down had outdistanced Lieutenant Bundy and his advancing men; and the barrage had raced far ahead of them, and beyond the German line. The German soldiers had plenty of time to let the smoke clear and to then rise in their trench line, sight-in on the advancing British line, and cut it to pieces. "Suddenly . . . an appalling rifle and machine gun fire opened against us and my men commenced to fall. I shouted 'down' but most of those that were still not hit had already taken cover. I dropped in a shell hole and occasionally attempted to move to my right and left, but bullets were forming an impenetrable barrage, and exposure of the head meant certain death. None of our men was visible but, in all directions, came pitiful groans and cries of pain."[41]

To most Americans, this was none of their business. President Woodrow Wilson had declared that the policy of the United States was to be strict neutrality. When German submarines began a campaign of unrestricted warfare, he declared that he would hold them "to a strict accountability."[42] Germany then backed off and called off the submarine offensive. Wilson even offered to act as a mediator for both sides, but the combatants wanted none of his interference.

After two years of war, Wilson faced reelection in the United States and ran on a campaign that declared, "He has kept us out of war,"[43] and won a second term. But Germany saw that the defeat of Britain could only be brought about by strangling her supply line, which included U.S. shipping, and its submarine force was the weapon that could do the strangling. In January 1917, Germany again declared a policy of unrestricted submarine warfare, and by March four American ships had gone down from German torpedoes.

Almost simultaneous with this new German offensive, British intelligence intercepted, and then revealed to the United States, a shocking German telegram—to be known as the Zimmermann telegram. In a *Most Secret* communication, Germany sought an alliance with Mexico whereby Mexico would enter the war against America if the United States were to declare war against Germany. As a reward for this alliance, Germany would ensure that Mexico would be restored part of the lands lost to the United States in 1848 after the Mexican War. There was even a suggestion to draw Japan into the war on Germany's side.[44] The shocking events of declared submarine warfare and the Zimmermann telegram forced an outraged President Woodrow Wilson to ask Congress for a declaration of war, and on April 6, 1917, the United States was at war with Germany and her allies.

At that time, Brigadier General Lejeune was serving as the assistant to the Commandant of the Marine Corps in Washington, D.C. On April 26, just twenty days after the declaration of war, senior officials from all the allied nations came to Washington. This included France's Marshal Joffre, Viviani of Italy, and Britain's Lord Balfour. It was at this Washington conference that John Lejeune first saw the desperation seeping out from the facade of the European leaders' diplomatic faces. "Never shall I forget the faces when I first saw them," he said. "They bore the marks of the agony of soul they had endured. . . . Soon we learned of the great gravity of the situation. . . . France was bled white and could not resist much longer without assistance. England was in grave danger of starvation because of the submarine warfare. Russia was on the brink of an internal explosion."[45]

Britain's Foreign Minister, Lord Balfour, introduced a request for "500,000 untrained men, at once to our depots in England to be trained there, and drafted into our armies in France."[46] But that request fell on deaf ears. Seeing that the Americans were not going to tolerate being used as a manpower replacement pool, Marshal Joffre suggested that an American division be sent to France right away, where the French could teach the Americans trench warfare behind the lines.[47]

The President announced that a division of the army and a regiment of Marines was to be sent to France immediately, ahead of an American Expeditionary Force (AEF) that was to consist of a million and a half men. Major General John J. "Black Jack" Pershing was selected to command this AEF—chosen over five other major generals senior to him. On October 17, 1917, he was promoted to the four-star rank of general, completely bypassing the three-star rank of lieutenant general.

For the Marines, the commitment for a regiment was exhilarating, but there was a big problem facing General Lejeune and Commandant Major General George Barnett. Although the Naval Appropriations Bill of 1917 had dramatically increased the Corps to fifteen thousand enlisted men—there had only been eighteen hundred in 1898[48]—they were not all in one convenient place where they could be easily assembled. In fact, Marines were scattered all over the Caribbean and other far-flung places. There were no organized regiments located in the United States.

"All of our organized forces were in Haiti, Santo Domingo, and Eastern Cuba," said Lejeune. From these scattered units would be recalled the men who would become the nucleus of the 5th Marines. "A number of companies were brought home from the West Indies. . . . The first Battalion was assembled at Quantico, and the second and third at Philadelphia. . . . Each company had a Captain and one Lieutenant. The remaining Lieutenants were fresh from college."[49]

Finally, the 5th Marines were brought up to combat strength of 2,600. But from the beginning they felt like stepchildren to the army. Pershing had pushed for a completely distinct and independent American Army, not to be part of some British or French command—nor to include Marines. In fact, it was President Wilson's order to General Pershing that demanded American independence from European command. He explicitly ordered Pershing to "cooperate with the forces of the other countries employed against the enemy, but . . . the forces of the United States are a separate and distinct component of the combined force."[50] But General Pershing also seemed to want to keep his army units separate and distinct from Marines. At every turn he conveyed the idea that he had no desire to include Marine Corps units in it. "The Fifth Regiment was not part of the First Division," observed Lejeune. "It was in addition thereto, and since complete divisions only were being sent to France, there would be no place for it except in the rear areas."[51]

Generals Barnett and Lejeune, seeing the Marines being pushed to the rear, prevailed on the Secretary of the Navy, Josephus Daniels, and his able, fiery assistant, Franklin Delano Roosevelt, to approach President Wilson to approve a brigade of Marines to be deployed instead of just a regiment. Without hesitation the President agreed, enthusiastically issuing an order to expand the Marines to a brigade.

Thus, came into being the Fifth and Sixth Marine Regiments—called the 5th and 6th Marines—along with the newly formed Sixth Machine Gun Battalion.[52] These three units would form the newly designated 4th Marine Bri-

"First to Fight" poster
(U.S. Marine Corps)

gade. The 5th Marines would still sail first, with the army's already fleshed-out 1st Infantry Division, and the 6th Marines would continue recruitment and formation.

Lejeune desperately wanted to command the Marine Brigade that was to fight in France. He had keenly felt the pangs of always being just left out of combat, and now his new place was to command the Marine Barracks at Quantico. That was where the enormous job of preparation, training, and embarkation was to be done. It was an important job, but it was not combat. He was again destined to be "*in the rear with the gear.*" He had gone to the Commandant and said, "I now have one desire and that is to go to France."[53] But it was Colonel Charles A. Doyen, commanding the 5th Marines, who would be promoted to Brigadier General and given command of the brigade.

> All in all, it was a heartbreaking winter for those of us who were left behind. Unit after unit entrained and departed for the mysterious land where battles raged and between which and us, there seemed to be an impenetrable curtain hung.
>
> All of those left behind stood along the railway tracks and shouted three lust cheers for their departing comrades, while the bands played until their instruments froze.[54]

And then came the late May declaration that the Marines would have to find their own vessels for movement to France. Secretary of War Newton Baker wrote personally to Commandant General Barnett that there would be no room for the Marines in the first convoy.[55] If this was a last-minute surprise by General Pershing to remove the fifth-wheel Marines from his already formed four regiments of the 1st Division is unknown, but such a move had been anticipated by the ever-vigilant navy team of Secretary Daniels and Franklin Roosevelt. Earlier that month, Secretary Daniels had approached President Wilson about two German warships interned in the Philadelphia Naval Yard. Why not "change *Kron Prinz Wilhelm* to *Von Steuben*, and *Eitel Frederick* to *DeKalb*?"[56] Von Steuben and DeKalb were two honorable German officers who had helped in the American Revolution. With the stroke of his pen, President Wilson solved the Marines' shipping problem.

Perhaps Pershing's irritation with Marines was justified to an extent. His twenty-thousand-man army contingent, preparing to embark to France, was completely ignored in a *New York Times* headline that announced in bold print, "2,600 Marines to Go with Pershing."[57]

On June 13, 1917, Pershing and his advance party arrived in France, and ten days later Colonel Charles A. Doyen and his 5th Marines docked at St. Nazaire off the Bay of Biscay. On July 9 it was assigned to the Army's 1st Division, which was already complete with its own four army infantry regiments. As feared by Lejeune and the Commandant, the Marines were indeed pushed to the rear. Pershing parceled the regiment into small units to conduct guard and provost duty.

According to Colonel Albertus Catlin, future commander of the Sixth Marine Regiment:

> The Marines did all the provost work for the American forces. They functioned as military police in various places, policed the villages and cafes, had

> charge of American camps and debarkation ports, guarded the lines of communications and the various bases, and helped to keep in hand the flood of incoming troops.
>
> We had Marines doing provost duty all over France—at Harve, Tours, and a dozen other places, even at Southampton, England.[58]

This rear area duty obviously miffed most Marines, and complaints found their way back to the Commandant. Pershing obviously was aware of the Marine unrest, and in a December letter wrote a personal note to General Barnett that can best be described as patronizing. Noting that the Marines had been under his command for "almost six months," he wanted to explain why they had been assigned to this communications, policing, and provost duty: "the Marine Regiment was an additional one in the division and not provided for in the way of transportation and fighting equipment in case the Division should be pushed to the front. When the service of the rear troops and military and provost guards were needed . . . it was the Marine regiment that had to be scattered."[59] He concluded, "It is a great pleasure to report on your fine representatives in France."[60]

But if guard duty seemed fitting to keep the Marines out of Pershing's hair, he offered no evidence that it did. By September, the Sixth Marine Regiment had been trained and was ready to embark to join the 5th Marines, as the President had authorized. But General Pershing again moved to thwart the Corps. On September 1, he had cabled the War Department that it should send no more Marines to France. He offered the War Department the same reasons for exclusion that he had often repeated: the army had filled out the divisions with its own regiments; the Marines had insufficient transportation, or required support not included in the army's makeup; or a variety of other gripes. But this time Pershing did not prevail, and the War Department's answer to him clipped his wings: "Referencing your cable . . . the President has directed an additional regiment (the 6th) to be sent and is now impossible to change the arrangements. This regiment combined with the one already with you will form a brigade and become part of the 2nd Division . . . one brigade of regulars [Army] and one of Marines. The Marine brigade will be commanded by a general officer of Marines about to be nominated and while serving in France will be part of the army under your command."[61]

✢ ✢ ✢

Pershing's American Army consisted of four infantry divisions: the 1st, 2nd, 26th, and 42nd, each numbering about twenty-eight thousand men. Of these four, the 2nd Division was unique. Its four infantry regiments included both army and Marine regiments: the Army's 9th and 23rd, and the Marines' 5th and 6th. The division was commanded by the Army's General Omar Bundy, but Bundy had not impressed Pershing and was in his crosshairs. Pershing had confided to his diary that he would "relieve him at the first opportunity."[62]

The situation along the Western Front was filled with doubts and misgivings. The Germans seemed to be getting the upper hand. The Russians were suddenly out of the war on the Allied side. The November 1917 Bolshevik Revolution would murderously dispose of Czar Nicholas II and his royal family, and the Russians signed a separate peace treaty with Germany. And the Austrians climactically defeated the Allied Italians.

Suddenly trainloads of German soldiers and equipment were on the move from the now peaceful Eastern Front to the west, and soon outnumbered the combined Allied divisions. By March 1918, Germany fielded 3.5 million men on the Western Front in 192 divisions facing the Allies' 169.[63]

On November 11, 1917, at a military meeting in Mons, Belgium, German field marshal Erich von Ludendorff laid out the plans for a spring offensive that would win the war for Germany. It targeted the British army. Ludendorff concluded that if the British were defeated, they would be more likely to sign a separate armistice since they were war-weary and not fighting to defend their country. He code named the entire offensive Operation Kaiserschlacht (The Kaiser's Battle).[64]

That offensive began at dawn on March 21, 1918, with a massive artillery and heavy mortar barrage. For five hours this bombardment pummeled the British sector of the Allied line with more than a million rounds. The ferocity of the attack hit the British lines like a wrecking ball. Private Jim Brady of the 3rd Field Ambulance said, "The barrage fell on us like thunder and lightning, causing the dugout to shiver and quake and stout beams to groan under the shock of direct hits and the waves of blast, which roared down the stairway."[65] And then, through the early morning fog, came the ground attack. "We saw them at about 300 yards," wrote Private S. S. Taylor. "Their gray uniforms didn't show up too well against the mist. There were thousands of them, a blanket of men coming straight at us."[66]

Nothing seemed capable of stopping the German juggernaut. The British Army fell back, attempting to make periodic stands to slow the Germans down. But the German pursuit was relentless. Captain R. Chell of the Essex Regiment wrote: "New lines sprang up and were mown down again until one almost sickened of the slaughter . . . they were behind us, and only a narrow exit corridor remained. As it was, it was a miracle that any of the battalion got out at all, but we left many, many behind."[67]

Britain's Field Marshal Sir Douglas Haig issued a dire order to his crumbling army. "Fifth Army must hold the Somme at all costs. There must be no withdrawal from this line."[68] But the army was not able to hold at the Somme. The Germans pressed on and wedged fourteen miles into the British lines, capturing forty thousand prisoners. In Germany, the Kaiser was ecstatic: "The English have been utterly defeated!" Finally, on April 5, Ludendorff called a halt to the German attack after having penetrated forty miles. Over 100,000 reinforcements had arrived from England to help stem the tide, and the Germans needed time to lick their wounds. The two-week savage campaign had cost them 240,000 casualties.

In Quantico, as the winter turned to spring, the running news of the German offensive shocked an already depressed General Lejeune. "We did not know all," he said, "but we knew enough to realize that the situation was a desperate one, and that Field Marshal Haig and his gallant men were fighting with their backs to the wall."[69]

The war was passing him by again! "The very depths of my soul were filled with an indescribable yearning for service overseas. I had sent battalion after battalion from Quantico to France and I felt I owed to every officer and man . . . to follow them and share their fate."[70] Then in May came unexpected news. General Pershing had relieved General Doyen from his command of the Marine Brigade—ostensibly for physical disability. He had replaced him with his own Chief of Staff, Army Brigadier General James Harbord. Was this more Pershing skullduggery?

Lejeune caught the next train to Washington to meet with Commandant Barnett, and Barnett warmly greeted him and told him he had not forgotten his promise to send him overseas—but doubted he would now want to go under the present circumstances. He handed Lejeune a copy of Pershing's cablegram to the War Department. "In substance it said that General Doyen had been ordered home because of physical disability and that Brigadier General James G. Harbord . . . had succeeded him . . . and if a Brigadier General of Marines were sent to replace Doyen, he would not be assigned to command the Marine Brigade, or any other front-line unit, but would be detailed to duty behind the lines."[71]

Lejeune accepted these harsh predictions as a welcome opportunity to at least get closer to the battle line than he was at Quantico. At the end of the month, the troop carrier *Henderson* set sail for France with him aboard. On June 8, 1918, he landed at Brest. He immediately inquired on the progress of the war and was given a file containing *New York Herald* and *Chicago Tribune* articles revealing what the folks back home already knew. The German army had reached the Marne River, at Chateau Thierry; and Rheims was surrounded and threatened; and Soissons had fallen; and the Germans had crossed the Ourcq River. Paris was seriously threatened and daily shelled with a unique German gun that could reach the city from a distance of sixty-two miles. He then went to Navy HQ to report in before joining the AEF.[72]

The only silver lining to the dark military clouds was that "the Second American Division . . . west of Chateau Thierry and north of the Marne . . . had stopped the advance of the German troops in that sector, and that the Marine Brigade, in a brilliant attack, had succeeded in taking the southern part of Belleau Woods and had captured several hundred prisoners."[73]

The next day was most depressing. John Lejeune learned of the obstacles he would have to clear and the hoops that he would have to jump through to even visit General Pershing at his headquarters at Chaumont. He also learned that any chance to convey his own ideas and the ideas of his Commandant, concerning the Marine Corps, were at best remote. Pershing would have no time for any of that.

All casual officers, including Lejeune, were not to be sent to Chaumont, but to Blois—a place to avoid. Blois quickly earned its reputation as a dead-end limbo, where officers languished for weeks at a time, waiting to receive orders to places no one wanted to go. Lejeune then played his ace in the hole. He carried a personal letter from an officer's wife, and that officer was at Chaumont. He called the officer and told him of the letter, and then used the opportunity to confide that he had a great need to come to Chaumont to "confer with General Pershing on matters of great importance to the Marine Corps." That officer abruptly signed off and hung up the phone.

Next, he telegraphed another contact at Pershing's HQ and explained his situation, and this time received permission for him, and his small entourage, to come forward. They motored for two days to reach Chaumont, and Lejune finally was able to visit Pershing's Chief of Staff, who promptly informed him that "The Marine Corps has no authority over you while you are on duty with the A.E.F."[74] The quick-thinking Lejeune acquiesced and said, "I would not have it otherwise . . . my chief reason for asking to be ordered to Chaumont . . .

was to urge most earnestly the increase of the Marine Corps force in France by a division." The Chief of Staff's response was that he "will bring the matter to General Pershing's attention when he returns."[75]

That evening he was invited to dine with the general at his quarters. Lejeune would write, "We had a very good dinner and the General was most agreeable." There was small talk and civility, but the visit ended when Pershing abruptly excused himself declaring that "he had a great deal of work to do."[76] Lejeune had not had an opportunity to broach his appeals or suggestions. Three days later the Chief of Staff informed him that "the subject of the Marine division had been discussed."[77] And Pershing never seemed to be in.

And then, that same afternoon, Pershing was suddenly in; and received Lejeune to inform him that he could not approve the assignment of a Marine Division. "It would interfere with his plan for a homogeneous army." It was too complicated. He was already having trouble replacing Marine casualties with other Marines. But Black Jack said that he would be more than happy to have Marine Corps officers with the AEF if they were "assigned to, or commissioned in the National Army."[78]

Lejeune was distraught. Upon his departure from Pershing's office he made one last-ditch appeal to the Chief of Staff and asked if there was any possibility for a change. The answer was emphatically, "No, it's final."[79] He then requested to command the 4th Marine Brigade, now being commanded by an army general, and was told no. He would report as an *observer* with the Army's 35th Division. "I was depressed and disappointed," Lejeune lamented. But John Lejeune never made it to the 35th Division. On July 5 he motored to the 32nd Division headquarters only to be handed orders to command the Army's 64th Brigade. "We were jubilant!" he exclaimed, "at last my period of probation had ended."[80]

Lejeune settled in and gave his first combat orders for action: a raid to be conducted by one of the regiments, "but before it took place, orders for the relief of the entire Division . . . were received, and [my] instructions for the raid were countermanded."[81] In fact, Lejeune's command of the 64th Brigade was to barely last three weeks. Shortly after he celebrated Bastille Day (July 14) with his French Allies, German General Ludendorff launched yet another attack in his offensive that had begun in March. Lejeune's new brigade was immediately entrained in those famously cramped boxcars that were marked "*huit chevaux ou quarante hommes*" (eight horses or forty men).

"We had no inkling of our destination," he said. After thirty-six hours, the train screeched to a halt in the woods of some unknown forest. He reported to his division headquarters, only to learn that his luck was forever bad and he was no longer commanding the 64th. But as he received his new orders his spirits soared. Beyond his wildest dreams, somehow, he had been assigned to command the 4th Marine Brigade, consisting of the legendary 5th and 6th Marines, heroes of Belleau Wood.[82]

It was a long drive to the 2nd Infantry Division's HQ, the parent command of the 4th Marine Brigade. It was made longer by the gigantic traffic snarl occasioned by the entire Allied Army being on the move. This juggernaut moved in the densest of fog, made worse by the swirling dust raised by marching columns. It was a snail's pace. "It was a continuous procession of motor trucks, wagons, officers' cars, motorcycle riders dashing ahead at a mad pace, camions [trucks] filled with men, troops marching, field artillery guns, caissons, and . . . long trains of captured, camouflaged cannon being towed by trucks."[83]

On July 26, an ecstatic John Archer Lejeune issued his first order: "I have this day assumed command of the Fourth Brigade, U.S. Marines."[84]

If John Lejeune had built a reputation of always being a man on the move, on the outside looking in, just in time to be too late, always searching for a place to call home, nothing changed with his new dream assignment to command the 4th Marine Brigade. Two days after assuming command of the brigade, he was no longer in command. On July 28, his day had hardly begun when his 2nd Division commander, army Major General James Harbord, frantically called him to his office.

Harbord had just returned from a grueling, twenty-four-hour conference with "Black Jack" Pershing, who was not at all satisfied with the state of supplies. Pershing had summarily relieved the commander of the Service of Supply of the AEF and told Harbord to get over there and take charge. General Harbord looked very grave as he rushed around his office. He had asked Pershing which officer would relieve him to take command of the 2nd Division. Pershing had long ago proven to be no champion of Marines, or even a casual friend; but now, without hesitation, he instructed Harbord to turn command over to General Lejeune. After all, he was the senior brigadier on duty.

It was just that simple. President Woodrow Wilson then appointed Lejune a major general, and he was suddenly a Marine general in command of an army division. "To say I was surprised, is putting it too mildly," said Lejeune. "I was stunned."[85] He became the first Marine general to lead an army division into major combat. On July 29 he took command. Some historians have opined that Pershing's move to appoint Lejeune to army division command at the start of the American Army's first independent action reflected respect for Lejeune's strength of character. But while that was a nice tribute to the general, the obvious reasoning for his appointment was revealed in Pershing's own pragmatic words—"*He was the senior Brigadier on duty.*"[86]

Upon assuming command, Lejune was ordered to move his entire division to an unknown destination, two days away by train. That turned out to be the town of Nancy, which he described as "a miniature of Paris."[87] And then it was on to Marbache, "a very small, squalid, sad looking town,"[88] wrecked and of no significance other than it fronted the south face of the formidable St. Mihiel salient that the Germans had held and defended since the beginning of the war. There was no fighting now. It was deemed a "a quiet sector."

But as if to preview that something "unquiet" was afoot, the 2nd Division was relieved and sent behind the lines for intense training. On August 27, the American First Army, under Pershing's command, went into the battle line opposite the St. Mihiel salient. It was sandwiched between two French armies and tasked to attack the south and west faces of the massive salient, on a fifty-mile front. The Germans defended with at least twelve divisions.

The 2nd Division was positioned in the very center of this attack, dubbed Operation A. Lejeune had twenty-seven thousand soldiers and Marines under his command, reinforced with sixty-three tanks to engage the machine gun nests and breach the barbed wire.[89] His own infantry order reflected the tactics he had learned many years ago while studying General Emory Upton's tactics while at LSU. He would not attack on a long, vulnerable line, but as Upton had done—in swift-moving columns. The division would smash into the German defensive line while configured in a column of brigades, with its regiments abreast and those battalions in columns. H-hour was 5 o'clock, September 12, 1918.

With the rolling barrage preceding them, the soldiers of the 2nd Division, with the Marines in reserve, struck the German line with a shattering blow that drove the German defenders back. By 10 o'clock the division had smashed through the German main line of resistance, and then swept away a reserve regiment attempting to make a last stand. By early afternoon Lejeune and his 2nd Division occupied its objectives, without even committing the Reserve Marines.[90]

Two days later, the Marines probed and made sporadic contact and repelled several feeble German attempts to counterattack. The retreating German army was making its way back to its final line, the Hindenburg Line, seven miles to the rear. The St. Mihiel salient was no more. Lejeune's division had routed the enemy, taking 3,300 prisoners and capturing 118 guns. His own losses had been light: 195 killed and another 1,000 wounded.

On September 20, the division pulled out of the line, supposedly for rest, but three days later it received orders to entrain again to move to yet another unknown destination. The only known was that it would be another ten-hour ride in those damned 40 x 8 boxcars followed by more marching. One Marine described the movement. "After many hours of crowding and quarreling over mashed toes and bruised ears, we reached Chalon . . . and made another of those long marches . . . to an area near the Champagne front not far from the city of Rheims."[91]

While the men of the 2nd Division might not have known the significance of the terrain as they detrained and assembled on September 30, the French knew it well as another powerfully held German line called Le Massif du Blanc Mont. Like St. Mihiel, Blanc Mont Ridge had been impregnable and afforded the Germans the ability to maintain a stranglehold on Rheims and Champagne and bombard it with their dominating artillery from the mountains.[92] It was even rumored that during the July German offensive the Kaiser himself had watched the launching of the attack from Blanc Mont.[93]

General Henri Gourand, Commanding General of the Fourth French Army, well explained the significance to General Lejeune. "If I could take this position by assault," he said, "and hold the ground against the counterattacks which would be hurled against my troops, the enemy would be compelled to evacuate Notre Dame des Champs and Les Monts."[94] Looking at the map, Lejeune could readily see that the German army would then be forced to retreat nineteen miles, all the way back to the Ainse River. Between Blanc Mont and the Aisne there was no place to establish a new defensive line. "But," continued General Gourand, "my divisions, however, are worn out . . . and it is doubtful if they are now equal to accomplishing this difficult task."[95]

Lejeune delivered a surprising answer: "General, if you do not divide the Second Division, but put it in line as a unit on a narrow front, I am confident that it will be able to take Blanc Mont Ridge, advance beyond it, and hold the position there."[96] And so, the 2nd Division entered the front lines on the night of October 1, 1918. The entire area was a panorama of four years of war and desolation. There was not a stone left upon a stone. Even the topsoil had been blown away, revealing the underlying white chalk that gave the massif its name. No trees, no bricks to even give evidence that a building had previously been there. All that had been replaced with trenches and tangles of barbed wire. Lejeune described it. "The debris of battle was still lying about—broken cannon and machine guns, rifles, bayonets, helmets, parts of uniforms . . . fragments of human bodies were often found. Arms and legs thrust out of the torn soil, and unrecognizable long-buried human faces, thrown up to the surface of the ground by exploding shells were frequently visible. The fearsome odors of the battlefield, too, were always present. . . . It was not a home, but horror."[97]

The first order of business prior to the 2nd Division's attack to seize Blanc Mont was to sweep any remaining enemy from the trenches immediately to the division's front. Then, at 0550 on the morning of October 3, the assault began. Lejeune employed his cherished Emory Upton tactics, with a rolling barrage sweeping the ground in front of his closely following infantry. The Marine Brigade was on the left with the Army Brigade on the right, each in a fast-moving column of regiments. The flanking French support infantry lagged behind the advancing Americans, who drove up the mountain and were able to signal back to Lejeune before 0900 that Blanc Mont was in their hands.[98]

As expected, the Germans counterattacked furiously for the next two days, desperately trying to regain the high ground, but they were systematically beaten back by the soldiers and Marines. The division continued to push the German army back toward the Aisne River, and, after much hard fighting, drove them from the Hindenburg Line. By October 10 it was all over. The enemy had given up ground and positions that it had held since 1914. But the cost to the 2nd Division was extremely high—five thousand casualties.[99]

Everyone could feel that the end of the war was close at hand. But being close was not the same as being over. The British had advanced toward Cambrai; the French continued to advance around Rheims; and the Americans were poised to launch the offensive that should break the German's back—the Meuse-Argonne. Lejeune was told that his 2nd Division was to be the "point of the wedge for a great attack."[100]

That final attack began on November 1, at 0530. An earlier two-hour artillery barrage plowed the land in front of the German line, and again Lejeune configured his attack to emphasize speed of advance. Three regiments of his division jumped off, each in columns of battalions along a mile and a quarter front. The seasoned army and Marine veterans quickly cozied up behind the creeping artillery wall of exploding steel, as if they were simply following a line of rolling vehicles in a parade. It was called "leaning on" the rolling barrage, and the very nearness sometimes caused a few friendly casualties.[101]

The attack surged ahead. The defenders in three small towns were swept aside, and by 1100 the 2nd Division had captured and occupied all of its objectives. The overrun enemy was streaming to the rear. Lejeune's lightning attack was the source of jubilation at both Corps and Army headquarters, and it drew a commendation from Fifth Army Corps: "The Division's brilliant advance of more than nine kilometers, destroying the last stronghold in the Hindenburg Line . . . may justly be regarded as one of the most remarkable achievements made by any troops in this war."[102]

For the next nine days, the allied armies pressed on to reach and then cross the Meuse River. Finally, at 1100, on the eleventh day of the eleventh month of 1918, the Germans quit. The war was over. The Armistice was signed.

After leading the division on the march into Germany following the armistice, General Lejeune remained in command of the 2nd Division until August 1919, when it was demobilized. He returned to the United States to again command the Marine Barracks at Quantico. On July 1, 1920, he became the 13th Commandant of the Marine Corps, serving until March 1929. He is often called the "greatest of all Leathernecks." Perhaps the greatest military tribute given him came from France's Marshal Philippe Petain, who called John Lejeune "a military genius who could and did do what the other commander said couldn't be done."[103]

After thirty-nine years of service, Lejune retired from the Marine Corps to pick up the reins as superintendent of Virginia Military Institute, which he held for eight years. He died on November 20, 1942.

CHAPTER 8

Graves Blanchard Erskine

BELLEAU WOOD LIEUTENANT, IWO JIMA GENERAL

The electric news of President Wilson's call for a declaration of war against Germany in 1917 arrived on the LSU campus like a thunderbolt. Like all young men anxious to answer the call to arms, nineteen-year-old senior Graves Erskine strained at the leash. Despite his young age, he was already a veteran of General "Black Jack" Pershing's Pancho Villa Expedition along the Mexican border. "I was a senior at LSU," he said. "I had the previous year gone to the border with the National Guard, the 1st Louisiana Infantry, and we were late getting back, and when war was declared, I was still a member of the Guard."[1]

Erskine was a perfect example of an extremely poor boy having an opportunity to rise from the family's dirt farm, to pursue a college degree that could lead to a career in medicine. He had been born near a place called Perrins Island, near Columbia, Louisiana, slightly south of Monroe. He had gone to school in a one-room schoolhouse in Death Ridge where schoolmasters taught class and maintained discipline with corporal punishment. "I went there with great fear and trepidation of what might happen. . . . They did not spare the rod."[2]

Graves aspired to be a doctor. His uncle was a well-known surgeon practicing in Monroe, and the youngster helped out around the operating room as his assistant. "I became fascinated with the idea of being a great surgeon some day."[3] But now those plans were pushed to the rear. He was anxious to fight, and since he was still enlisted in the 1st Louisiana Regiment, he expected to be called up. Indeed, some of the Regiment were called up, but not him, and he was outraged.

"My grandfather had always told me—he was an old Civil War soldier—that when your country was in trouble, you volunteer; and I felt that I would not have been true to him if I didn't go to war."[4] His grandfather's lawyer just happened to be the colonel in command of the 1st Louisiana Regiment, and Graves must have felt that that relationship qualified him, as a sergeant/trumpeter, to send a telegram to the colonel telling him "We had a war on and I wanted to know when the hell we were going?"[5]

An army training camp opened at Plattsburg, and many of the LSU cadets were signing up, but Erskine was too young—not yet twenty-one. Undaunted by this latest setback, he marched down to see the Adjutant General of the Louisiana Guard himself. "He gave me pretty short orders and practically threw me out of the office, and reminded me that I was only a sergeant, and sergeants did not come to see the Adjutant General."[6]

His next step was to go to see the Commandant of Cadets at LSU, Captain Sanderford Jarman, who listened to his complaint and told him about a request for ten cadets from the LSU Corps to apply for reserve commissions in the Marine Corps. So, he went to New Orleans to try his luck there. The board looked over the ten applicants and, although Erskine was underage, gave him a waiver and selected him for one of the probationary commissions. He gleefully returned to the LSU campus, expecting to be called away at any time, but found yet another roadblock. It was April, and he would not be the required twenty years old until the end of June.

A visit to see the LSU president, Thomas D. Boyd, solved nothing. In fact, it brought more problems. He had to have a diploma, but the school year was not out and graduation was in June. President Boyd was very understanding, and promised to see if there was anything he could do to have the faculty board give him credit for the courses he was now taking. He had high marks, and the board approved, except for one course in political science—"which was really my downfall,"[7] Erskine lamented. The political science professor gave Erskine an emphatic "No!" Despite all pleadings, he told the young cadet, "I don't think you've opened a book. . . . You haven't answered a question in the entire year."[8]

Erskine was desperate. His chance for military service in combat hung in the balance. After several meetings with the adamantly negative professor, he meekly admitted that the professor's charges against him were all true, and then followed with the lamest reasoning, hoping to change his mind. "I thought I was absorbing enough to have a passing grade." Still the answer was no. But Erskine persistently visited the professor until he finally relented. "I'll give you a passing grade to get you out of my class."[9] And just that quickly, Graves Erskine had his LSU diploma, and on July 2, 1917, received his Marine Corps orders.

He reported in at Quantico, and one of the first people he met there was the base Commanding General, John A. Lejeune. Erskine had been detailed for duty as the Fire Patrol Officer, and he was to report to General Lejeune during his shift. The general had a tiny office, and it was a rainy and muddy day, and Lieutenant Erskine made sure to shine himself up, spit and polish, to report in. "I looked real snappy when I got there, and he let me stand in the doorway for five minutes before he spoke to me. He was looking at me, and I was putting on my best military brace, and finally he said, 'All right, Napoleon, come in and tell me what you want.'"[10]

The next Officers Training Course convened for several hundred new lieutenants. The instructions began with the very basics: lectures on the customs of the service, drill, and the manual of arms, all of which was old hat to Erskine. He was soon bored stiff. "I felt that I was a veteran, having been to the Mexican border, and three years at LSU, which was rated pretty high as a military school; so, I was pretty disgusted that I wasn't learning anything that I couldn't figure out. . . . I sent in my resignation, because I felt it was my duty to go to war. . . . I didn't see any chance of getting to war in that school."[11]

His resignation made it all the way to the head of the officers' school, Colonel Presley Rixley, and he soon found himself standing before him to explain. Rixley gave him a stern lecture on patriotism and duty to country, and after the severe berating, asked, "Do you have anything to say?"[12] The salty, ramrod-straight second lieutenant immediately answered, "Yes [sir], that's why I'm getting out of this outfit; I haven't learned a damn thing since I've been here, and I'm getting pushed around."

When the colonel asked about his future plans, Erskine snapped back, "I'm going to Canada and joining the Black Watch."[13] With that, his resignation letter was set aside and his request summarily rejected. He was sent to take command of the 2nd Platoon of the 79th Company of the 6th Marines (Regiment).

For the rest of 1917, much to the chagrin of the young lieutenants eager for combat, it was days upon days and weeks upon weeks of training and preparation. Colonel Albertus W. Catlin, commanding the 6th Marines, said: "Through the summer and fall of 1917 we drilled 'em and we drilled 'em, until they were fit to go up against any foe on earth. We taught them to shoot straight and use the bayonet, and we had them mopping up trenches and cutting wire; we hardened them with hikes, and we got them to handle machine guns like baby carriages. We filled them full to bursting with the spirit of the Corps, and then we shipped them off to France to fight."[14]

Catlin was himself an inspiration to his own Marines. His towering combat credentials were awe-inspiring, ranking with other legends of the Corps. As the Marine Detachment commander, he had survived the 1898 explosion and sinking of the battleship *Maine* in Havana Harbor. In 1914, he received the Medal of Honor for his actions as a battalion commander in Vera Cruz. President Woodrow Wilson had ordered the Marines to Mexico to prevent a German shipment of machine guns and ammunition,[15] and Catlin brilliantly and heroically led his battalion in that operation. He had faced an enemy battalion of Mexican troops, heavily armed with machine guns, and a large contingent of armed convicts freed by the Mexicans. The machine guns controlled all the streets of Vera Cruz, and the rooftops bristled with snipers. Any advance seemed suicidal.

Using long-ago learned Indian tactics, Catlin avoided the deadly enemy fire sweeping the streets and pushed the Mexicans steadily back by advancing, untouched, through the walls of the side-by-side houses. At the end of one block, they crossed the street and continued their advance through the walls of the next block of buildings. "Our chief weapon was the pick-ax."[16] By 8 o'clock that evening, the bristling Mexican defensive line was no more and the outwitted Mexicans had been driven from the town. Catlin's losses were five men killed.[17]

His 6th Marines now lagged behind the already-deployed 5th Marines, who were first to embark to France with Black Jack Pershing. When his regiment was finally deployed, it was an annoying piecemeal embarkation—a battalion at a time. The 1st Battalion arrived at the port of St. Nazaire on October 5, 1917, after an eighteen-day Atlantic crossing. Five weeks later, on November 12, the 3rd Battalion showed up at Brest. It was not until January 5, 1918, that the 2nd Battalion, with Graves Erskine, finally steamed into St. Nazaire. Together, the 5th and 6th Marines completed the 4th Marine Brigade that became half of the Army's 2nd Division.[18]

The Marines were assembled in a farm area near Verdun, fifty miles behind the front lines. For the next two months they underwent a severe training regimen under the direction of the French 77th Regiment. "They were drilled constantly in trench organization, signal systems, and all the details of trench organization as it existed at the front. And all this in addition to the routine drill of the Marines," wrote Catlin. "It was winter, cold, and often stormy, but the weather made no difference. The training went forward every day, and manoeuvers were executed in snowstorms. . . . But they learned . . . and became hard as nails."[19]

On March 15, 1918, the Marines began entraining to the front. It took five trains to move each regiment: fifty cars per train, one hundred cars to move the Marine Brigade. They were taken to a sector on the heights of the Meuse River southeast of Verdun and relieved the French forces in the trenches. The rule for service in the hated trenches was twenty days in, and twenty days out.

It was during this time that news of a tremendous German offensive reached the ears of the trench-bound Marines. "Every man of us knew all about it, for we picked up the German and French communiques by wireless,"[20] said Colonel Catlin. But knowing about this juggernaut offensive and being able to do anything about it were two different things. The Marines were restless and frustrated. "For a month we looked for orders to move, but none came,"[21] complained Catlin. Finally, on May 14, after two months in the trenches, orders came to move out. "We rejoiced," confided a gleeful Catlin, "for the order meant a rest from trench digging, relief from the nightly peril of No man's land, [and] a fond farewell to the mud and rats and cooties."[22]

The German offensive that had begun on March 21, had swept across the Somme River and advanced in an unstoppable wedge. The first check of this irresistible force was delivered not by the French or British armies but by the American 1st Division, which captured the city of Cantigny on May 28.[23]

German General Erich von Ludendorff was undaunted. Despite this minor setback, he began a new drive. Seventeen German divisions advanced and swept aside the 21st and 22nd Divisions of the French Sixth Army defending at both Chemin de Dames and at the Aisne River near Soisson.[24] The defeated French general confided to Prime Minister Georges Clemenceau that "there was nothing to stop the Germans but '*de la poussiere*,'—mere dust."[25]

"On they came," said Catlin, "sweeping everything before them, demoralizing the French army opposed to them, and heading straight for the Paris of their dreams."[26] The Metz-to-Paris road was wide open. Forty divisions, four hundred thousand of Germany's best, had advanced in the Ludendorff Offensive at a rate of six to eight miles a day, equipped with an enormous assortment of tanks, machine guns, and poison gas projectors. The retreating French inflicted significant casualties on the assaulting Germans, but nothing had stopped them. Now they approached Chateau Thierry, only thirty-five miles from Paris.

The city of Chateau Thierry was built on both the north and south sides of the Marne River, the north side being the larger. Two bridges spanned the river: The stone bridge first crossed from the south bank to an island in the middle, with a second span continuing to the northern bank. The second bridge, to the east, was a railroad bridge.[27]

Confident of victory, the Germans approached Chateau Thierry from the north in columns of four, their rifles slung over their shoulders and singing songs of victory. They anticipated a quick river crossing and a continued march to Paris, pushing the retreating, demoralized French defenders in front of them. The war seemed over! But as the Germans approached, an unexpected relief column had arrived on the south bank to cover the withdrawal of the steadily retreating French. Part of this arriving force was an American machine gun battalion that had been in the rear.

A desperate call had gone out: anyone who could come to Chateau Thierry should come on the double, before the war was lost. The 7th Machine Gun Battalion, commanded by army Major James Taylor, had been sixty-two miles to the rear, and raced to board anything that rolled. In a wild overnight movement, the lead elements arrived at Chateau Thierry on May 31. They deployed some guns on the north side of the mid-river island and on the south side of the adjoining railroad bridge. The bridges were all that blocked a German march to Paris. The Machine Gun Battalion had arrived, but without any support. By 4 o'clock they were dug in as French engineers prepared to blow both bridges. During the night, the rest of the 7th arrived and took up additional positions in the houses and gardens facing the river.[28]

At 4 a.m., the German 231st Division began their infantry advance through the north of the city, and by dusk, they were facing the bridges. At 10:30 p.m., the American machine guns opened fire, raking the north bank and drove the surprised Germans to cover. French demolition teams blew up the north span of the road bridge.[29]

At dawn on June 2, the Germans unleashed a tremendous artillery barrage against the annoying Americans. They maneuvered to force a crossing of the railroad bridge. "Then came the Germans," wrote Colonel Catlin, "a long, gray flood of them, streaming down to the bridges. The Americans opened upon them a fire so furious and accurate that the advancing columns hesitated, wavered, and then faltered behind the barrier of their fallen comrades. Then they came on again."[30] The surprised Germans increased the fury of their attack in an attempt to sweep the defenders from their rock-solid positions on the south bank and to suppress the galling fire from the machine gunners that constantly thinned their ranks. But the south bank defense held, and with superior fire stopped the German thrust in its tracks.[31]

The German command, seeing that their attacking army now faced a major river crossing without a bridge, and facing a formidable defensive line on the south side of the Marne, reevaluated its plans. It would now simply sidestep

Chateau Thierry and advance to the west, along the northern bank of the Marne, and find a better route for their march to Paris. As the Germans began their sidestepping, the American 2nd Division, containing the two regiments of the 4th Marine Brigade, moved into position to defend the Metz-to-Paris Road.[32] General Omar Bundy placed his division directly across the new path of the advancing Germans.

Second Lieutenant Graves Erskine had been at the army school at Gondrecourt, taking a course in automatic weapons and musketry, when he heard that the Marine Brigade was moving to confront the German advance. He had been in Gondrecourt for a month, and on May 31 promptly checked himself out of school to join in the impending action. He worked his way to the headquarters of the 2nd Battalion, 6th Marines, at the town of Meaux, halfway between Paris and Chateau Thierry, and reported to his 2nd Battalion commander, Major Thomas Holcomb, who asked why he was not still in school.

"School's out, Major!" the cocky lieutenant replied, and moved to rejoin his 2nd Platoon already deploying to the west of Chateau Thierry. Erskine had formed his own opinion about the French fighting style. He recognized their spirit and combat *élan,* but engaging the enemy left a lot to be desired. "I didn't think they had any idea of marksmanship with a rifle. They wanted to throw grenades at everything, and we wanted to kill at 500 yards away, a little bit beyond grenade range."[33]

The 6th Marines arrived at the battle line exhausted from their thirty-hour ride in the French camions. But those who rode were actually the lucky ones. Some had gotten lost and shouldered their sixty-pound packs and hiked to the battlefield. "When they arrived," said Colonel Catlin, "they were gray with dust and hollow-eyed with fatigue. They looked more like miners emerging from an all-night shift than like fresh troops ready to plunge into battle."[34] They were without tanks, gas shells, or flame projectors, and the French military hierarchy was hesitant to throw the 2nd Division into the line. But General Harbord, who commanded the Marine Brigade, was the image of confidence and told the French, "Let us fight in our own way, and we'll stop them."[35]

Colonel Catlin sent two battalions of the 6th Marines up the Metz-to-Paris road to within a half mile of the defensive line being formed. From there, they marched in on foot. Lieutenant Erskine marched his 2nd Platoon of the 79th Company into the battle line, forming the extreme right of the 2nd Battalion, and linked up with the Army's 9th Regiment, whose line then swept further to the right. "The men were told that orders were to hold the line at all cost,"

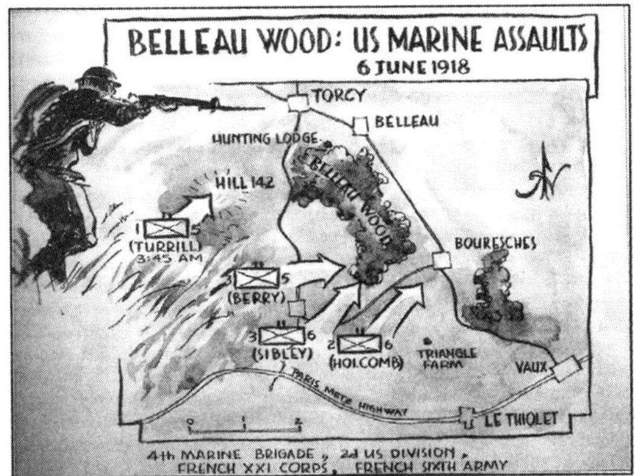

Belleau Wood
(U.S. Marine Corps map)

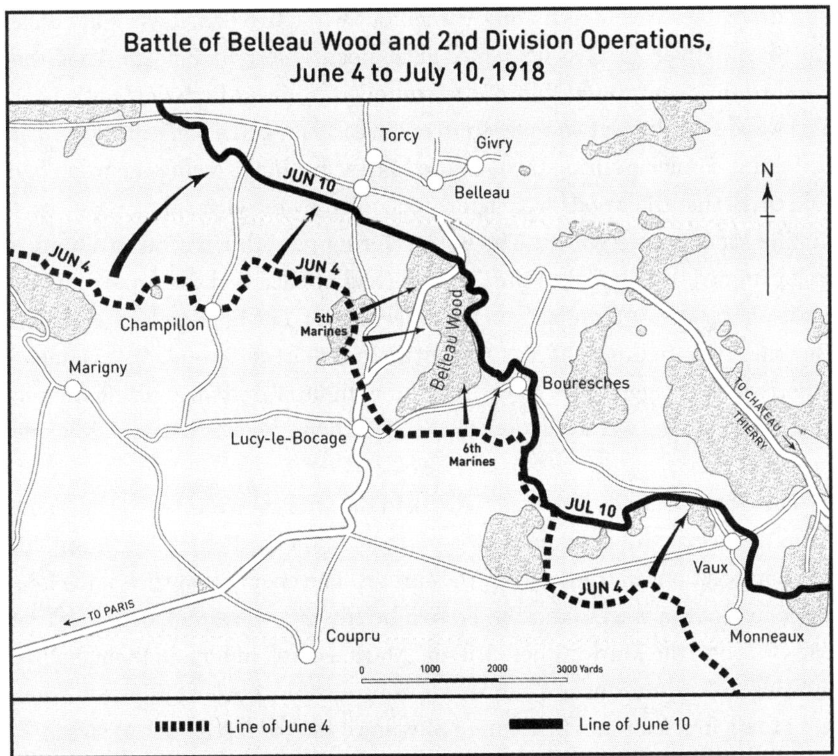

Battle of Belleau Wood and 2nd Division operations (U.S. Marine Corps map)

wrote Colonel Albertus Catlin. "They knew that there were few, if any, troops in their rear, for they had just come that way. They knew that their own thin, determined line lay between Paris and awful destruction."[36]

The Marines' defensive line ran generally along a northwest axis: from the Paris/Metz highway at Le Thiolet to Triangle Farm, then skirted the southwest edges of Belleau Wood and Hill 142 and beyond. It was a front of almost four and a half miles. An officer in the 2nd Division described the countryside as having seen no trenches and no war. "The area we were in was untouched by war. It was farming country. There were fields with half grown wheat and occasionally fairly large stone farm buildings with large courtyards. . . . There were small undulating hills and scattered hamlets. About every mile, there would be a small patch of woods."[37]

Just before midnight on June 2, the German army occupied the line from Torcy through Belleau to Bouresches. The plan was to enter Belleau Wood and drive the steadily retreating French out while two divisions would cross the Marne River—a crossing denied at Chateau Thierry.[38] Belleau Wood was perched on high ground that was surrounded by chest-high fields of wheat. The wood was almost two miles from north to south and a half mile deep. The retreating French pouring through the Marine positions facing the wood had told them that the woods were lightly occupied by the Germans.

The French were wrong. The entire German 237th Division was hidden in its tangled density.[39] But the Germans had no idea that the American 2nd Division was across their path on the other side, and its Fifth and Sixth Marine Regiments were dug in facing the entire length of the wood. "We watched [the French] stagger back to the rear . . . with utter despair written on their war-weary faces," wrote Colonel Catlin. "To them the war was lost, life held no hope."[40]

The German attack began at 5 o'clock in the afternoon on June 3, and came from the north and northeast, sweeping down on the French forces still located in positions in front of the 5th Marines. The enemy came down the long slope in platoon waves, and the French briefly met the attack before falling back through the Marine lines. Still the Marines, well hidden in their shallow fighting holes, or crouched behind the several stone walls of a French farm, held their fire. Lieutenant Lemuel Shepherd had ordered "Battle sights,"[41] and the metal Springfield sights clicked to three hundred yards—point of aim was point of impact. The distinctive metallic sound of rifle bolts chambering .30-caliber rounds rippled down the line, and the silent Marines fixed their eyes upon the German advance "across wide wheat fields bright with poppies that gleamed like splashes of blood in the afternoon sun. . . . On they came," observed Colonel Catlin, "never wavering, never faltering, apparently irresistible."[42]

As the German soldiers in the front assault ranks closed toward the Marine line of resistance, watchful eyes tracked their every step through Springfield rifle sights. When that gray line stepped up to the 300-yard killing line, the Marines opened fire—reminiscent of the shock of volley fire. The front ranks collapsed, and recently arrived artillery joined in with machine guns to hammer the Germans to the ground. "Shrapnel, machine gun and rifle fire was poured into these advancing lines," wrote Catlin. "It was terrible in its effectiveness. . . . Three times they tried to reform and break through that barrage, but they had to stop at last."[43]

This first day's success, on June 3, in no way portended the savage fight that was to become the Battle of Belleau Wood. On June 6, the Marines launched their own attack to seize Hill 142, the western edges of the wood, and the town of Bouresches. In similar fashion to the previous German attack, the Marines crossed those terrible wheat fields almost in parade formations and were mowed down by quick-firing German Maxims. Extraordinary heroism brought limited success, and as night fell they had only a tenuous hold on Hill 142, the southern end of the woods, and most of Bouresches. The cost? Almost eleven hundred dead and wounded.[44]

Lieutenant Graves Erskine had led his 2nd Platoon, 79th Company of the 6th Marines' attack to seize Bouresches. "We went over the top at 5 o'clock in the afternoon. I had about 900 yards to go through this wheat field. We had to crawl most of the way because fire from the Germans was pretty darn intense. . . . We got a hell of a lot of machine gun fire after we started out, and it just cut us to pieces."[45] Pinned down, Erskine sent a runner back to the company commander to tell him that he couldn't move. The captain's unsympathetic reply was, "Get going goddammit!"[46] What was left of Erskine's platoon finally shot its way into the town well after dusk. On the way in they captured a machine gun along with the German gunner:

> This fellow was still firing. I walked up behind him and kicked him on his shoe; he fired a few more bursts, and he finally got up, and slung his gun over his shoulder. I was holding my pistol on him—a great big husky German, and my hand was shaking.
>
> Before I knew it, he had turned me around and was drinking out of my canteen. [I] only had five or six men left at that time out of fifty-eight.[47]

The 6th Marines consolidated their hold on Bouresches, and Erskine met up with the company commander, Captain Zane, who asked how many men were left. Erskine replied that he only had five. The captain ordered him to take whatever he had and go out and locate the Germans. Erskine was incredulous, and the salty lieutenant shot back that he knew exactly where the Germans were, and when Captain Zane asked, "Where?" he said, "They seem to me to be all over!" But the captain repeated his order, and the subdued lieutenant moved toward the front mumbling "this is my last trip."[48]

At the end of the patrol he took his few men along with fifteen or twenty engineers and was ordered to defend part of the line to the northeast of the town. He dug in behind a long hedge. They were able to fire under the hedge, but its height prevented the attacking Germans from lobbing grenades into their position. Grenades that cleared the hedge fell far behind Erskine's line. The Germans launched three such attacks against him, all to no avail.[49]

For the next five days it was a continuous slugfest for control of Belleau Wood, with high casualties from artillery, gas, and heavy machine gun fire. The Marine Brigade was depleted and exhausted. The Germans were no better off. By June 13 the opposing German 237th Division, which had begun the fight with over thirty-two hundred men, was reduced to fewer than fifteen hundred.

In Bouresches, the 79th Company was hard hit. In addition to Erskine's Marines absorbing the three German infantry attacks, gas attacks had caused numerous other casualties and Captain Zane had been seriously wounded. Lieutenant Graves Erskine suddenly found himself to be the only surviving officer and took command of the forty men left of the two hundred-man 79th Company. He was in command when the 79th was pulled out of the town and moved to the extreme northern edge of the wood and placed in front of a hunting lodge south of the town of Belleau.

As Erskine dug in, a sharply dressed, middle-aged gentleman in a French helmet and a green and white armband approached him, looking every bit the newspaper correspondent. He asked who was in command. The sharp-tongued Erskine, having never mellowed from his early days at LSU, snapped back, "Who the hell wants to know?" The gentleman with the green and white armband introduced himself: "I am your brigade commander." Erskine did not know that Marine General Charles Doyen had been replaced as brigade commander by Army General James Harbord now standing in front of him. "Don't give me that. I know my brigade commander, and it's not you! I don't know what goddam army you're in, but you better pull out of here pretty damn quick. We don't tolerate strangers up here."[50] Finally, Erskine made General Harbord show him identification, and when he had proven who he was, Erskine was still not chastened and said, "You don't know how close you came to really getting it up here." He would later say, "It would have been terrible if I had shot my own brigade commander."[51] Erskine led the general around his defensive position, and Harbord was impressed.

The struggle at Belleau Wood lasted until June 26. The final Marine attack to drive the Germans, once and for all, from this battleground began in the late afternoon. A battalion of the 5th Marines followed a rolling barrage against German positions grudgingly holding the northwest corner. That final attack inflicted 450 more casualties on the Germans, at the cost of 250 Marines. The Marine commander sent General Harbord the simple message: "Belleau Wood now U.S. Marine Corps entirely."[52]

Erskine went on to fight at Soissons, where an artillery round knocked him flat with a severe concussion. He finished his war in the attack on St. Mihiel, where machine gun bullets shattered his leg, and he was carried from battlefield. At the field hospital he begged them not to amputate. They reluctantly agreed, passing him through three French hospitals before he was sent home for a long recovery involving nine surgeries in the Portsmouth Naval Hospital.[53] For his heroic action he was awarded the Silver Star.

After two years of recovery, Erskine begged for a return to duty and, in 1921, was ordered to Haiti to take command of the 54th Company. But he had only been there for eight months when, in September, he was ordered to return to the United States for special assignment. On March 4, 1921, Congress had approved a resolution "*calling for the return to the United States of an unknown American soldier killed in France and his burial with appropriate ceremonies in a tomb to be constructed at the Memorial Amphitheater in Arlington National Cemetery*."[54] Erskine had been ordered "for some reason to organize the Guard of Honor for the Unknown Soldier and go aboard the USS *Olympia* and pick up the body."[55]

On September 25, Erskine landed in Norfolk after leaving Haiti. It was only forty-seven days until November 11, the scheduled date for the interment of the Unknown Soldier at Arlington Cemetery. He met Colonel William McKelvey, commanding Marine Barracks, Norfolk, who was most concerned: "How are we going to do this overnight?" the colonel asked.[56]

The order was to form a thirty-eight-man Honor Guard to receive and transport the special casket from France to Arlington for the ceremonial reburial. McKelvey promised to make available anything that Erskine needed

General Graves Erskine, USMC (U.S. Marine Corps photo)

and assigned the thirty-eight Marines from the Sea School to the guard. "The old ship [*Olympia*] didn't have any regular quarters," said Erskine. "They hadn't had any Marines there for a long time, and we didn't have very much in the way of facilities."[57] *Olympia* was indeed an old cruiser, but in 1898 had been the legendary flagship of Commodore George Dewey during his climactic victory in the Battle of Manila Bay. It was a fitting selection to now transport the honored remains of the Unknown Soldier back to the United States. *Olympia* sailed from Norfolk, first to the British port of Plymouth and then to Pier d'Escale in Le Havre.

On October 22, the caskets of four unknown soldiers were exhumed, one from each of the four American cemeteries, and taken to Chalons-sur-Marne for a secret selection process. That selection ensured that even the cemetery from which the chosen body had originated would remain unknown.[58]

Shortly after 1 p.m. on October 25, the special funeral train bearing the casket of the Unknown Soldier from Chalons-sur-Marne arrived. "Thirty French soldiers removed the floral pieces from the train and took position in the column for the procession to the docks. The American body bearers then carried the casket from the funeral car and placed it on a waiting caisson while the band played '*Aux Champs*' and French school children showered the caisson with flowers."[59]

At Pier d'Escale, Captain Graves Erskine's Honor Guard formed in two ranks next to the ship's band. As the caisson arrived, the casket, draped with the American flag, was carried forward by six army body bearers. The Honor Guard snapped to "present arms," and *Olympia*'s band alternately played the national anthems of France and the United States.

Chopin's *Funeral March* was the slow dirge as the procession of six Sailors and two Marines relieved the Army bearers to carry the casket across the gangway. *Olympia* was absolutely level with the dock, so there was not the slightest incline. The casket passed through the rails and onto the ship, and was piped aboard. French and American dignitaries followed in procession.[60] The casket was placed on the stern deck in a prepared, decorated spot, and floral tributes already arranged there were enhanced with additional flowers brought aboard by local schoolgirls. At 3:20, *Olympia* cast off, and accompanied by an escort of nine ships, put out to sea with a final 17-gun salute.[61]

But it was as if this booming salute was a preview of an upcoming battle to be waged between *Olympia* and an Atlantic Ocean storm now boiling with the fury of two hurricanes. Hurricanes #5 and #6 had both churned across the Florida Peninsula and out into the Atlantic. They tracked to the northwest—directly into the path of homeward-bound *Olympia*.

A wooden weather case protected the casket from damage, but its exposed position on the stern deck was a dangerous place. The ship's captain, Henry Lake Wyman, ordered the case moved to a higher spot, away from the anticipated destruction of crashing waves that were sure to wash over *Olympia*'s stern. Going up was the only choice since the case was too large to pass through the bulkhead doors and passageways.[62] The highest spots on the ship were the two bridges.

"The *Olympia* had two bridges," said Captain Erskine, "the forward bridge was for the navigation and operation of the ship; the after bridge was where they gave all the signaling; all the signal lights were there." There was a small, flat spot between the ship's port and starboard funnels, at the highest point of the ship, and it was there that the Marines chose to secure the casket and case. "We lashed this fellow down with everything we could tie on him."[63] One member of the Honor Guard said that the lashing-down was with "enough line to secure the battleship *Wyoming*, fore and aft, with breast and spring lines to boot."[64]

On October 31, the ship passed the Azores, and "Near gale force winds ripped into *Olympia*. . . . Green water smashed into the bow of the ship and washed down her wooden decks like a mountain stream in spring. When the bow rose on a growing wave, the fan tail dipped low with green water gushing across the place where the Unknown Soldier previously rested."[65] "There were times we thought we might not make it home," said Captain Erskine. "The chaplain and the captain got together and held a special service praying to God that the ship wouldn't sink."[66] A terrible thought haunted the young Marine captain. "What if the Unknown Soldier—the hero America is waiting to honor—is washed overboard? I knew if such a thing happened, I might as well jump over with him."[67]

After ten frantic days fighting the rampaging Atlantic weather, with Marines guarding the casket strapped to the ship's bulkheads and railings, the seas finally calmed. On November 9, in a steady rain, *Olympia* sailed up the Potomac River, trading salutes with other military posts. It docked at the Washington Navy Yard, which was jammed with a legion of dignitaries forming the delegation to receive the nation's Unknown Soldier. In that legion were General of the Armies John J. "Black Jack" Pershing and Major General John A. Lejeune.

Two of Captain Erskine's Marines and six sailors from the ship's company carried the casket to the gangplank. *Olympia* began firing a 1-gun salute each minute, the boom echoing over the hushed crowd. The band again played Chopin's *Funeral March.* A solemn silence followed the last notes. Those gathered on the dock focused on the flag-draped casket, now borne by the ramrod-straight sailors and Marines standing at the head of the gangway.

Dignitaries held their hats over their hearts, and the military saluted. The overbearing silence was suddenly pierced by the boatswain's shrill whistle to pipe the Unknown Soldier ashore. Eight army soldiers from the 3rd Cavalry met the ship's bearers and bore the casket to the waiting black-draped caisson.[68] Two days later, the solemn interment of the Unknown Soldier at Arlington Cemetery added another page of honor to American history. Graves Erskine and his thirty-eight Marines had been the first in what was to become a long line of trustworthy soldiers guarding the body of this unknown warrior.

⚜ ⚜ ⚜

Casket carrying the Unknown Soldier is brought down the gangway of the USS *Olympia,* 1921 (National Archives)

Unknown Soldier with escort (National Archives)

On the twentieth anniversary of the burial of the Unknown Soldier, Captain Graves Erskine was a senior colonel teaching Marines at Quantico. A month later, Japan attacked Pearl Harbor.[69] Recognizing his enormous talent, war planners immediately made him Chief of Staff, Amphibious Force, Atlantic Fleet, to teach and supervise the training of the Army and Marine Corps personnel in the art of amphibious warfare.[70] But that lasted only a short time before he was sent to Pearl Harbor to assume the duties of Chief of Staff of the Amphibious Corps, Pacific.[71]

Through the Pacific campaigns of 1943 and 1944 to seize Tarawa, Kwajalein, Saipan, and Tinian, Erskine was Chief of Staff to the Amphibious Force commander: the crusty and hard-nosed General Holland M. "Howling Mad" Smith. In August 1943 Erskine became the youngest brigadier general in the Marine Corps and pioneered amphibious warfare doctrine; and was General Smith's chief planner of those four operations.[72]

In 1944, after the Tinian Operation had successfully concluded, General Erskine had accompanied General Holland Smith to the navy flagship and was surprised when Smith bluntly announced: "I am going to see that you get promoted to major general, and you are getting the 3rd Division!"[73] The 3rd Division had just completed the operation to recapture Guam, and was earmarked for the next operation, code named "Detachment." It was to be the attack to seize the island of Iwo Jima in a three-division assault. Iwo was only six hundred miles from the Japanese mainland. For the first time in its history, the Marine Corps would actually fight as a corps—V Amphibious Corps—that included the 3rd, 4th, and the newly formed 5th Marine Division.

"The concept was to land the 4th and 5th Divisions abreast on the east side of the island," said Erskine, "and my division was put in general reserve. . . . The 5th Division was on the left and the 4th Division on the right."[74] The landing beaches were on the southwestern side of the island. The 5th Division landed by the narrow neck that pinched off the towering height of Mount Suribachi and drove straight across the half-mile narrow neck to the opposite coast. One regiment of the division was to break off and then surround and seize the volcanic mountain itself. The 4th Division was to land to the right of the 5th and immediately turn to the north to drive up the eastern side of the island.[75]

The barrenness, bleakness, and hostility of the terrain of this desolate island was well-described by historian Joseph H. Alexander: "Mount Suribachi dominated the narrow southern end, overlooking the only potential landing beaches. To the north, the land rose unevenly onto the Motoyama Plateau,

Underground Japanese tunnel system
(U.S. Marine Corps sketch)

falling off sharply along the coasts into steep cliffs and canyons. The terrain in the north represented a defender's dream: broken, convoluted, cave-dotted, a 'jungle of stone.' Wreathed by volcanic steam, the twisted landscape appeared ungodly, almost moon-like . . . something out of Dante's Inferno."[76]

To command the defense of this hellish place, the Japanese High Command had selected a fifty-three-year-old samurai warrior, Lieutenant General Tadamichi Kuribayashi. Reflecting Japan's desperation in the face of a collapsing military, he was told: "Only you, among all the generals, are qualified and capable of holding this post. The entire army, and the nation, will depend on you."[77] It was a thinly veiled order to fight to the death.

The twenty-one thousand Japanese defenders dug in and swore to take ten of the enemy with them before they died. They dug underground tunnels, sixty to ninety feet deep, and made side rooms, sleeping quarters, dining rooms, resting areas, and fighting positions; and connected them all to vertical shafts and other horizontal tunnels.[78] "The key point of position construction," wrote Japanese major Yoshitaka Horie, "was to make a cave from which we could snipe, and to link these caves with underground paths. Everybody, including the general, staff officers and men, made his position—and his tomb—by himself. . . . General Kuribayashi even published his 'Cave Digging Discipline.'"[79]

Colonel Takeichi Nishi, a Japanese tank commander, famous as a gold medalist in the 1932 Los Angeles Olympics, wrote, "We are going to dig the land 20 meters deep and make underground streets; then we won't worry about the enemy's one-ton bombs."[80] Indeed, Iwo Jima had been pummeled by air bombardment since August 1944, and on December 8 the heavy bomber horses of the 7th Air Force joined the naval aircraft hounds to hammer the desolate island, hopefully into submission.

B-24 heavy bombers from the Marianas began a seventy-four-day aerial campaign that rained down a cascade of bombs and explosives. It became the greatest aerial effort conducted in the entire Pacific Theater.[81] But the effectiveness of this massive aerial offensive, as revealed in a Japanese defender's letter home, was minimal. "The enemy air raids come more than 10 times a day," wrote Lieutenant Colonel Nakane to his wife, ". . . We can get some fish because whenever the enemy makes air-raids, many fish come to the beach, being killed by their bombs. . . . So, if they do not come, we miss them. . . . We're gladly waiting for the enemy."[82]

None of this was known except to the island's defenders. When the name of Iwo Jima was revealed to the embarked Marines, it only drew blank stares, and the inevitable question: "Where's that?"

So the twenty-one thousand Japanese defenders on the surface of Iwo Jima slowly disappeared, one by one, gun by gun, into the island; 361 artillery pieces, 65 mortars, 33 large naval guns, and scores of large-caliber antiaircraft guns were swallowed into the earth. Kuribayashi's own command post, in the extreme badlands of the northern sector, was seventy-five feet deep and connected with more than five hundred feet of tunnels. "One installation inside Mount Suribachi ran seven stories deep."[83]

On February 17, Transport Squadron 11 weighed anchor in Guam to carry General Erskine's 3rd Marine Division and all attached units to the battle area. Two days later, the ships anchored eighty miles off the Volcano Island of Iwo Jima in a spot called the Transport Area—nothing more than an imaginary corral for ships. It was simply a point on the map identified by longitude and latitude map coordinates. Here, in division reserve, the 3rd Division would wait for its call to battle. Erskine wrote, "The ships had steamed in, as dark, phantom shadows: USS *Cape Johnson, President Adams, Callaway, Frederick Funston,* and *Napa.* Just before midnight, the last of them arrived."[84] On February 19, D-Day, the assault regiments of the 4th and 5th Marine Divisions stormed ashore on the black sand beaches; but no one in the Transport Area

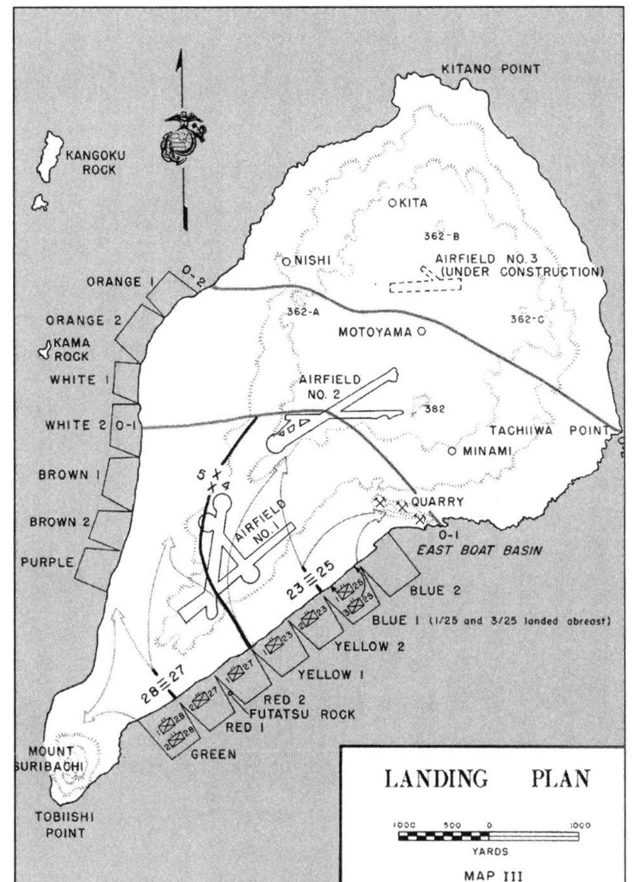

Iwo Jima landing plan, Map III

(U.S. Marine Corps map)

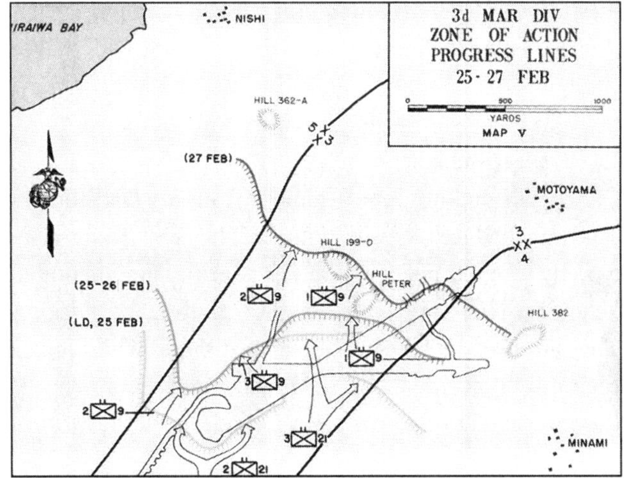

3rd Marine Division Zone of Action

(U.S. Marine Corps map)

knew anything of the attack. Their job was to simply wait, and if called upon, to hurl the weight of their attack into the battle.

Every Marine in the invasion force had carefully studied the models and terrain maps while on board the ships. They realized it would be another typical island invasion: strong Japanese beach defenses, backed by a main line of resistance, defended by an enemy who would, in final desperation, hurl himself in the human wave of a banzai charge. There would be few enemy survivors, and when it was over there would be the next island.

The call to action for the 3rd Division was not long in coming. The ships suddenly weighed anchor and on D+1, February 20, moved to within fifteen thousand yards of the smoking beaches. The klaxons and boatswains' pipes called the Marines to their debarkation stations at the ship's railing. All eyes strained to catch a glimpse of the objective, but Iwo Jima was only a blur in the distance, shrouded in explosions, smoke, and haze. Only the very top of Suribachi, thrusting its peak into the blue sky, was visible. One artillery observer described this chaotic, smoking scene as "Hell with the fire out."[85]

Erskine's command would not land as a division. Its commitment to the battlefield was to be piecemeal. The 21st Marines would go in first. Congestion on the beach did not allow his other two regiments to land. So, in practiced formation, the 21st Marines climbed over the sides, four at a time, and began the difficult descent down the rope cargo nets into bobbing landing craft. Their feet engaged the horizontal strands of the net, and then felt downward for the next strand. Their hands gripped the vertical strands as if they were the sides of a ladder. To mistakenly grip the horizontal strands was to invite the boot of a descending Marine above to step upon and crush the fingers, dislodging the man for a life-threatening downward plunge. One Marine described this maneuver: "Climbing down rope nets with full gear and weapons required the agility of a monkey, combined with the skill and daring of a trapeze artist, to avoid being slammed against the side of the APA or being crushed by the LCVP as it pitched and rolled."[86]

"They brought in one regiment from my Division," said General Erskine, "and attached it to the 4th Division. They were put in a very hot sector and had a pretty rough time."[87] Those 21st Marines landed under heavy enemy fire on Yellow Beaches 1 and 2 and fought with the 4th Division for the next two days. They attacked up the sliding terraced slopes while Japanese gunners, deep in their dug-in positions, thinned their ranks. On the 23rd, after bitter fighting, they managed to claw their way to the edge of the Motoyama 2 airfield. There, they waited for the supporting tanks. But the twelve tanks moving up from the south end of the airfield to support them were quickly neutralized by concealed mines planted along the airstrip, and by the heavy-caliber Japanese guns that had registered every square yard of the airfield. Hundreds of pillboxes delivered covering fire over the entire strip. First Lieutenant Raoul Archambault, looking out over the flat runways, said, "This is fighting on a pool table."[88]

In a heroic effort to suppress the Japanese defensive fires from those almost invisible pillboxes, Corporal Herschel Williams from West Virginia manned the last flame thrower in his company and conducted his own personal war against the buttoned-up Japanese gunners. Carrying the seventy-two-pound weapon on his back and protected only by four riflemen to keep the Japanese pinned down, Williams charged his first target. With his left hand he squeezed the handle that lit the spark igniter, and then, bracing himself against the anticipated recoil, squeezed the firing trigger with his right. The jellied gasoline leaped from the nozzle, creating a stream of roaring flame. The occupants in that first pillbox were incinerated, and when a Japanese soldier leaped from his concealed position Williams turned and ignited him. Five more pillboxes fell that day to Williams and his flamethrower. He received the Medal of Honor for his actions.

Despite the heroics of Williams and the 21st Marines, the regiment could not hold its hard-won position, and was forced to give it up and back down off the runway. Succeeding attempts with more armor were also repulsed. Major Holly Evans, commanding the 3rd Tank Battalion, said, "On the second day of fighting for Motoyama 2, only 19 of my original 46 tanks were operative. We were hit by anti-tank weapons, machine guns, cannons, five-inch guns and 150 mm mortars, as well as mines of all descriptions. Most of our men and equipment were lost on the airfield—ten tanks were knocked out on the strip within a few hours."[89]

General Graves Erskine and his 9th Marine Regiment were finally brought ashore on D+5, February 24, and rejoined his 21st Marines. His final regiment, the 3rd Marines, remained on board, the last of the reserves. "The Division was committed as a division on D plus 6, between the 4th and 5th Marine Divisions in the vicinity of Airfield No. 2."[90]

Sandwiched between the other two Marine divisions, and attacking to the north, Erskine's 3rd Division discovered that it was testing the Japanese main line of resistance. This formidable line, bristling with firepower at every point and covered by artillery, stretched out along a line from coast to coast, through

Motoyama 2 airfield, and was invulnerable to any flanking attack.[91] To breach the Japanese line would require a step-by-step, yard-by-yard, grinding frontal assault.

On February 25, the attack began. At 0930, Erskine's 9th Marines jumped off, supported by twenty-six tanks. The tanks crawled onto the southernmost runway of Motoyama 2 and headed north with three tanks in the lead: *Ateball, Agony,* and *Angel.* But they did not even make it to the second runway crossing before Japanese fire brought them to a halt. *Angel* and *Agony* exploded in flames, and *Ateball* was stopped when a high-velocity shell blew off a track. Japanese gunners, with perfect observation, were able to fire unobstructed down the long, flat runways.

For the next four days, the two regiments of the 3rd Division battered against this formidable, rock-solid line, with the full power of massive artillery, naval gunfire, and air support. Slowly and grudgingly, the Japanese line bent. On February 28, D+8, part of the line broke. But Marine casualties had been ghastly. A horrified veteran Marine correspondent, Jim Lucas, who had covered the previous savagery at Tarawa, now wrote of Iwo Jima: "It was bad there [at Tarawa], but it was all over in seventy-two hours. This bastardly battle just goes on and on from one ridge to another. When will it end? And will anybody be alive when it does?"[92]

On March 8, Erskine's battle line occupied the cliffs in the north, overlooking the sea, and, by his order, a reconnaissance patrol descended to the shoreline to scoop up a canteen of water—proof that it had come to the end of the island. That canteen was happily sent back to V Corps commander Major General Harry Schmidt with a note: "For inspection, not consumption."[93] General Graves Erskine had fought his division, with only two of his regiments, to the opposite end of the island, across the most heavily defended central highlands. Along the way he had begged to have his 3rd Marines landed to relieve his worn-out men. "They wouldn't let me bring them in," he said. "I'd asked the question of [Admiral] Kelly Turner, and [General] Holland Smith, and the usual answer was, 'You got enough Marines on this island now; there are too damn many here.' I said some of these people are pretty tired and worn out, so take them out and bring in some new ones. They practically said, 'you keep quiet; we've made the decision.' . . . I didn't get all of my artillery ashore either!"[94]

But General Erskine was also miffed by the actions of the 5th Marine Division, which had all three of its regiments. "The 5th on our left had been stopped dead. . . . Their lines were way back from ours, maybe 500, maybe

1,000 yards. We were mopping up when the acting chief of Staff for General Harry Schmidt called me: 'We want you to start making plans to join up with the 5th Division to clean out the Kitano Point area.'"[95] Kitano Point was at the very end of the island in the 5th Division's zone. Erskine was incredulous and yelled into the field phone that he wasn't going to do it, and just "why the hell can't they take it?" and "if you want it taken, give me the job; those guys [the 5th] are finished."[96]

General Schmidt himself got on the phone, and Erskine told him the same thing. He was not "bailing out" the 5th Division or "conducting any joint operation with them," and if General Schmidt wanted to take Kitano Point, "he could take it in one day." Schmidt shot back, "The 5th has been trying for 5 days," and the sharp-tongued Erskine informed the general: "Hell, I'm talking about the 3rd Division, not the 5th!"[97] The next morning the 3rd Division's Kitano Point operation kicked off, and at 1300, Erskine reported to General Schmidt, "the island is secure."[98] His division had accomplished the mission in half a day.

General Graves Erskine continued his Marine Corps career, first begun at LSU, until retirement in 1954. In 1953, as a lieutenant general, he became Commanding General, Fleet Marine Force, Atlantic. Upon retirement in 1954 he was promoted to the four-star rank because of his record of extraordinary heroism in combat. He died on May 21, 1973.

CHAPTER 9

Arthur Abramson

SINKING THE RISING SUN

Arthur Abramson was in the second grade and gloomily sat in a classroom at a table cluttered with construction paper, scissors, glue, and pegs of all shapes. It was the special classroom reserved for the mentally deficient, and for students whom the teachers deemed to be uncontrollable. It was 1929 in the city of New Orleans, and his well-conceived plan to escape this "God-forsaken" school had gone astray.

After two years of what he deemed to be confinement in a school he hated, he had decided to run away to protest his forced attendance. But running away as a second grader was not for the faint of heart. It required great planning, not to mention bravery and determination, and seven-year-old Arthur Abramson knew he had both. After morning formation on the day he had chosen for his great escape, the teacher picked up the handbell and rang it loud and long. That was the signal to the students that the singing of "America the Beautiful" was over, as was the Pledge of Allegiance, and the school day was about to commence with a march to the classrooms. To young Arthur, that ringing bell was his signal to break ranks and bolt for the front door and to his freedom on the other side. He would not spend one more day in this accursed place.

He ran as fast as his legs could carry him and hit the door in stride, only to have his way temporarily blocked by some meddlesome teachers intent on barring his flight to freedom. He would have none of that, and he steeled himself for the fight. His small fists flailed away, making periodic contact with some beefy bosoms, and soon he broke free and was in full flight for home, four blocks distant. He had allies there. Arthur was now the perfect model of an ancient Roman maxim: "Fortune Favors the Bold."[1]

An unexpected ally was there—his father, who had returned home to retrieve something he'd forgotten, and he scooped up the running boy just as he entered the house and asked what was wrong? Arthur enthusiastically explained the whole situation: his brilliant plan, his daring escape, his flight from the tyranny of the school. But his dad did not seem to understand the situation no matter how Arthur explained it. Ignoring the great bravery his son had just displayed, he unceremoniously packed him in the car and returned him to school.

The betrayal was humiliating, and the teachers, especially the two who had tried to stop Arthur from leaving, immediately took him into custody and banished him from his classroom. They concluded that Arthur was in some way mentally deficient and placed him in the special classroom, and, thus, he now was spending the rest of the year cutting and pasting, and putting the geometrically shaped pegs into the proper geometrically cut holes.[2]

Fortunately for Arthur, his dislike for school gave way to better habits, and he finished high school at sixteen, in the top of his class. That was in 1938, one year before Adolf Hitler invaded Poland. Arthur went to work, mostly as a "gopher," for $25 a month. At that time, the service seemed very attractive, and he had this desire to one day be a navy pilot. War was not in his mind, but flying was, and he was so enthralled with the prospect of it that he mentally planned a career as a naval pilot in the peacetime navy.

In 1940, he had saved some money, and entered the fall semester at LSU and was in the cadet program. He excelled in the military regimen and was recommended for the advanced ROTC program. But he was most attracted to a government program called CPT (Civilian Pilot Training). It would teach flying, and upon the successful completion of the course the student would receive a civilian license. Additionally, this CPT program issued the students a set of khaki coveralls, trimmed with LSU purple, and a set of gold wings. That flashy uniform made Arthur feel "nine feet tall." While he wore his LSU cadet uniform on drill days, he was free to wear his aviation uniform at any time. Since he only had one change of civilian clothes, at every opportunity he donned his "aviator's uniform" and walked the campus with the other cadets, including the new female cadets. No one else had a uniform with wings of gold. The girls loved the uniform, and Arthur loved the girls.[3]

On the day that Japan bombed Pearl Harbor, Abramson and his friends gathered to discuss the situation. Like most other young Americans, there was an immediate desire to sign up and answer the call to arms. Some said they were going to join the paratroopers; Arthur opted for the navy. Six months later he quit the university, enlisted in the navy, and became a seaman 2nd Class assigned to the Naval Air Station in New Orleans. He opted for flight

LSU female cadets
(LSU Archives)

training, went to Pensacola, and upon completion of flight training in January 1943 was commissioned as an ensign in Naval Air. Whenever asked to list his first three choices of aircraft to fly, he filled in all three blanks with "Fighters, fighters, fighters." When his supervisor asked what he would do if he was not selected to fly fighters, he candidly replied, "I would want to be released from the navy!"[4]

Advanced flight training in February took him to Chicago and the freezing cold weather whipping off Lake Michigan. Advanced carrier training was on USS *Wolverine,* a poor excuse for an aircraft carrier. It was an old paddlewheel boat that had been converted to include a flat deck laid on top. Each hopeful pilot was required to make eight takeoffs and landings. Training on *Wolverine* was unusual at best. To make the required landings, the aviators were once again placed in their advanced trainer aircraft, but these aircraft had no tail hook with which to catch the arresting wires on deck. Naval ingenuity, however, had solved that problem. A hook had been attached to a rope and was run from the tail of the aircraft, up over the top of the tail assembly, to the cockpit, and inside the aircraft. To accomplish this, the cockpit had to be left open. This was not unusual since the cockpit was left open for safety purposes, but what was unusual was that the rope, with the hook, was placed in the cockpit. There was no place to hang the hook, and the pilot couldn't just hold it, so he draped it on one of the unused knobs on the controls.

After takeoff, with the cockpit open and the frigid Chicago air whistling in, he would circle the ship for the inevitable landing. As the plane was in the act of landing, he would throw the hook out of the open cockpit hoping to snag an arresting wire. Amazingly, and despite some anxious moments, the hook worked well.

But it was not the main problem for Abramson. The stabbing cold, with the wind whipping through the open cockpit, was unbearable. He had never been issued any winter clothes, so he and the rest of the pilots flew in their thin khaki uniforms which they had used in Florida. Later they would joke how it had been perfect training to deploy to the Pacific.[5]

Fifteen months later, on June 19, 1944, Lieutenant (junior grade) Arthur Abramson sat in the cockpit of his F6F *Hellcat,* waiting to launch off the aircraft carrier USS *Yorktown.* It was a dawn launch in the Philippine Sea and was no different from the thousands of other carrier launches conducted by ships of the United States Navy during the two and a half years of war since Pearl Harbor. On this day, *Yorktown* turned its bow into the wind along with the other three carriers of Task Group 58.1 and prepared to send its fighters and bombers into the sky.

The mission today was to support the Marines attacking the islands of Saipan and Tinian to wrest control from the Japanese. Saipan was the larger island and most heavily defended, and this strategic, fourteen-mile-long stronghold in the Southern Marianas was defended by thirty-two thousand determined Japanese soldiers, who had proven themselves formidable foes. Early progress reports of this landing indicated that the struggle was as savage as had been the battle for Tarawa. Twenty thousand Marines and army forces landed on June 15 on the western beaches, and by nightfall more than 10 percent of the American invasion force had been killed or wounded. By June 19, the two Marine divisions, having suffered frightful casualties, had joined up and began to swing to the north, a move that would crack the Japanese defensive line and bring the campaign to an end. But there was still hard fighting ahead, and that required support, particularly air support, and some of the aircraft from Task Group 58.1 were detailed for that mission.

Task Group 58.1 was massive: four carriers, four cruisers, and eleven destroyers. But it was only the tip of the enormous iceberg known as Task Force 58. This monstrous naval force was an armada of four carrier groups and one battleship group. The force counted fifteen carriers, seven battleships, twenty-one cruisers, sixty-nine destroyers, and 891 carrier aircraft. The mission was to capture Saipan, Tinian, and Guam as well as defend against any naval attacks from the newly formed Japanese naval force known as the Mobile Fleet. The

Mobile Fleet was centered on nine carriers, five battleships, thirteen cruisers, twenty-eight destroyers, and 430 carrier aircraft. This prowling Japanese force had recently been sighted operating three hundred miles to the west of Saipan.[6]

To Arthur Abramson, who sat in the cockpit of his F6F *Hellcat,* this grand picture of opposing, maneuvering fleets was completely unknown. He was not concerned with grand strategy. The only thing he knew was that the wind was whistling down the deck, the launch officer, known as Fly One, was to the right-front of his aircraft with his flag snapping in the wind, and he had just given him the go. He pushed the throttle fully forward, and it was now his job to get the two thousand horsepower, twelve thousand-pound armed and loaded *Hellcat* into the air. His mission was to bomb and strafe targets on the island of Guam, and, like the hundred other combat missions he had flown since becoming a naval aviator, that was all he needed to know. Thirty other aircraft were on that mission to Guam.[7] Once his mission was completed, he was to return his aircraft safely, be debriefed, and stand by for his next mission. That was life in the air war in the Pacific.

Flying with Abramson were other aircraft from his own squadron, VF-1, as well as separate flights of aircraft from the three other carriers of Task Group 58.1: *Belleau Wood, Bataan,* and *Hornet.* These "newer" versions of *Hornet* and *Yorktown* were namesakes of the original carriers sunk by the Japanese. But they carried the same names to confuse the enemy and deny him the knowledge, and satisfaction, that he had sunk the originals. The missions assigned to the aircraft from Task Group 58.1 were varied. They included supporting the troops fighting on Saipan, bombing and strafing the enemy airfields at Rota and Guam to the southwest, and conducting air strikes to "soften up" Guam, next on the invasion list. The remaining aircraft, still on the carriers, provided task force defense.

After launching, Abramson flew close to the water to avoid radar detection as he began the two-hundred-mile flight to the target. He was twenty-two years old, but his youth belied his experience. He was a veteran of a hundred combat missions and was flying with an esteemed fighter squadron known as the "*High Hats.*" As the strike force flew on to Guam, young Abramson studied his instrument panel and navigational data, and finally reviewed his target assignment. He carried twenty-four hundred rounds of .50-caliber ammunition, which fed the six machine guns in his wings; and today he also carried a 250-pound bomb under his right wing, which he hoped to deliver to a particular antiaircraft position at Agana, on the Orote Peninsula, on the western side of the island. He would unleash the .50-caliber ammunition on the fuel and repair facilities on the Japanese airfield, and, if lucky, on a Japanese *Zero* that might venture up to challenge the force. He had a score to settle with the Japanese and their *Zeroes*.

It had been seventeen months since commissioning, and this combat mission to Guam was a long way from home, college, family, and especially the girl he had met at LSU while wearing his "aviator's uniform." Flying close to the water, Abramson watched the whitecaps pass rapidly beneath him. The war in the Pacific was a seducing war. He had gazed out over the miles upon miles of open water which was deep blue and spectacularly beautiful. He marveled at God's work in the breathtaking beauty of the cloud formations that jutted into the sky. It was hard to believe that a war was taking place. From the air, there was no sign of war like on the land. No shattered buildings and bombed out villages. There were no craters and twisted railroad tracks or burned vehicles pushed to the side of the road—lifeless relics of earlier battles. There was no field with rows of white crosses interspersed with Stars of David; there was just water, the sun, and the crystal-clear majesty from twenty thousand feet.

Even in aerial combat, he realized that killing was sanitized. The killing field was in a small, cylinder-shaped space rising from the water to fifteen thousand feet. In that space, dueling aircraft would swirl and dive in a ballet of death. Planes and pilots plunged inside this cylinder, often to a flaming fate; but when the dogfight was over, the area was again peaceful and serene—no broken aircraft or bodies littering the battlefield—just the Pacific Ocean in all its beauty.

Guam came into view, and Abramson rose to attack altitude and swung his *Hellcat* to approach the target area, easily identified by the Orote Peninsula and the town of Agana. As he had done so often before, he put the fighter in a glide-bomb run and lined up the area of the suspected AA position in his sights. At 425 knots, his right thumb pressed the red bomb release button on the tip of the control stick. The 250-pound bomb released, telegraphing its departure with a slight uplifting of the aircraft, and Abramson pulled out of his glide. The bomb would continue along the glide path; and as he gained altitude and banked to the left, he looked back to see the explosion. A dirty cloud was all that appeared from the spot of the impacting bomb. His second, third, and fourth passes were strafing runs, and his .50-caliber wing guns shredded storage, maintenance, and fuel areas. Fires sent black smoke into the air marking his path. Other aircraft engaged Japanese *Zeroes* that rose to challenge the attackers, but those engagements were brief. He had no opportunity

Escort carrier much like the *Nassau*
(Naval History Archive)

to settle that old score with the accursed *Zero* aviators. Out of ammunition and with fuel running low, he headed for the rendezvous point where his squadron would gather for the return flight to *Yorktown.*

What a difference this mission was compared to his first combat mission seven months earlier. He had entered the war zone on the small jeep carrier *Nassau,* which had never been designed to handle the *Hellcat.* It was so small that it looked like the training ship *Wolverine* on Lake Michigan. The only difference was that he didn't have to throw the tail hook out of the cockpit to land. Flying on that tiny carrier was nothing like being on the big fleet carrier *Yorktown.*

The escort carrier (CVE) had been a brilliant invention born out of great necessity. In the early days of the war, when carriers were in great demand and very short supply, industrialist Henry J. Kaiser promised President Franklin Roosevelt that he could turn out smaller carriers in just three months instead of a year. He was already turning out Liberty ships in a month. With the President's approval, Kaiser forged ahead. Using the existing hulls of oilers, cruisers, and cargo ships fitted with flat deck tops, he turned out scores. Kaiser's carriers were lightly armored and much slower than the fleet carriers.

Looking like baby carriers, they were mocked and ridiculed as "Kaiser's Coffins," "Woolworth Flattops," and "One Torpedo Ships." The running joke was that their CVE designation stood for Combustible, Vulnerable, and Expendable.[8] Since they lacked the speed to run with the fast carrier task force, these baby carriers often sailed with the amphibious landing force to provide air cover and close air support for the attacking Marines.[9]

Such was the case at Tarawa. VF-1's twenty-four *Hellcats* were embarked

Hellcat landing
(U.S. Navy photo)

on USS *Nassau,* CVE-16, and the mission was to bomb and strafe the island in support of the 2nd Marine Division. The navy pilots were green; for most it was their first combat mission. They anxiously examined their maps to find the battle area in the vast Pacific Ocean and finally gave up. When told to look in the Gilbert Islands, they asked, "Where the hell is that?"[10] Their fingers finally traced the triangular shape of Tarawa Atoll, formed by fifteen islets, and the tiniest of those islets was tucked into the southwestern tip of the atoll. That was the target: Betio—one half mile wide and three miles long.

In the days leading up to the Marine invasion, VF-1 flew air cover to protect against intruding Japanese aircraft, but none appeared. Then they participated in prelanding bombardment, joining the battleships in pounding every square yard with shattering explosions. Despite these massive prep fires, the invasion was stopped at the water's edge by a fierce Japanese defense. The battle was to become three days of slaughter, best described by one historian as "Utmost Savagery."[11] On D+1, Arthur Abramson and his fellow *High Hats* were pressed into service to provide close air support for embattled Marines facing an enemy determined to die to the man. Rockets, napalm, and gunfire pounded Japanese positions. On D+2 and D+3, it was more of the same. Finally, on D+5, VF-1 was ordered to launch from the deck of *Nassau* and land on the newly captured airfield on Betio. It was to be their new home. The last remnant of the Japanese force was still fighting on the extreme tip of the island when the *Hellcats* roared in to land.

They landed in a scene straight from Dante's *Inferno.* If the young aviators of VF-1 had thought life on the jeep carrier left something to be desired, they soon discovered that their new life on Betio would have them begging to return to the *luxuries* left behind on *Nassau.* Ironically, it was Thanksgiving Day, November 25, 1943, when Arthur Abramson and VF-1 touched down on the compacted coral runway. They were greeted like saving angels. Abramson had hardly brought his *Hellcat* to a stop when he was surrounded by more than a hundred bedraggled Marines cheering his arrival on the terrible island as if he was a conquering hero. But the celebration was short-lived, and the cheering Marines were soon flat on the ground when the whine of snipers' bullets zipped overhead. Abramson found himself alone standing upright until a bellowing Marine sergeant ordered him to get his "f--ing ass" down.[12]

The following day, with part of the island still unsecured, Abramson and another pilot dared to go souvenir hunting. The first souvenirs they found were the ghastly remnants of bodies strewn all over, in various stages of disfigurement and decomposition. Their wanderings brought them to the main Japanese block house, bombed and burned out, and piled three-high with bloated corpses The stench was overwhelming. They returned with no souvenirs except decomposed human remains stuck to the bottoms of their boots.

With Betio secured, VF-1's mission was to now fly combat patrols to the north in preparation for the next leap forward into the Marshall Islands. These were monotonous four-plane patrols from the Tarawa airfield, day in and day out, never encountering one Japanese plane rising to confront them. Monotony became their biggest enemy, until it was joined by something worse: dysentery and dengue fever. One by one the pilots succumbed to the ravaging diseases, which were the by-product of the exposed, unburied, and rotting corpses baking in the tropical sun. And their close companions, the blue-green flies, happily flitted everywhere. In death, these former Japanese soldiers were still inflicting casualties on the Americans.

Life on Betio was devoid of any redeeming qualities. Comfort was measured by new standards. This new comfort included keeping most of the swarming blue-green flies off of one's food and having your canteen of distilled seawater slightly hot, and the hope that the water intake was not close to a decomposing body bobbing in the lagoon.

To the parched and sunburned Marines on the island, this precious water was swallowed despite its awful taste in long gulps to slake the never-ending thirst. Happiness was dreaming of a cool sip of ice water. But what started out only as wishful thinking soon became a reality, and the pilots of VF-1 found themselves the answer to the Marines' dreams. It started slowly: first one

VF-1 carrier launch (U.S. Navy photo)

Marine approached a pilot, and then another. Soon there were long lines of Marines, each holding one canteen wrapped in a water-soaked cover, waiting to hand it to be placed in the recesses of the *Hellcat*'s fuselage. The extreme cold at the high altitudes would chill the filled canteens to near freezing, and upon landing, the pilots passed the precious cargo out to the same waiting line. It was one step from heaven.

But there were few heavenly things about Tarawa, and sickness and dysentery became a way of life. Sickness did not stop the necessity to patrol; sick or not, each pilot flew his turns. The flesh melted off their bones, and after two months Arthur Abramson's meager 155 pounds had shrunk to 123. The parachute riggers teased him and said, "Mr. Abramson, you'd better never jump out of your plane, because you don't have enough meat on your butt to deploy the chute."[13]

When the army completed taking nearby Makin Island in the Tarawa Atoll, VF-1 began to fly northward to escort bombers to the next island, and then to the one after that. Each one had an equally unfamiliar name: Kwajalien, Eniwetok, Wotje, Mili, Maloelap, and Jaluit. The dysentery and dengue fever never let up, nor did the combat patrols. Oh, for the clean sheets and tablecloths aboard ships, especially the jeep carriers, instead of the rot of the hellhole of Tarawa Atoll. The daily scene at the flight line bore a contorted resemblance to a dusty bar in a wild west saloon. But instead of a bartender serving whiskey in shot glasses to worn-out trail hands, two navy corpsmen served the exhausted and emaciated fliers shots of paregoric and bismuth in paper cups, ladled from gallon jugs set up on a makeshift table. Like the trail hands of old, the pilots were red-eyed and gaunt-faced and tossed down a couple of "jiggers," hoping that the magic elixirs would quell the boiling rumbling in their intestines. It usually didn't work, but like punch-drunk fighters they lined up every morning to swallow the shots before staggering off to the flight line. If during the flight the pilot suffered gut-wrenching cramps and diarrhea, his only relief was on his cockpit seat. It was not uncommon to see a cockpit canopy slide open and the pilot's shorts fly out into the air. As the weeks and months went by, each man finally accepted his fate, and that was the realization that he would spend the rest of the war not as a glorious carrier fighter pilot on a majestic ship, but as a wasted warrior in some hell like Tarawa.[14]

On one of the high-cover missions protecting army bombers flying to the Marshall Islands, Abramson finally got his chance to face a Japanese *Zero*. It was not what he expected. He was flying as wingman for the squadron's executive officer, a man not possessing great fighter pilot skills and who was notorious for riding herd over the junior officers as if they were recruits. On this mission, with no warning, young Abramson saw the exec's external fuel tank fall away, which is the maneuver prior to going into combat to lighten the weight of the aircraft. But there was no alarm of enemy in the area. Bending down to release his own tank, he hesitated, and after straightening up to scan the clouds, he found that he was alone. No sign of the exec. He had no idea of even where to look, so he continued his patrol to provide cover at twenty-five thousand feet for the bombers flying below. Suddenly two Japanese *Zeroes* came across his front from his right. They were at least two thousand feet above him, which put his *Hellcat* in a very disadvantageous position. No one had ever told him what to do in this situation when he faced the enemy alone. He thought for an instant, and then did what he'd been trained to do in the face of the enemy. He attacked.

His attack forced him to try to climb the additional two thousand feet and engage the speeding enemy in a full deflection shot. His other option was to attempt to get behind for a zero-deflection shot. Despite his bravery and resolve, the maneuver he was attempting was impossible. He was too green to recognize that the Japanese planes were decoys, flying this tempting pattern in the hope of luring an American airplane into this exact maneuver.

Abramson took the bait. As he began firing upward in his steep climb, his aircraft began to shudder, and he knew instantly that he was being shot from behind. Two other *Zeroes* had zoomed down upon him from above. Their 20-millimeter cannon fire smashed his right aileron, and he felt the stick between his knees go dead. The *Hellcat* did not respond to his command. He looked in his rear mirror and the tail section seemed okay, but a quick look to the right revealed a soccer ball-size hole in the wing.

With the aileron shot away, the plane went into a deep spiral, and the *Zeroes* followed, pouring cannon fire into the stricken *Hellcat.* Abramson placed both feet on the right rudder and pulled the stick all the way back to get the plane to respond, but it continued to fly in its deep, descending spiral. In his rearview mirror, he could see the two *Zeroes* and even see the faces of the grinning enemy pilots, as they hammered him. He radioed a *Mayday* as he watched more holes appear in his wings, as if some giant were stabbing it with a huge ice pick. His instrument panel was shattering before his eyes, and he could feel the thuds as the rounds impacted the armored plate in the back of his seat. He hunkered down as far as he could. He pushed his entire 123 pounds against the rudder, wondering how long the aircraft could remain up with the pounding it was taking. Just when he thought it couldn't get any worse, one of the Japanese pulled off for an oblique shot and sent rounds into the propeller. There were zinging sounds like ricocheting bullets in a Wild West movie.

With the propeller nicked, the aircraft began severe shuddering and vibrating, and a panicked Arthur Abramson could now see through more holes in the wings. He had endured ten minutes of battering and had lost altitude. But luckily, at the lower altitude with the denser air, the mechanics worked a little better, and he was able to slow the spiraling. He flew in a flat turn, not going anywhere but at least not going down. A glance at his watch told him it was time for rendezvous. The aircraft of VF-1 would assemble to return to Tarawa, and he'd be left alone.

The sickening thuds against his seatback reminded him that the enemy was still there. He wondered why they hadn't tried to make a high side run, a maneuver that would kill him and cut his plane in half. But they seemed content to stay behind and pound him, like a cat with a mouse, continually tossing it in the air before killing it.

Abramson knew the plane was finished and thought of bailing out, but knew that the Japanese enemy would strafe him, so he stayed put. Even if he made it to the water, he knew the possibility of making it to some speck of land in the vast ocean was remote, and the land out here belonged to the Japanese, and capture meant torture and most likely an unpleasant death. The thought of digging his own grave and kneeling while some sword-wielding Samurai bastard lopped off his head and then kicked his scrawny ass into the hole behind his still-rolling head kept him in the cockpit. But minutes later he changed his mind. Despite the grim possibilities of jumping, one fact was sure: death was certain if he stayed in the *Hellcat.* At the first lull he would jump. They must be close to being out of ammunition and low on fuel.

He went through the checkoff for bailing out. All straps and cords were free. It would be a disaster to jump and suddenly find yourself snagged on the doomed aircraft like a cowboy, fallen from his horse, with one foot stuck in the stirrup. The aircraft would fly around, dragging him outside, until it made its final dive to the ocean with him in tow. Straps free, he said to himself: breathe deeply from the oxygen bottle; jettison the canopy hood; make sure the survival kit is in the backpack and stand up.[15]

"I would jump out the left side since that was the direction of the turn," he testified. "When I had one foot out, I felt for the wing, just as the Jap fired again, sending tracers whistling past me, and I fell back into the cockpit. Blood was running down my back. I reached back around to check my wound, and my fingers discovered the fluid was clear and tasted like water." His survival pack had absorbed the bullet, which pierced his bottle.

He now resolved to die with his aircraft. He carved out his plan for victory, even if his victory would be a small one. He would deprive his attackers the thrill of returning to their officers' club and bragging how they had killed an American. It wouldn't change the outcome of the war, but it would be his own private, personal victory. He would keep his plane flying until they were forced to leave him. *He* would determine when he would die, even if that was only a minute after they left him.

With that resolve, he felt a sudden calm come over him. He marveled at how the *Hellcat* continued to fly against the odds, and his thoughts now turned to his girl, Toni; and thoughts of dating and dances at LSU and how that was all thousands of miles away. She would never know what happened to him. He retrieved her picture encased in a leather holder and placed it on the shattered instrument panel. He would relax and spend his last moments with her.

At the lower, denser altitude, having regained some control over the aircraft, he flew just above the water to keep the enemy from making a high side run on him. Flying just twenty-five feet off the surface made that maneuver impossible. Finally the *Zeroes* broke off. He'd sent them home without a confirmed kill. He'd won. They'd drink no saki over him tonight.

Limping along, he found his handheld compass and plotted a reciprocal course for Tarawa. It was something to do. His radio periodically emitted static as the jiggling wires occasionally made contact, and although there was some comfort in that, he knew that his comrades were long gone. He resolved to continue his one-man war.

Suddenly a voice broke in: "Hold on, I'm coming." Someone had heard his original *Mayday,* and he looked up to see if his ears were playing tricks on him. The welcome sight of another *Hellcat* appeared up on his left, and the pilot was giving him a big thumbs-up. It was his closest friend, Lieutenant Norman Duberstein, who had joined the rendezvous and, finding him missing, had broken from the group to search. The two plotted their course home and flew on for forty-five minutes, like a seeing-eye dog leading a blind man; but now, Abramson knew that when he had to ditch, someone could report his position.

His hopes soared when the island of Makin came into view, but this joy was short-lived when the control tower informed them that a crashed aircraft had closed the runway. Seventy-five more minutes to limp to Betio. He expected his aircraft to disintegrate along the way, but it miraculously stayed aloft. With Toni's picture still on his dashboard, Abramson finally made his approach, more crablike than straight, to the fifty-foot-wide, twenty-seven-hundred-foot-long airstrip.

His wheels were down, but he didn't know if they were locked, and he could not get the wings level. With both feet on the right rudder and the stick jammed all the way to the right, he came in listing 30 degrees. On either side of the narrow strip, other aircraft were parked like dominoes. He knew that he would smash into one side or the other, careening and exploding into the parked aircraft, and he would not survive to enjoy the end of his long, miraculous flight. To his direct front he could see men running to escape the path of his wildly careening *Hellcat.*

The first part of the aircraft to touch the runway was indeed his dipping left wing tip. But instead of grabbing and slinging the plane to the side and into the parked aircraft, the contact flipped the plane up onto its two wheels, and Abramson suddenly found himself rolling straight down the coral strip. He came to a halt at the very end and struggled to a standing position in the cockpit on legs of jelly. Vehicles and men raced toward him as he stepped out onto the wing and lowered himself to the ground.

He began an inspection of his pile of junk, just as if he were going to return it to the navy for future use. In the space behind the cockpit he found more than thirty 20-millimeter cannon bullets that had bounced off his armored seat and collected behind him. The ground crew began counting holes in the aircraft and stopped counting at 275. A bulldozer came and pushed the wreck off the strip.[16]

✶ ✶ ✶

Abramson shuddered as he flew home from his Guam mission and reflected on his narrow escape from death seven months earlier. In the distance he could see *Yorktown* on the horizon. What a joy to call the big carrier home. After three months, flying missions from Tarawa, his time in hell had ended as abruptly as it had begun. One day the squadron commander called all his pilots together and happily announced that they had been relieved and would go to Hawaii for a week's leave. The jumping and jubilation knew no bounds, and then he further announced that they would be assigned to a fleet carrier. They left Tarawa hoping never to see the place again.

At the end of their week's leave in Hawaii, they saw a huge carrier pull in and anchor at Ford Island. It was *Yorktown,* and no one had to tell them that this was their new ship. VF-1 had been assigned to The Fighting Lady, and Abramson and his friends went to the dock to admire the monster 872-foot ship. *Nassau* had only been 550 feet. He knew, at last, his lifelong dream was to be fulfilled.

Within weeks he was at sea on The Fighting Lady, and now, seven months later, his flight of four aircraft returning from their morning strike on Guam flew home. The *Hellcats* flew down the starboard side, spaced in an echelon column in preparation to land. Arthur Abramson was third in line, and as the first plane passed the ship, the pilot peeled off to the left to bring himself astern to approach the flight deck. Several seconds later, the second plane broke from the column to make the same stern approach, and then Abramson broke off.

He could see the first plane had landed and was rolling forward while folding its wings. The second pilot made his approach, correcting his path and attitude from signals from the Landing Signal Officer (LSO), who stood to the left in full view. First his arms were level, then tilted, then level again as he mimicked the plane's attitude and signaled it to the pilot on his approach. At the last moment, the LSO whipped his right paddle across his chest indicating to cut power, and the second pilot dipped the aircraft's nose, pulled it back up, and slammed the *Hellcat* down hard, catching one of the arresting wires.

Abramson approached as the crew fought to clear the second plane from the deck. The deck of the 33,000-ton carrier pitched and bobbed, and was anything but stationary, but he had the LSO in view and adjusted to his arm movements. At 105 feet, and gliding perfectly for the deck, the LSO gave him a sudden wave off, Abramson rammed the throttle forward and the engine roared to gain air speed. He pulled the nose up and flew low over the flight deck, where he saw that the second plane had not been freed from the barricade wire. He banked to the left for a second approach, and now watched the fourth plane land. A minute later, he too was down and walked toward the ship's island for debriefing. When he arrived below deck, the whole ship was buzzing. Everyone was cheering and slapping each other on the back and creating an atmosphere like New Year's Eve. What had happened? What was all the fuss? When he was able to get one of his friends to calm down, he learned that he'd missed the greatest naval victory since Midway.

While on his mission to Guam, the Japanese Mobile Fleet had launched four attacks against the entire Task Force 58. Each enemy attack had not only been defeated but annihilated by the American planes and ships. Those few enemy aircraft that had managed to break through the defensive phalanx of fighters had met the withering fire of the ships' antiaircraft guns. The navy had lost only thirty aircraft while shooting down over 350 of Admiral Ozawa's *Zeroes* and torpedo aircraft. In his excitement, one young American pilot quipped that "this was like an old-time turkey shoot." A member of the press picked it up, and the battle was forever called "The Marianas Turkey Shoot."[17]

Arthur Abramson and the thirty aviators from the Guam mission were delighted with the victory, but inwardly very disappointed to have missed the greatest of air battles. They had been part of the force from the beginning, and had been left out of the glory, and the lost chance to have struck a great blow against the enemy was more than frustrating.

The next day, the crew of *Yorktown* settled into a very light schedule. Morning flights were flown by some of the new replacement pilots. They were off at 0400, after a full breakfast of steak and eggs, which was a navy tradition for the early flights. The early strikes usually had the heaviest casualties, so the sumptuous breakfast came to be jokingly called the condemned man's last meal. The veteran pilots now found spaces to hang out and relax. Some wrote letters home in the ready room, others read in little corners or worked crossword puzzles. Most enjoyed the light day and were convinced that the war in the Pacific would go on forever. There were thousands of islands between them and Japan, and they had seen the Japanese fight for every inch of terrain, down to their last soldier; and when the island hopping was over, there was the Japanese mainland itself to invade.

The favorite hangout on this light day was the officers' wardroom. Most pilots described it as an oasis where they could return to civilization. The room gleamed with clean, starched, white tablecloths, polished silver, napkin rings, and sparkling crystal. Meals were served by white-jacketed Filipino stewards with white gloves. The food was delicious, the coffee hot, and the tea ice cold. It was a place of civility and elegance in a world that included hell holes like Tarawa.

At 1400 on this lazy day in the Pacific, June 20, the lounging men were jarred back to reality by the grating voice over the P.A.: "All pilots and crews, report to the ready rooms." Most everyone figured that they would get an update on details of yesterday's climactic battle and shuffled off with their unfinished letters and crossword puzzles in hand. Ten minutes after assembly, the ship's intelligence officer came on. His announcement was electric. "The long-range search planes have located Ozawa's Fleet fleeing northwest for Japan."[18] It was escaping to fight another day. There was no way that Task Force 58 could catch it. They were already 240 nautical miles away and increasing the distance, and the pilots lamented that they would have to fight them again.

The word quickly spread that, with the enemy in full flight and out of range, Task Force 58 would cruise slowly and refuel. That was a promise of, perhaps, a few more days of light duty. The *High Hats* went back to their letter writing and crossword puzzles. But at 1545 the P.A. again howled to life. "The Task Force Commander has decided to launch a maximum attack against Ozawa's Fleet. Strike in forty-five minutes. Stand by for General Quarters!"[19]

In the ready rooms, there was dead silence. Only the clicking teletype broke the pall. No one had to tell the squadron that if an attack was launched at 1630, it would be dark three hours later. And no one had to announce that the extreme distance to the target meant that the strike force could not return to the carrier after the attack. Still the aviators scribbled on their boards, figuring out the data necessary before the strike.

The squadron leader, Lieutenant Commander Bernard "Smoke" Strean, finally spoke. "Okay, gentlemen, we have our orders. I'll lead the entire strike. We'll be first in to find the target."[20] The only noise was the scribbling on the

Yorktown pilots after briefing on Ozawa strike: Arthur Abramson (*foreground*), George Staeheli (*left*), Ralph Wines (*right*), M. M. "Mad Dog" Tomme (*background*).
(U.S. Navy photo)

plotting boards. Slowly the names of those who would go were called and marked on the blackboards on the walls of the ready room. Abramson's name was assigned to one of the two standby aircraft that would be ready to replace any aircraft with mechanical problems. No one asked the ancient question, "Why me?" They already knew the ancient answer: "Because you are here!"

Hellcats would be rigged with an external 150-gallon fuel tank to bring the fuel capacity up from 250 gallons to 400. The bad news was that instead of running the engines lean, which would consume the precious gasoline at a rate of thirty-five gallons per hour, the force would have to catch the retreating Japanese fleet, and that would require a high-power setting that consumed close to 100 gallons per hour.

Abramson rechecked his calculations. No matter how he figured it, the result was always the same. His night splashdown in the ocean would still be 150 miles short of home. Twelve pilots from VF-1 were designated to go. Four hundred aircraft would launch from the entire force of Task Force 58. The chance to annihilate the fleeing enemy fleet overrode the concern for the safety and well-being of the men involved in the attack. The importance of the mission made them expendable. As if to mention the obvious, the squadron commander instructed the selected men to refresh themselves on the techniques for night bail out and water landings. It was all matter-of-fact, no drama, no emotion. It was warriors following orders, and training and discipline took over.

Those officers not flying the mission went topside to watch the eventual launch. The others donned their flight gear. Abramson approached his aircraft, and the three members of his deck crew were standing waiting for him. They were three navy enlisted men: a radioman, an ordinance man, and a mechanic. They had a special relationship with their pilot and considered the plane theirs—they only lent it to the pilot for a mission, and they expected him to return it to them intact.

As with most deck crews and pilots, the relationship was informal. No stiff salutes or anything like that. In fact, a breezy touch of the cap was more in order, and when the aircraft went to the line the deck crew usually gave the machine a loving pat as if it were a pet. Today was different. They helped Abramson into the aircraft and made sure he was buckled in. The tug on his straps seemed extra caring. With that final act, the deck crew jumped from the plane. With goggles down, the young pilot watched the three men straighten to attention, snap a salute, and hold it. Tears welled in his eyes, and when they cut the salute away, he was glad his goggles covered his eyes. They knew this was goodbye.

Engines coughed to a start, and the whole deck became a scene of whirling propellers. An officer on deck flashed a chalkboard with the latest flight data up to the pilot, and the first aircraft was off at 16:27. Abramson waited, wondering if he would be needed, and realized that the 240-mile estimate to the enemy fleet was wrong. The flight would be longer. *Yorktown* had turned into the wind to launch, and that was east. The Japanese fleet was fleeing to the west. He also realized that if by some miracle they could make it back, most of the pilots had never made a night carrier landing. He had made a couple, and it was sheer terror. First it was necessary to find the ship in the dark, and the only way to do that was to fly to the spot where it was thought to be and look for the telltale phosphorescent trail which all vessels make as their props churn the water. After finding the biggest trail, indicating the carrier's path, the pilot had to make an approach from the stern in the pitch dark, trusting to fate that the ship was there. If he found the LSO in his pink and yellow, fluorescent suit with his fluorescent paddles glowing in the dark, he would look like some ghost or underworld specter. When he gave the signal to cut power, the pilot had to slam the *Hellcat* down, and only when the aircraft hit the deck was he certain that the carrier was beneath him.

But why worry about a night landing? There was no way to make it back. He sat in his cockpit with his engine idling and watched his friends roar down the deck and into the air. He checked once again that his flaps were lowered to the proper position, just in case. Only a few planes left to launch—all on a one-way trip.

The banging on his fighter's fuselage startled him. It was his signal to launch. He sent his *Hellcat* racing, full throttle, down the flight deck and into the air to become part of the first wave of a strike force of eighty-five fighters, seventy-seven bombers, and fifty-four torpedo planes.[21]

With the first wave off, the deck crews brought the second wave aircraft up from the hangar decks. On board *Lexington,* Admiral Marc Mitscher's flagship, a radio operator rushed to the admiral and handed him a message. Mitscher read it and was dismayed. He gave orders to stand by to recall the flight while he studied the chart. When he was finished, he canceled the second wave launch but did not recall the aircraft now flying west, toward the Japanese fleet. The message had come in at 16:05, but by the time it had been decoded and reached the bridge, the last of the first wave was gone. There was a mistake. The corrected report placed the enemy fleet sixty miles farther away. It was not a 240-mile flight; it was 300 miles. Two-hundred and sixteen aircraft flew into the setting sun.[22]

Shortly before sunset, the navy aircraft from Task Force 58 spotted the first ships of the fleeing Japanese fleet. Two oilers plowed along in the ocean far below, and several planes attacked them, sinking both. Forty miles later, the main body came into view. Commander Strean led a diving attack on a large carrier, and the *High Hats* placed three bombs on the frantically maneuvering ship.[23]

In the darkening dusk, the Japanese antiaircraft fire took on beautiful colors. It became streaks of red, yellow, and orange. Also in the bright sun, nasty, black puff appeared, but these exploding rounds made the sky over the battle area look like fireworks. In fifteen minutes the attack was over, and the aircraft from the strike force turned for home. All external tanks had been dropped, and fuel gauges dipped below half.

The force flew to the east and into the night. Every man was on his own, and most of the flight flew in small groups. The pilots leaned the fuel mixtures back until the engines coughed. Abramson teamed up with Ensign M. M. Tomme. Everyone liked him, but no one knew what M. M. stood for, and he insisted that everyone just call him, "Tommy." Yesterday had been a red-letter day for Tomme. He had scored his first two kills in the Turkey Shoot.

Abramson and Tomme flew at fifteen thousand feet and set their flight path on a slightly descending angle. That would ensure the greatest conservation of fuel and allow gravity to stretch out the distance covered. Radio silence was broken, and there was chatter among the planes as they "held hands" on the return flight. After an hour, the first aircraft began to call out longitude and latitude in preparation for ditching. These were the aircraft that were damaged, or those that had burned the most fuel in high-power settings in the target area or had carried more weight in bombs. Sometimes the radio crackled with, "I'm going in." As the miles increased, the calls increased. There were ingenious techniques for ditching. Some decided to ditch together as a group rather than crash individually, and the radio crackled with, "Here we go" or "The Navy's losing its best fighter pilot, I'm going in."[24]

The *Hellcat* fighters had a great advantage. They had been able to carry extra fuel without having to carry additional ordinance. The torpedo bombers and dive-bombers were not afforded that luxury because they were ladened with bombs and torpedoes. As the miles increased, the bombers splashed into the ocean.[25]

Tomme and Arthur flew on. More and more planes ditched, and Abramson suggested they recalculate their fuel. It was 2115 and pitch dark as the two aviators scribbled separately in their cockpits. Abramson calculated twenty-two minutes left. Tomme figured twenty-three. They decided to take Tomme's estimate. Their altimeters showed they were at four thousand feet. They had glide-flown to stretch every mile, but they knew the end was near.

Both men were silent, engrossed in their own thoughts for the next fifteen minutes. Abramson revisited Toni. Her picture was again on the dashboard in front of him, just as it had been during that hopeless situation off Tarawa. This time he was confident that someone would tell her he was missing and presumed dead. But where on the ocean could they point to his grave? At 2130 an excited voice crackled over the radio and shook the two pilots from their melancholy thoughts.

"Lights! Lights, at 10 o'clock!"

Anxious eyes snapped to that bearing, and sure enough on the horizon was the glow of lights reflecting in the night sky as if a city were there. Every aircraft headed for the illumination. After Admiral Mitscher had launched the flight in the afternoon, he had turned the task force to the west and ordered a maximum speed pursuit. For four hours Task Force 58 had been racing at flank speed to try and shorten the return distance to save the pilots and planes.

Now he chanced lighting the fleet with total disregard for the possibility of a Japanese submarine attack.[26] The fleet was lit as if in daylight. Running lights, truck lights, and searchlights stabbed into the sky. Star shells were fired from guns to mark the way. The carriers were attacked like a swarm of hornets by the frantic aircraft parched for fuel.

The first arrivers landed easily, but the ensuing stampede caused the LSOs to wave off many. Some aircraft, so short of fuel, could not stay airborne with a wave off and were forced to ditch. Other ships moved to rescue them. Pilots landed on the first carrier they could find. Some pilots committed the court-martial offense of ignoring the wave off and crashed onto the decks, resulting in destruction and death. On one carrier, two planes landed simultaneously, miraculously not crashing into each other.[27] Some aircraft coming in on their final approaches discovered at the last moment that the ship they were approaching was not a carrier but a cruiser or destroyer that had its truck lights burning. Some of those who pulled off their mistaken approaches then ran out of fuel trying to find the carriers.[28]

Tomme and Abramson found *Yorktown* and made their approaches. They promised to meet in the ready room as soon as they landed for some of the medical officer's "prescription," brandy, and Abramson went in first. He could already taste the brandy. He was never so glad to see the LSO signaling him that his wings were level and he was on course. When he gave the cut engine signal, he cut his engine and put the *Hellcat* on the deck.

In a moment he had taxied forward, while crewmen folded the wings. He followed parking instructions, signed the yellow sheet as if he were returning a rental car, and jumped to the deck. He ran to the island, looking to see Tomme set his aircraft down and catch a wire. In the next moment, he, too, was rolling past with a big grin and a big thumbs-up. Abramson stepped through the door and down the ladder to the ready room. Doc's "prescription" was almost in his hand. But when he had almost descended the steps, he heard the telltale sound of a crashing airplane, and wondered who it was. Was it a VF-1 plane, or a plane from a squadron off another carrier who had found *Yorktown?* After he had checked in, he burst into the ready room to wait for Tomme so they could be debriefed, but the assembled pilots there were not smiling and there was no exhilaration. Something was wrong.

And then they told him Tomme was dead. Was this some kind of joke? He had just waved to him on deck, and he'd be down in a minute for a drink; come on fellows. But the shaking heads confirmed it was no joke. An aircraft pilot, out of fuel, had ignored a wave off and the 13,000-pound *Hellcat* crashed on top of Tomme's plane. The prop had ripped into the cockpit and killed him instantly. Abramson sank into a chair and buried his head in his hands.

The next day, Arthur Abramson served as one of eight pallbearers for Ensign M. M. Tomme's funeral. Tomme's body was wrapped in a canvas bag and lay on a board next to the edge of the flight deck. On either side four of his friends faced each other. Each man's somber stare spoke of deep sorrow. They wore their flight suits and vests. The chaplain prayed, a bugler played *Taps,* and the board, under the American flag, was tilted until Tomme's body slid off and down into the deep. The eight men then manned their *Hellcats* for the next mission.[29]

⚜ ⚜ ⚜

The Battle of the Philippine Sea, 19–20 June 1944, was a crushing defeat for the battered Japanese navy. Admiral Ozawa lost over 350 aircraft and, more importantly, the pilots flying them. He also lost three carriers, and two others were damaged. The Mobile Fleet was forever driven from the Marianas, and the occupation of the island chain was ensured. The juggernaut American Navy, led by the Fast Carrier Task Force, now set its sights for the next chain, the Bonin Islands, and its gemstone, Iwo Jima: a close base to begin the air attacks necessary for the eventual invasion of the Japanese homeland. It was the last classic carrier battle of the war. The Japanese could not replace the planes or pilots or ships to venture out as a naval air threat to the American Navy.

The day following the dramatic night mission, ships from Task Force 58 steamed westward along the path of the return flight, searching for downed airmen. Almost eighty aircraft had ditched into the ocean, and only half of the crews were recovered.[30]

In the battle of June 19–20, Ensign Tomme had downed two enemy aircraft. In later action, Abramson would shoot down four *Zeroes.* When his combat tour ended, he was credited with four kills and two probables, but those probables would deny him the official status of ace. He was awarded the Distinguished Flying Cross for his combat action while a member of VF-1.

CHAPTER 10

Major General Troy Middleton

TIGER IN THE ARDENNES

In December 1944, the mood inside the luxurious Hotel Trianon in Versailles was as ugly and miserable as the cold and wet outside. The beautiful hotel had been converted into the very nerve center of the war effort against Nazi Germany. On September 20, General Dwight D. Eisenhower had moved the Supreme Headquarters Allied Expeditionary Force (SHAEF) from southern England to Versailles.[1]

Back then, the world was euphoric following the success of the June Normandy Invasion and the subsequent breakout from the beaches. The Allies had raced across France in hot pursuit of the fleeing German army, retreating so rapidly that they couldn't catch them. Paris exploded in joy. The images of the massive victory parade down the Champs-Élysées filled the front pages of the world's newspapers, and theaters showed the intoxicating news clips to gleeful audiences. The war was almost over, certainly by Christmas if not before.

But that was then, and now it was December and the end of the war was nowhere in sight. In fact, the Allies were bogging down. Their long supply line, running all the way from Cherbourg in France, and a failed British airborne operation through the Netherlands, brought their seemingly invincible offensive to a complete halt. In concentrating on the airborne assault through Holland, the British Army, on the left of the advancing Allied line, had failed to take and open the vital port of Antwerp. Because it could not break through the German defenses during Operation Market Garden.[2]

Tensions ran high between the Supreme Commander, General Dwight Eisenhower, and British Field Marshal General Bernard Montgomery. Montgomery always seemed to have an excuse as to why failures did not involve him. While the British Army advanced ploddingly to the northeast, the Americans advanced rapidly on a relentless eastward track. The Third Army under the fiery General George Patton had actually crossed the Meuse River, one hundred miles east of Paris, and was now equidistant to the Rhine River and Germany.[3]

On December 7, Eisenhower visited the American front stretched out along the Belgian/German border. The American First Army under General Courtney H. Hodges was deployed to the north of the Ardennes Forest, while Patton's Third Army was to the south. As Eisenhower rode along, he noticed how spread out the line was through the forest. He questioned General Bradley on its vulnerability, and Bradley informed him that he could not reinforce the Ardennes without stripping forces from Hodges and Patton, and thereby weaken their abilities to attack. But he assured Eisenhower that any enemy attack foolish enough to try a breakthrough would quickly be counterattacked on each flank long before it could reach the Meuse River.[4] And as late as December 15, General Montgomery echoed the same sentiment, and proclaimed the weakness of the enemy. "The enemy is at present fighting a defensive campaign on all fronts; his situation is such that he cannot stage any offensive operations."[5]

So, the thinly held Ardennes front was more like a *ghost front* whose fifty-five miles[6] was defended by the three infantry divisions of Major General Troy Middleton's VIII Corps and the recently arrived 9th Armored Division. Actually, VIII Corps' entire front stretched for over eighty miles, from Losheim, southeast of Aachen, in the north, through the Ardennes Forest, and then south into Luxembourg. And one of those defending divisions was brand new to the war. The 106th was freshly arrived from the States. The other two were the veteran 28th and 4th Divisions, resting and refurbishing after months of hard fighting in Huertgen Forest—a fight that had inflicted a combined nine thousand casualties upon them.[7] On December 10, a heavy snow cloaked the entire forest and converted the Ardennes into a picture-perfect winter wonderland. A scene rivaling Currier and Ives artwork.

General Middleton's defensive assignment was not enviable. Everyone, including the Germans, knew that he was stretched thin. Historian Charles B. McDonald called the sparsely manned front "the nursery and old folks home of the American command,"[8] because its defending divisions were either green or beat-up and recovering. To make matters worse, the three divisions were new to the command of VIII Corps. The 28th had been assigned on November 19, the 4th Division on December 7, and the 106th on December 11.[9] Middleton lamented, "In November, I had the Eighty-third, the Eighth, and the Second Divisions. All three had been with me out of Brest. All were well-led. Then all three were moved out." The Germans had placed their own assortment of second-rate troops into their line opposite VIII Corps, confident that no Allied offensive would begin from there. But these lesser troops were still perfectly capable of patrolling, infiltrating, and scouting to keep a finger on the pulse of the Ardennes.

The German army had actually stopped retreating. They had managed, as the Allies slowed their pursuit, to also stop, and began to reinforce and rebuild with both hands. Rolling stock and railroad losses, miraculously, had been replaced through systematic "looting" of the occupied countries of the Reich. The freight yards held every type of European locomotive and rail car known to man. The reconstituted *Reichsbahn* began to run with the efficiency it had shown when the war first began, and under the cover of bad weather, darkness, and the foliage of the dense forest it began to move, undetected, the equivalent of sixty-six divisions to the front opposite the Americans occupying the Ardennes.

That gathered force now sat in a rock-solid defensive position with short-

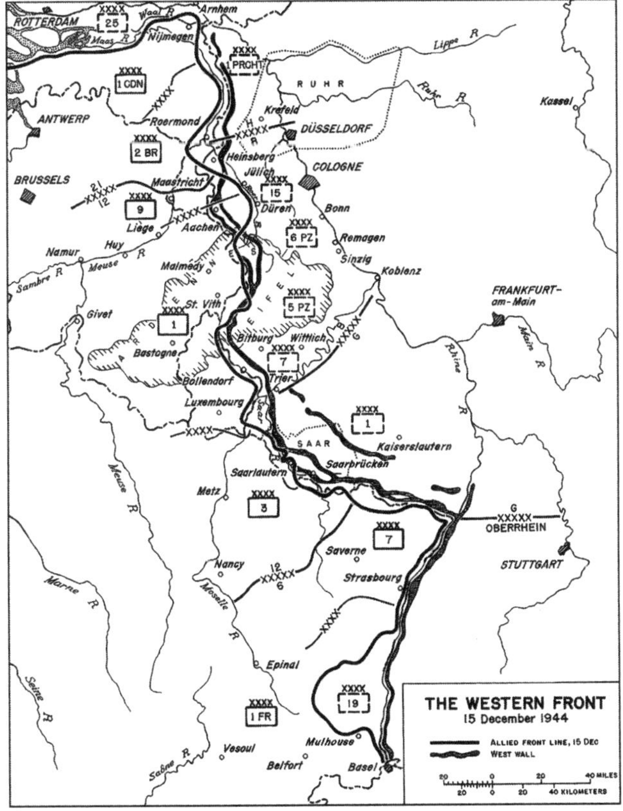

The Western Front, December 15, 1944

(U.S. Army map)

Colonel Troy Middleton in World War I
(National Archives)

ened lines of supply and communications back to the Fatherland. And the Rhine River, and the final protective Siegfried Line, was still intact to their rear as a fallback position. Although they had fewer tanks and artillery than the Allies, the Germans had superior numbers of infantry; if they seized the initiative and chose the battlefield and the time of attack, they could mass those fewer tanks and artillery into an initial advantage.

General Middleton was most aware of his compromised position and had discussed it with his boss, General Bradley. But Bradley called it a calculated risk. To strengthen the axes of his upcoming multiarmy attacks, he had to weaken VIII Corps. "No one, at any level of command, made a great fuss about the risks being run, risks being routine in warfare."

Middleton's VIII Corps Headquarters was at Bastogne, twenty miles to the rear (or west) of the front lines, marked by the Our River. That strategic river ran through the Ardennes Forest. Middleton was no stranger to combat or adversity; in fact, he had been in the army longer than General Eisenhower, having enlisted as a private in 1910. He obtained a commission in 1912 and almost immediately became a combat veteran. He had served during the Mexican Revolution in 1914 and skirmished against the forces of Pancho Villa in 1916. During World War I, Troy Middleton proved his military brilliance in the field, and had greatly impressed Black Jack Pershing's Chief of Staff, Colonel George Marshall. Middleton first led a battalion in heavy combat, and then the 39th Regiment in the Meuse-Argonne Offensive. In Middleton's military file, Marshall had written, "This man was the outstanding infantry regimental commander on the battlefield in France."

When World War II began, Middleton was at LSU and came out of a comfortable five-year military retirement. He had been assigned during his army career to LSU as Professor of Military Science and Commandant of the Cadet Corps that numbered five hundred. When he left after six years, Middleton's Cadets were the pride of LSU and mustered over seventeen hundred.[10] He felt it was his duty, even after twenty-seven years of military service, to again offer himself to his country in this new war and was quickly made a brigadier general and embarked upon a path of continuous combat. He commanded a combat division in North Africa in 1942, and then as a major general in the invasions of Sicily and Italy in 1943; on June 6, 1944, he led VIII Corps across the beach in Normandy.

General George Patton had specifically demanded Middleton's services. The fiery Patton had inquired who was commanding VIII Corps for the upcoming invasion of Normandy and was told a name he did not recognize. "Never heard of him," he roared. "What's his combat record?" When told that the officer had none, that was enough for Patton. "Get rid of him. I want a man who knows how to fight. Get me Troy Middleton."[11]

Indeed, Patton and Eisenhower demanded their best team for the invasion, and in Troy Middleton there was none better. Years later, Eisenhower would refer to him as "his mentor," having been a student under Middleton in the 1925–1926 class at the prestigious Command and General Staff School in Leavenworth, Kansas.[12] Now, five months after Normandy, Troy Middleton's VIII Corps was deployed in the changing weather in rural Belgium. November always brought the rainy season and the first blasts of penetrating cold. Middleton had seen it all before, and on November 11, Armistice Day, the irony of his déjà vu presence in Belgium, prompted him to write to his wife: "In 1918, on this day, I was south of Metz on the Moselle River when the Armistice was signed. Little did I then think that I would be somewhere in Belgium in 1944 and at war with the same enemy."[13]

His mission for now, however, was not to attack. It was passive and clandestine. He was "to attempt to deceive the Germans by active patrolling, and to prepare for an eventual attack to the Rhine River."[14] Middleton's attempts at deception were sometimes bold and sometimes clever, but all proved to be ineffective. On October 20 he had one of his divisions put on three river crossing demonstrations so the Germans would fire upon them and reveal their positions and concentrations. But the Germans did not take the bait. Then, on November 11, he had a division in the line unleash a tremendous volume of fire, as if it were the prelude to an infantry attack, but again the Germans revealed nothing. In early December he had special troops wear the uniforms and shoulder insignia of the 75th Division to hopefully convince the observing Germans that a new division was joining VIII Corps.

These troops drove their marked vehicles all around, and occupied command posts, and broadcasted the sound effects of a multitude of vehicles. And within this hustle and bustle, their radiomen transmitted fictitious messages and phony supply requests. It was all a grand show. Later, captured German documents revealed that the Germans were somewhat fooled and entered the insignia of the 75th on their tactical maps. But after several days they figured it out and dismissed the 75th Division as the grand hoax that it was.[15]

Still Middleton was not completely blind to the hidden build-up assembling in front of his corps. On December 14, two prisoners said that an attack was coming in his area. And on that same day, "A German woman came across the *Our* [River] and reported that a great deal of bridging material was being moved in at night near her home. She also told us that other residents talked of seeing troop concentrations coming into the area every night." He dutifully sent this intelligence to General Bradley's headquarters in Luxembourg. "But it was not taken seriously," Middleton wrote. "All I could do was to alert all my troops."[16]

Ever so quietly, on the east side of the Our River, those second-rate German troops that had faced VIII Corps' lines slowly disappeared from the Ardennes front. And in their place, the soldiers of a massive German juggernaut slipped into those vacated positions. Over two hundred thousand men formed the ranks of three massive armies—the Fifth SS Panzer Army, the Sixth SS Panzer Army, and the Seventh Infantry Army. "They had traveled by night," wrote historian Frank Price. "They never got out from under the plentiful cover of the Eifel hills during the day. They maintained perfect quiet and all but perfect deception."[17]

> Thirteen infantry and seven armored divisions were ready for the initial assault. Five [other] divisions . . . were on alert or actually en route to form the second wave, plus one armored and one mechanized brigade at reinforced strength. Approximately five additional divisions were listed in the OKW [Wehrmacht High Command] reserve. . . . Some 1,900 artillery pieces—including rocket projectors—were ready to support the attack. The seven armored divisions in the initial echelon had about 970 tanks and armored assault guns. The armored and mechanized elements of the immediate OKW reserve had another 450 to swell the armored attack. If and when the Fifteenth Army joined in, the total force could be counted as twenty-nine infantry and twelve armored divisions.[18]

The total strength of VIII Corps was only 68,288.[19]

Sunday, December 16, 1944, was dreary, cold, and rainy in Versailles. But it was a day full of cheer and great celebration at SHAEF headquarters. The U.S. Senate had announced that General Eisenhower had been confirmed to his new rank of five stars: General of the Army. That rank had previously been conferred upon only two other officers: Generals George Marshall and Douglas MacArthur; and it was the equivalent of the British rank of field marshal.

But for two enlisted personnel at SHAEF, it was an extra special day. Pearlie Hargrave, one of General Eisenhower's chauffeurs in the Women's Army Corps [WAC], and his orderly, Sergeant Michael McKeogh, were married. Their wedding ceremony took place in the historic Marie Antoinette's Chapelle Royale at the Palace of Versailles. Pearlie and Michael were the first couple to be married there in 174 years. In 1770, the ill-fated Louis XVI, the final French monarch, and his bride, Marie Antoinette, had also been wed in that famous chapel. Ike briefly attended Pearlie and Michael's ceremony, and then hosted a champagne reception for the happy couple at his house at Saint-Germain.[20] It was replete with a wedding cake, and Eisenhower made sure it was well photographed.

Eisenhower at wedding on December 16, 1944
(National Archives)

After the festivities, in the late afternoon, General Bradley dropped in. Eisenhower's military strategy to continue to advance against the German army on a broad front was not without problems. Although the tactic forced the German army to defend everywhere, it had placed an enormous manpower and supply strain on the American Army. All but one of the American divisions were now committed to the attacking line. Bradley reported that the flow of replacements, necessary to flesh out these massed divisions, was not keeping pace with manpower needs. While they were discussing these logistical problems in the Supreme Commander's office, they were interrupted by the arrival of British General Kenneth Strong, SHAEF's Chief of Intelligence (G-2). Strong had been Ike's personal pick as intelligence chief,[21] and he carried urgent news from the front.

The three men gathered around a large map spread out on a table. Strong pointed to the VIII Corps area and the front line running through the Ardennes Forest. The German army had attacked in force at 0530 that morning. Bradley examined the map but quickly brushed off the significance of the reported attack. It was a "spoiling attack," initiated by the Germans to draw off parts of Patton's army from its assembly position near Verdun, where it was poised to launch its own attack against the German line. That attack was scheduled for December 19.

Eisenhower did not agree. He saw it as much more. The Ardennes offered no strategic or tactical advantage for the Germans if they captured it. Looking past the Ardennes front, to the west, across the Meuse River, the poor terrain ended, and became good tank-operating country. A German thrust there would put it on the obvious path to the critical port of Antwerp. And an attack there would split the British and American armies and threaten the British with encirclement.

"That's no spoiling attack," he said, instantly recognizing the threat. "I think you had better send Middleton some help."[22] They continued their map study and Eisenhower directed Bradley to get 7th and 10th Armored Divisions moving toward Middleton. Bradley again objected, remarking that Generals Hodges and Patton would be upset to lose those divisions which were parts of their respective armies. But "with a touch of impatience, Eisenhower overruled Bradley."[23]

The group was joined by Eisenhower's personal assistant, General Everett Hughes, and they settled down amid the leftovers of the champagne reception. Hughes opened a bottle of Highland Piper Scotch, and they passed anxious hours playing bridge.[24] General Strong would have important details in the morning.

But Strong's details on December 17 offered no encouragement and were as dire as his original report. Eisenhower immediately signaled General George Marshall in Washington on the situation and returned to his high-level council. His only two reserve divisions were the 82nd and 101st Airborne, both in Mourmelon, France, preparing for an anticipated airborne operation, and he ordered them to begin to move immediately toward the battlefield. By December 18 the two airborne divisions were on the way: "11,000 trucks carried 60,000 men, plus ammunition, gasoline, medical supplies, and other materiel into the Ardennes."[25]

The mode of transportation was mostly open-air cattle cars, and the men stood in the freezing wind for the 107-mile trip. "These trucks had no benches, and damn little in the way of springs. Every curve sent men crashing around, every bump bounced them up into the air. . . . They drove with lights blazing until they reached the Belgium border."[26]

⚜ ⚜ ⚜

Troy Middleton had been asleep in his van in Bastogne on December 16 when the opening artillery barrages crashed into the town at 0530. "By 10 A.M. I had word that elements of sixteen different German divisions had been identified in the attacking force," he said.[27] Knowing the magnitude of the German onslaught, he issued a "hold-at-all-costs" order for all units on the west side of the Our River, and he gave orders that the two regiments of the 106th Division on the German side should immediately pull back across the river. But that pullback never happened, and on December 17, those regiments were surrounded by the swarming German forces of the Sixth SS Panzer Army, cut off, and eventually captured.

To the right of the shattered 106th Division's line was the continuing line of the battle-weary 28th Division, manning an impossible twenty-four-mile front. Opposite that line, and ready to smash through it, was the German Fifth SS Panzer Army, commanded by General Hasso von Manteuffe; and the spearhead of this massive army was XLVII Panzer Corps, led by General Heinrich Freiherr von Luettwitz, who had a reputation for drive and audacity. His orders were simple: "cross the Our and Clerf Rivers, make a dash 'over Bastogne' to the Meuse, seize the Meuse River crossings near Namur by surprise, and drive on through Brussels to Antwerp."[28] Therein was revealed the entire German plan, code named *Wacht am Rhein* (Watch on the Rhine).

But General Luettwitz's attack, instead of overrunning battle-weary troops when it crossed the Our River, ran into a buzz saw. Instead of rolling over the American defenders and racing for the Meuse River, his force had its hands full just getting past the starting line. All day on the 16th the depleted regiments of the embattled 28th Division either hurled back the German offensive or fought it to a stalemate with the help of artillery and especially the devastating firepower of quad .50-caliber machine guns mounted on half-tracks. One historian wrote, "The meat-chopper quad-50s were quite efficient."[29]

On the 17th, General Middleton rushed Combat Command R (CCR) from his 9th Armored Division, widely scattered across his eighty-eight-mile front, into the fray to back up the most threatened parts of the 28th Division's defensive line.[30] Those regiments were to hold until reinforcements could be brought up from the west. They were set up in what could best be called island defenses: clustered in and around villages situated along the major north/south highway that paralleled the Our River. That major highway was known as Skyline Drive.[31]

Middleton particularly ordered two roadblocks: one at a road junction west of Clervaux, at Lulange, thirteen miles from Bastogne and on a direct east/west road. The other was set at Allerborn, even closer, only eight miles distant. These two positions were along Highway N-12, a critical road from the German border directly to Bastogne.[32] He ordered these two roadblocks to be held "at all costs."[33] It was not a meaningless death wish order. It was absolutely necessary. If either roadblock fell, the gate to Bastogne would be wide open and there would be nothing to stop the German army from racing through toward Antwerp.

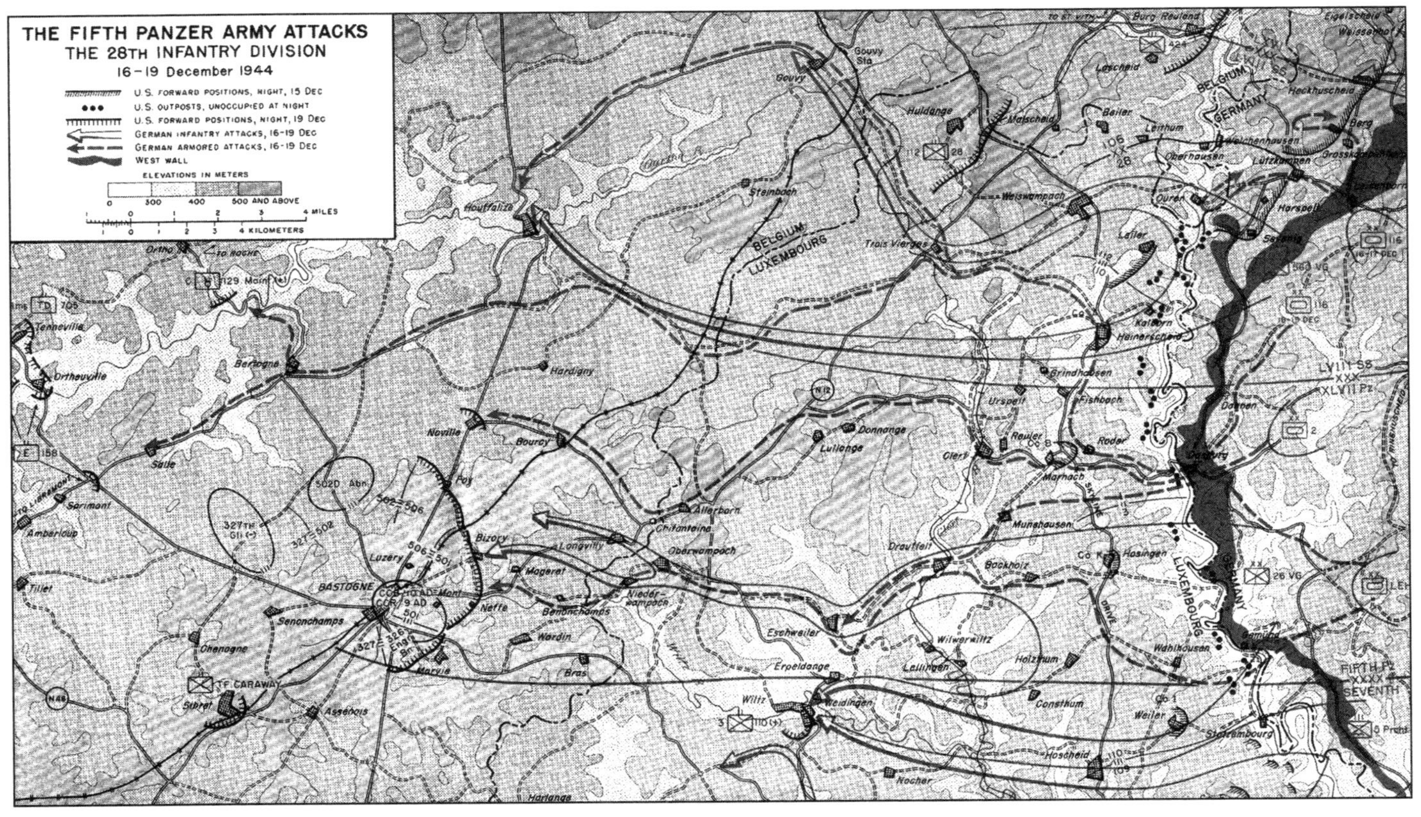

Fifth SS Panzer Army attack (U.S. Army map)

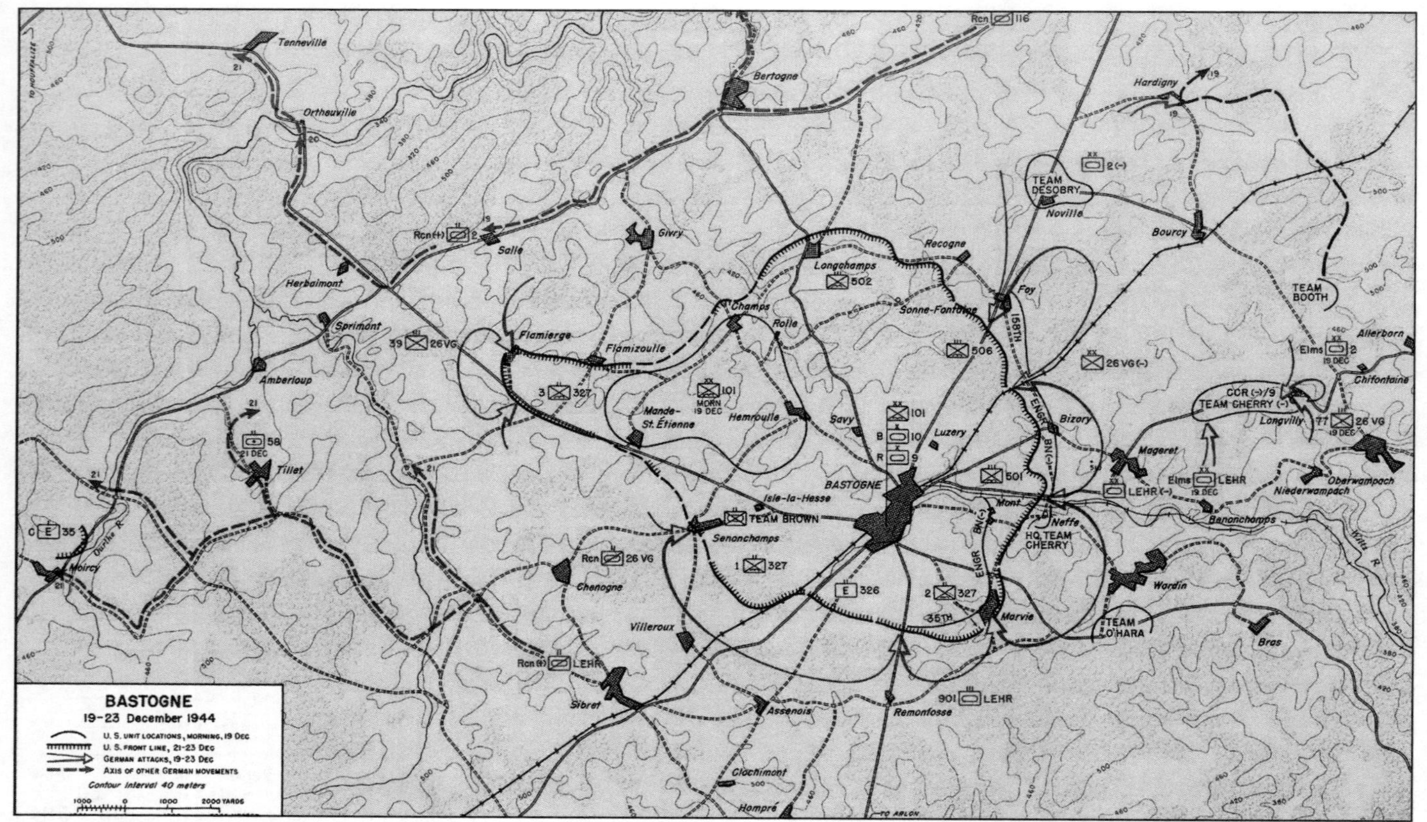

Bastogne (U.S. Army map)

"Without some armor to back up our roadblocks, we couldn't have stopped anything," said Middleton. He had thoroughly reconnoitered the entire area during the previous months, and he knew the terrain. It didn't take him long to figure out what the Germans were up to. The configuration of the land was a dead giveaway, and the crossroads town of Bastogne was the key to the whole German offensive. It was a classic hub town. All the roads leading to the west, the axis of the German attack, passed through it.

During the four years that Middleton had taught at the Command and General Staff School, his favorite course was the one that made the students think creatively, Tactical Principles and Decisions. "Working from the *Gettysburg* map, the student was required to make a decision and write an order supporting it."[34] Bastogne was every bit the critical hub town that was Gettysburg, and Middleton had demanded his hundreds of students to tackle critical tactical problems presented on that Gettysburg terrain. The 28th Division now fought on similar terrain. If the situation had been presented in the Command School as a tactical problem, the students would have been required to plan a defense of Bastogne and present a supporting argument. Middleton's decision was never in doubt. Bastogne must be held. Although defending it would make it a defensive island, surrounded by the onrushing torrent of rapidly advancing German armor and infantry, to not defend it would be absolute folly.

But not everyone agreed with Middleton, including a most senior general. In fact, without Middleton's uncanny perception, Bastogne would never have been defended. On one occasion, General George Patton, in an outrageous outburst, ridiculed Middleton's logic. He said, "Troy, of all the goddamn crazy things I ever heard of, leaving the 101st Airborne to be surrounded in Bastogne, is the worst."[35]

But Middleton was no pushover, and his answer to Patton, and to anyone else who simply would not see the obvious, was short and to the point; and it was the very answer he would have demanded at the Command School. "Bastogne sat astride all the important highways in that part of Belgium."[36] It was just that simple. "To hold Bastogne was just common-sense soldiering."[37] But holding Bastogne with a force that was already partially overwhelmed was not for the squeamish commander. It would require an absolute commitment by all to stand their ground against overwhelming odds, to not give an inch, and if necessary to die with their swords in their hands.

The tankers of CCR of the 9th Armored Division, backing up the infantry along Skyline Drive, knew exactly that. They also knew exactly what every

American tanker knew: they were no match for the superior German Tigers and Panthers in a head-to-head duel. "The Shermans were parceled out in and around the villages in two's and three's. Hidden by walls, houses, and hedgerows, or making sudden forays into the open; the American tankers stalked the heavier, better armed panzers, maneuvering under cover for a clear shot at flank or tail, or lying quietly in a lane until a Panther or Tiger crossed the sights."[38]

The relentless onslaught against the 28th Division's thinly held line eventually focused on the area of the two roadblocks. At the road junction coming from Clervaux, Task Force Rose, commanded by 9th Armored's Captain L. K. Rose, braced for the German attack. Rose had placed a platoon of infantry facing north of the junction, and another platoon facing east, all peering into the gloom and gray of the ominous winter weather. His fifteen Sherman tanks had dug in to present a low silhouette—the ground was not quite frozen—and they were three hundred yards behind the infantry. At 0830 on December 18 the defenders saw three enemy tanks, accompanied by infantry, feeling their way forward in the fog and snow. This force had just come from mopping up the last American defenders in Clervaux, and were now, the German vanguard, pressing on toward Bastogne.

The Shermans waited and tracked them and held fire until the enemy tanks had moved closer. Finally the order was given, and in quick succession the Shermans unleashed their 75 mm cannons and scored hits on all three German tanks. But then the American tankers' nightmare became a reality. The German tanks' armor, especially its thick front plating, merely shrugged off the hits. Only one was disabled, and the other two, unfazed, scurried off for cover. "Shortly after this initial engagement, four enemy tanks supported by infantry emerged from the woods just northeast of the American forward positions. A few minutes later an entire German tank column barreled down on Task Force Rose from the north. The enemy tanks first turned their guns on Task Force Rose's armored infantry posts, forcing the infantrymen to withdraw to the Shermans. As the lead panzer came into view, it was turned back by fire."[39]

But the attacking Germans then laid a smoke screen in front of Task Force Rose and decided to wait for the main German column to arrive. Fifteen minutes later the smoke lifted, and at 1100 Task Force Rose saw that it had arrived. The Germans chose to blind the Americans again. The arriving tanks laid down a second smoke screen that obscured them for ninety minutes. When it finally lifted, the German armor had maneuvered to within eight hundred yards of Task Force Rose.

Sixteen German panzers now bombarded the Americans, along with devastating fire from distant German artillery zeroed in upon their position. Task Force Rose was quickly dissolving, and two surviving Shermans beat a hasty retreat to the rear to get help from Task Force Harper at Allerborn. But no help was to come, as General Middleton nixed the request. A later request to pull the remnants of Task Force Rose back to join Harper was equally, emphatically refused, and the two tanks returned to their positions to face their fate.[40]

They were to fight, with swords in hand, where they stood. Such was the desperation of VIII Corps to buy time. No time would be bought by compromising either position. "Hold at all costs" meant exactly that. On December 18, at 1500, an hour after the final requests for help, CCR headquarters at Longvilly received the last message from the northern roadblock. Task Force Rose ceased to exist.[41]

The German column now moved forward toward the second roadblock at Allerborn and was in positions to its front by the late afternoon. But it did not attack or fire upon the American position until nightfall. Then the combined fire of German armor and infantry overwhelmed it. "In the opening onslaught, the Germans swept the first line of defense with machine-gun fire to clear out any infantry that might be protecting the tanks."[42] Then they turned their attention to the destruction of the tanks and the remaining infantry defenders. Their infra-red sights easily zeroed the Shermans, and the supporting infantry were picked off as they were silhouetted by the resulting fires. "By 9 pm, Task Force Harper ceased to exist as a unified, cohesive fighting unit."[43]

The road to Bastogne was now open, but by midnight on the 18th no German armor approached the wide-open town. General Luettwitz had pushed his XLVII Panzer Corps on to accomplish its mission, and his orders were to "make a dash 'over Bastogne' to the Meuse, seize the Meuse River crossings."[44] As such, he had turned off the road to Bastogne and pressed on for the bridges across the Meuse River. By nightfall on the 18th, the 28th Division was no longer a viable outfit, but it had delayed the German timetable by thirty-seven hours.[45] And the combined defense of the 106th and 28th, with the support of 9th Armored Division, had delayed the German offensive three days. The cost to VIII Corps? Extremely high.

"The 110th Infantry virtually destroyed; the men and fighting vehicles of five tank companies lost, the equivalent of three combat engineer companies dead or missing, and tank destroyer, artillery, and miscellaneous units engulfed in this battle." Historian Fred Cole further wrote, ". . . the Germans paid dearly for their hurried frontal attacks against stonewalled villages and towns—but the final measure of success and failure would be in terms of hours and minutes won by the Americans and lost to the enemy."[46]

✶ ✶ ✶

During the evening of December 17, General Eisenhower approved the order sending the two American airborne divisions to the Ardennes battlefield. Camp Mourmelon was near Reims and 107 miles from Bastogne. For an airborne division, that would be less than an hour via a C-47 aircraft. But this was not to be an aerial move, and the troopers dubbed it a "tailgate jump."

Somehow the Army Communications Zone in that area managed to assemble enough ten-ton open-air trucks and trailers, along with the workhorse two-and-a-half-ton trucks, to mount the two divisions for their move to Belgium. The 82nd would lead the way. As the convoy rolled through the freezing gray and fog, there was no immediate plan on how the airborne should be deployed, only that "Bradley and Hodges were agreed that they would be thrown in to block the German spearhead columns."[47]

General Hodges's First Army headquarters was most concerned by the dangers posed by Jochen Peiper's spectacular armored spearhead thrust toward Werbomont in the north, near Malmedy. Nothing seemed to be able to contain Kampfgruppe Peiper's battle group of the 1st SS Panzer Regiment that pressed the attack of the Sixth SS Panzer Army and made its mad dash for the bridges over the Meuse. So, Werbomont seemed to be the destination of the 82nd and 101st.

The first of the reinforcements to show up in Bastogne was at 1600 on the 18th. It was Colonel William Roberts, commanding Combat Command B of 10th Armored Division. General Eisenhower had recognized the German threat for what it was and had ordered General Bradley to get 7th and 10th Armored Divisions moving toward Middleton, and Roberts was the first to arrive. "When Bill Roberts came up to Bastogne on December 18," said Middleton, "I told him to break up his outfit into three task forces. Bill didn't like it at all. He told me, 'Troy, this is no way to use armor.' And I told him that I knew it as well as he did. But we weren't fighting any textbook war here."[48]

Roberts split his command into three teams and sent them out on three roads that were certain to confront the advancing enemy. Team Desobry, led by Major William Desobry, was detailed to take his fifteen tanks with a platoon of tank destroyers and proceed five miles north, to Noville, on the highway leading to Houfallize. There was no infantry to accompany him, but he was promised that infantry from the 101st would follow him shortly.

Desobry was no fool. He knew that he was being sent directly into the path of a massive approaching armored force. The idea was, as it had been for all VIII Corps defenders, to hold, and to continue to hold, to buy time for reinforcements to arrive from General Patton's Third Army at Verdun—whenever that might be. And he knew that he was out on the end a very long limb, and there was no air cover—or even a promise of air cover—because of the socked-in weather and the forecast for more of the same.

Everyone knew of the recent extraordinary heroism of 9th Armored, and Team Rose and Team Harper—and their "hold-at-all-costs" defenses. Perhaps that is what General Roberts of 10th Armored had on his mind when he briefed Desobry prior to his departure to Noville. "You're young," Roberts said, "and by tomorrow morning you'll probably be nervous, and you might think that it would be a good idea to withdraw from Noville. When you begin thinking that, remember that I told you it would be best not to withdraw until I order you to do so."[49]

With that, Major Desobry and his combat team shoved off to Noville and arrived in the eerie darkness and silence of the abandoned city. At 0230 he set up his last-stand defense. The two other teams from Combat Command B of 10th Armored were dispatched to the east. Team Cherry (Lieutenant Colonel Henry Cherry) hustled out to Longvilly, and Team O'Hara (Lieutenant Colonel James O'Hara) rolled to Wardin.

The next commander to report to Middleton was the Acting Commanding General of the 101st Airborne, Anthony McAuliffe, who came in at 1700. He had taken a detour from his route west of Bastogne. His intended line of march had been toward Werbomont, but before proceeding on he stopped to confer with Middleton for a tactical update.[50] But fate also intervened to redirect the 101st to Bastogne. At the village of Herbomont (seven miles west of Bastogne) two main roads extend southeast to Bastogne and northeast of Houffalize (en route to Werbomont). When the leader of the 101st column, Col. Thomas L. Sherburne, Jr., reached Herbomont about 2000, he found two military police posts some two hundred yards apart, one busily engaged in directing all airborne traffic to the northeast, the other directing it to the southeast. This confusion was soon straightened out; the last trucks of the 82d roared away in the direction of Werbomont and Sherburne turned the head of the 101st column toward Bastogne.[51]

So, 805 officers and 11,035 men of the Screaming Eagles rolled to their destiny in a small Belgium town. They had braved the icy rain and snow flurries on the road from Camp Mourmelon, and the trucks in the back half of the column made most of the run in the dark. All along the way, the entire column confronted the reverse traffic, sometimes panicked and despondent, streaming away from the battle area. The 501st Regiment, in the van, took eight hours and arrived at midnight. The others followed in trace, and by 0900 on 19 December General McAuliffe had all four regiments in Bastogne, and General Middleton had his promised reinforcements.[52]

Generals Bradley and Hodges approved the transfer of the 101st to VIII Corps and ordered Middleton to move his headquarters out of the city, seventeen miles to Neufchateau. He would be of no use to anyone if captured by the Germans. Upon his departure, General McAuliffe took sole command of Bastogne's defense. Middleton telephoned Bradley that "there was no longer a corps reserve." He had committed everything, and "the 101st might be forced to fight it out alone."[53] His parting words to McAuliffe were of hope: "you're going to be surrounded here before long . . . help is on the way from Patton. . . . There's a lot of artillery backing you up. It can put plenty of fire on any point around Bastogne."[54]

⚜ ⚜ ⚜

As the Americans rushed to circle the wagons around Bastogne, Team Desobry was the farthest out of the three armored teams waiting for 101st infantry to join them. In the dark he had set roadblocks east, west, and north at the crossroads at Noville. There was nothing friendly in front of him, and nothing behind for five miles. At 0425, December 19, his northern outpost of two tanks detected the distinctive sounds of German armor prowling around in the night. The plan was to pull the roadblock back into the Noville main defensive line just before first light, but that was still an hour and a half away.

In the dense fog, the Americans listened; and the sounds seemed to be closer, and then almost on top of them. They strained to see through the gloom, and still the sounds inched closer; and then stopped. In the swirl of the moving fog they vaguely detected the shapes of three German tanks stopped in a tight column. Perhaps the Germans had detected them too, but it was an American who got off the first shot, at point-blank range—and missed.

The German lead tank did not, knocking out both Shermans as the crews scrambled out and raced to the rear. A brief small-arms firefight broke out as they retreated to Noville. The other two roadblocks joined the line at daylight and were just in time to be targeted by a German 88 mm gun that smashed six vehicles. Simultaneously, two German Tigers bolted out of the fog on Desobry's right, but bazookas and supporting fire quickly put them out of action.[55]

Fog concealed the rest of the German force, but the sound of their maneuvering was easily heard. When it finally lifted, the defenders were treated to a sobering sight. The Germans were everywhere except behind them. To the west, a few tanks were only two hundred yards away. To the north, where the roadblock met them during the night, fourteen more were on the high ground. They were fewer than one thousand yards distant and began to roll in a great attacking line.

On they came, and as they closed the distance they fired round after round into Noville. The American tank infantry had taken up defensive positions in the buildings, but the German tank fire soon crumbled many of them. And then the fog came in and stopped the fire, but it was fleeting and soon lifted to the German advantage. The fighting, like the fog, swirled all morning. On one occasion, when the fog blinded both sides, four American tank destroyers arrived undetected from Bastogne to the south. When it lifted, the tank destroyers were presented with a long line of unsuspecting German tanks, strung out on a ridge to the north. In short order, the tank destroyers demolished nine German tanks and put the rest to flight behind the cover of the ridge. It was now that Major Desobry pondered Colonel Roberts's order from the previous evening. He could look toward Bastogne and see the small town of Foy, just two miles distant, and he was thinking that it offered a tempting defensive fallback position. The allure was just too tempting.

He called and suggested that he should pull out of Noville to a new position at Foy. Colonel Roberts was sympathetic, but the answer was no. But Roberts reassured him that an infantry battalion of the 101st was already on the march toward him. At 1330 it finally arrived, but not unnoticed by the surrounding Germans. "They came under heavy, well-aimed artillery fire, twenty to thirty rounds exploding on the village every ten minutes. Vehicles and buildings were aflame."[56] But with the arrival of the reinforcing infantry battalion, Noville's defense was greatly strengthened. Desobry had thirteen tanks, six tank destroyers, and one thousand American paratroopers.[57] Aligned against him was 2nd Panzer Division with its seven thousand men, eighty of its original eight-eight tanks, and a full array of supporting artillery and supplemental guns.[58] Unknown to Desobry and his defenders, Noville was the focus of the entire Panzer division.

That night, senior officers from the 101st huddled with Desobry to coordinate plans for night defense. He had made his headquarters in building #15 directly across the street from the bombed-out church, now missing its tall steeple. The airborne had dug in and occupied other positions in the rubble, and in building #15 the furniture had been stacked against the windows and a massive armoire was jammed against the main window that overlooked the highway just yards from the crossroads.

Suddenly a high-velocity shell from a German 88 exploded through that window, splintering the armoire and sending shrapnel flying about the room, killing the 101st battalion commander. Desobry was slammed to the floor with a painful, disabling head wound. But that shot proved to be only the initial shot of a massive artillery bombardment. The Germans pummeled Noville all through the night, and their only loss was two tanks that had advanced to the edge of the city.

Colonel Robert Sink and Major Charles Hustead, who took over for Desobry, concluded that they could not reasonably expect to hold out for the rest of the day under such an onslaught of fire. They called General McAuliffe in Bastogne and suggested that it was time to pull back; the Germans were on top of them. McAuliffe agreed, but when he sought approval from Middleton, he emphatically refused.

"You can't to it," chided Middleton. "We can't hold Bastogne if we keep falling back."[59] To Middleton, it was all a matter of time. It was a critical battle of delay. Bastogne was not the target. The Germans gained nothing by capturing Bastogne. It was all about the roads. Each minute delaying the German advance along those roads toward the real target was a minute closer to German failure and defeat. Already the attack was five days old, and the German army was nowhere near where it was supposed to be. And the reason the German army was stuck, in the cold and snow, trying to pass through a little, nothing town, had nothing to do with the tactics of the higher commanders. Patton certainly didn't believe in the value of holding the town. It was all due to an old general who was a master tactician, and who had taught those tactics to his seniors; and even if some didn't remember those tactics, he did. Middleton was determined to slow down the German armored onslaught by contesting every inch of those bleak, frozen highways upon which they had chosen to move. And he was going to hold on doggedly, like a bulldog with a bone.

"I simply didn't believe that the Germans could run the 101st out of Bastogne," Middleton explained. "Besides, I had Robert's Combat Command with tanks and guns, plus two corps artillery battalions, each with thirty-six, 155-mm guns that would reach anywhere around the defensive perimeter. I also had one and a half battalions of tank destroyers. We had a lot of firepower. . . . When you hold onto key points, you deny the enemy freedom of movement. They have to eat and be resupplied."[60]

Still, the fate of Bastogne was very much in doubt. Three German divisions were on the doorstep, and to the east, around Longvilly, desperate fighting at the armored outposts had already cost the Americans forty tanks.[61] At Noville, the badly wounded Major William Desobry had been readied for evacuation back to Bastogne. Despite the fact that German soldiers freely roamed the area, and the firefight was ongoing, he was bundled up and placed in one of the remaining ambulances. Hoping the swirling fog would provide concealment, the ambulance trundled off for the five-mile run to Bastogne. But luck was not with Bill Desobry, and along the way, the Germans captured the ambulance and he was taken prisoner.[62]

On December 20, the 2nd Panzer Division launched yet another attack against Noville. But chastened by the tremendous firepower demonstrated by the Americans, the Germans did not come in a headlong rush. They came in small advances from all three sides. Artillery from Bastogne held them at bay, but only for an hour. Noville had little left with which to resist. The tanks were out of armor-piercing shells, and just when it seemed the situation could not get worse, the Germans maneuvered around the flanks and cut the road behind them and were now to their rear. Any attempted withdrawal to their fallback position at Foy would now be heavily contested. Noville signaled, "All reserves committed. Situation critical."[63]

In the afternoon, the withdrawal began. The German blocking force to the rear bristled with armor and self-propelled guns. As the Noville defenders began feeling their way to the rear, the bad weather and fog did not conceal them. German tanks opened fire with deadly accuracy. The very first rounds knocked out the four Sherman tanks leading the column. Two American tank destroyers worked their way up from the rear of the column and opened fire upon the German blockade.

But now it was the Germans' turn to be trapped. The American paratroopers defending Foy were in their rear, and the combined fire from front and rear broke the enemy position and reopened the escape route. By 1700 the column reentered the defensive perimeter.[64]

The fight at Noville came at a terrible cost. Over two hundred men from the 101st battalion were dead, wounded, or missing. Team Desobry had no accurate numbers of its losses, but it reentered the Bastogne perimeter with little of its original force. Its fifteen tanks were gone, along with most of its other vehicles. Major Desobry was gone, and so were most of the tank infantrymen.

The 2nd Panzer Division lost thirty-one vehicles and half of its infantry.[65] But the real loss to the Germans was time. As if the delays experienced during December 16 to 18 were not bad enough, the American defense at Noville had cost them two more days (December 19–20) trying to take two miles of road.

On December 22, the Germans demanded that General McAuliffe surrender, and his famous "Nuts" answer has rightfully been enshrined in history. And then, disaster for the Germans. The bad weather hiding their attack from the eyes of the American air forces cleared. C-47 aircraft, 241 of them, each carrying twelve hundred pounds of resupply, dropped their cargo over Bastogne. Although not all drops hit their targeted area—some fell into enemy hands—enough hit the target, and the stranglehold on the encircled Americans was broken.

The next day, 160 more aircraft arrived. And on December 26, late in the afternoon, First Lieutenant Charles P. Boggess, commanding Company C, 37th Tank Battalion, rolled his tank through the American perimeter,[66] south of Bastogne. He was the first of General Patton's relieving Third Army. The German encirclement was broken. There was much hard fighting ahead, and most of the casualties from the Battle of the Bulge were yet to come, but when the encirclement was broken, the Germans had lost the Battle for Bastogne.

It would take another month to shove the Germans all the way back to the Our River, from whence they had begun the attack. But with each step backward the German army marched toward total defeat. On January 20, 1945, General George Patton, who had originally objected to defending Bastogne, wrote a very different tribute concerning Middleton's defense. He praised it as a "magnificent tactical skill," and as "a truly suburb feat of arms."[67]

"We made the Germans take the risks," said Middleton. "They had to have all the roads that ran through Bastogne. They had to try to take them away from us. That 'stroke of genius' George [Patton] later attributed to me for deciding to hold Bastogne was just common-sense soldiering."[68] That commonsense soldiering was in knowing where and when to fight. "I had to draw a line our troops couldn't drop behind in that battle. There are times you have

Eisenhower and Middleton at the Buchenwald Concentration Camp (National Archives)

to stay and fight, no matter what the odds. Little groups of Americans threw the Germans so far behind schedule . . . they were never able to get far along with their drive to the Meuse River."[69]

At this critical time in World War II, Middleton still had time for LSU. The Board of Directors still consulted him on important matters, emblematic of the high esteem in which he was held. Although he had departed the university after Pearl Harbor and was an ocean away fighting in every offensive in Europe, the administration always sought his advice. Historian Frank Price recorded just such an incident concerning an oil well lease, even at the height of the war. "So highly did the Board of Supervisors regard Middleton's opinion, they had forwarded to Europe a file dealing with the university's oil lease agreement."[70] He dutifully answered it on December 15, just hours before the German onslaught.

With the final German resistance broken, the Germans had, in reality, played their last card, and began to collapse. Ike's Army rolled on with Middleton's VIII Corps foremost in the battle line. And then the horror of the concentration camps and German death camps were discovered. On April 12, 1945, VIII Corps broke into the Buchenwald concentration complex at Ohrdruf, and General Eisenhower demanded that everyone see the horror. The Signal Corps was ordered to film it all because one day people would deny it had happened.[71] A Signal Corps photographer captured the moment when Middleton and Eisenhower strode through the infamous camp.

On June 5, 1945, Middleton was promoted to Lieutenant General, and in

August returned to LSU and resumed his duties as Comptroller, a position he had left in 1941. Five years later, in 1950, after a Board meeting that he did not attend, three of the members walked into his office and announced, "the board has elected you president of the university, and we are here to notify you."[72]

From February 1, 1951, until his retirement in February 1962, Troy Middleton presided over LSU. His tenure included its greatest years of academic and athletic growth and the difficult time of initial desegregation. He did not favor integration, but his first duty was to uphold the law, and the Supreme Court had ruled that educational integration was now the law of the land. He said, "The board believes in law and order and that it respects the decisions of the courts . . . it should set an example as a responsible, law-abiding body."[73]

Notwithstanding his personal beliefs, Middleton had always put the law first. In 1950, long before the *Brown v Board of Education* Supreme Court decision, the law was "separate but equal." Despite this, he had integrated the LSU Graduate School since there were no separate graduate facilities for black students. From 1950 to 1956, 633 black students enrolled in the Graduate School. In 1956, he wrote, "There is no dodging the fact that as long as the state of Louisiana makes no provisions whatever for Negro education beyond the bachelor's degree, Negro students who wish to pursue their studies further are *legally entitled* as American citizens to utilize the facilities at Louisiana State University."[74]

Upon his retirement, January 11, 1962, was declared "Troy H. Middleton Day in Louisiana." Accolades came in from around the world. Former President Dwight Eisenhower would later proclaim the qualities of Middleton's character: "First of all, he had moral courage, and he had common sense, and he took full advantage of all the experiences he'd had. You know, I like that saying attributed to Napoleon, that the genius in war is the man who can do the sensible thing . . . when everybody else around him was crazy. Well now, that was Troy Middleton."[75]

In 1965, Governor John J. McKeithen came to Middleton with a request to serve on a special commission. The pain of desegregation and the recently passed Civil Rights Act had elevated racial tensions, and McKeithen sought to ease them. His plan was to create a biracial Commission on Human Relations, Rights, and Responsibilities, and it was to be composed of twenty-one black and twenty-one white individuals. The governor needed strong leadership for such a daunting undertaking and chose two stalwarts of the community: Troy Middleton and Dr. Albert Dent, President of Dillard University in New Orleans.

When asked why he agreed to serve on such a "hazardous venture,"[76] Middleton said: "Some believe we serve no good purpose and that we are merely forty-two troublemakers . . . I have had nasty letters and uncomplimentary phone calls. I took an oath in 1910 to defend my country, obey the laws, support the president, and in short to try to be a law-abiding citizen. The Civil Rights Act is a law; to violate it would be to violate my oath. I shall never do that."[77]

The commission, cochaired by Middleton and Dent, served for five years and was a steady influence for compromise during tenuous times. His leadership was nationally recognized. In 1966, the National Conference of Christians and Jews selected Middleton for its annual brotherhood award. That same year, the Louisiana Association of Broadcasters honored him as the Louisianan of the Year for his accomplishments in racial peace-keeping.

Troy Middleton died on October 9, 1976. In 1978, LSU's board voted unanimously to name the library the Troy H. Middleton Library.

CHAPTER 11

Claire Lee Chennault

THE FLYING TIGERS

In 1937, Imperial Japan enjoyed its sixth year of military misrule. The army controlled all facets of government after its 1931 takeover by the radical young officers of the Kwantung Army stationed in China. They had staged a phony "incident" to justify an attack on China. The goal was to take over Manchuria and control its wealth of natural resources.[1]

This national lusting for the rich lands of China was nothing new. By 1927, the sense of peace and prosperity that had spread throughout Japan with the defeat of Germany in the Great War had eroded into a feeling of want and apprehension. The world's financial collapse in 1929 accelerated this into the reality of unemployment and desperation. The civilian government was completely feckless. It became the target of popular unrest and was mocked in vulgar terms. The generals who made up the *kodo ha* (action group) of the army called these despised officials "frogs in the well." And while the *frogs* might have overlooked the obvious solution to Japan's economic and social woes, the death-before-dishonor *kodo ha* had not. Their chant was "take over Manchuria!"[2]

But that was easier said than done. There had to be an incident, some sort of atrocity or affront on the part of the Chinese that could be used to trigger Japan's military might to seize Manchuria. The chances of China committing such an act were remote, so it would have to be concocted, and then blamed on China.

On September 18, 1931, the conspirators of the Kwantung Army blew up a section of railroad track and promptly accused the Chinese of the "atrocity." The telegram announcing the incident to Japan's higher command claimed: "a unit of the Northeastern frontier Defense Army of the Republic of China had dynamited part of the Southern Manchurian Railway and attacked the Japanese railway guards in their barracks northwest of Mukden."[3] Of course, the telegram's conclusion was, "the second battalion of the Mukden garrison had raced in to annihilate the enemy."[4]

While the obviously staged incident caused much bluster among the diplomats of the western powers and the journalists of the free press, it proved to be just that: bluster. The United States and Great Britain did nothing. The *New York Mirror* printed the obvious: "wily Japan is doing her stuff." And the *Chicago Tribune* saw through it all: "Japan will cover her action with mollifying formulas and emerge from the scene with substantial profits from her military coup."[5]

By the end of 1931, the army was in total control of Japan's military and political life. And then on July 7, 1937, there was a second incident. More of the same: contrived and concocted. This time it was at the Marco Polo Bridge, outside of the Chinese city of Beijing. The Japanese army conducted war exercises just outside the city. One of the soldiers went missing, and the Japanese demanded to search the city. That led to an eventual attack and a full-scale war.

The Chinese government abandoned appeasement, and this time stiffened their backs. On July 30, the Japanese army continued its attacks and seized the city of Tientsin, and began its all too familiar regimen of cruelty, looting, burning, and rape. Generalissimo Chiang Kai-shek quietly ordered a general mobilization of China's armies, and what became known as the Second Sino-Japanese War began.[6] The Japanese forever shunned the word "war" and simply preferred to label it the "China Incident."

They set their sights on Shanghai and the Chinese capital of Nanking. General Iwane Matsui, commanding this new Japanese expeditionary force, departed Tokyo promising victory and proclaiming that his attack would capture the Chinese capital, bring China to her knees, and bring Chiang Kai-shek to his senses.[7] And the Japanese propaganda machine cranked out bizarre explanations to attempt to explain away the obvious: Japanese cruelty was misunderstood; the army always had "peaceful" intentions. Foreign Undersecretary Kensuki Horinouchi was a master of the art of attempting to put a better face on war crimes, and to deny the undeniable. He even went on CBS radio to explain to Americans their obvious misunderstanding of the current Japanese offensive, and to somehow justify Japan's very presence in China.

"We must emphasize first, that the expeditionary forces of Japan now in China have been sent there for no aggressive purposes, and secondly, that we have no territorial designs. Our forces are in China to safeguard our legitimate interests and protect our rights, and to secure the safety of our nationals.

These forces will be withdrawn the very moment that their presence is no longer needed."[8]

But the reality of Japan's attacks on the cities of Shanghai and Nanking proved to the observing world that Horinouchi's nonsensical proclamations were utter bombast. Millions of viewers actually witnessed in unedited news film horrific scenes of utmost savagery unleashed upon the vanquished people of China. The world was convinced that "Japan was a nation of beasts gone mad."[9] While the Japanese press printed not one photograph of this savagery, the rest of the world saw it all, particularly the damning film footage of a little child sitting in the burning wreckage of the bombed-out Shanghai railway station surrounded by dead bodies.[10]

And just when the world thought the wanton savagery could not possibly get worse, they witnessed the "Rape of Nanking." On December 12, 1937, the Japanese army broke through the final defenses of the Chinese line in Nanking and moved to surround the entire Chinese force. A few of the defenders managed to break out of this encircling ring, but the vast majority, one hundred thousand, did not. They were trapped and at the mercy of the rampaging Japanese army. To the Japanese, there was no Geneva Convention, not even basic rules of war, and the systematic slaughter and blood lust began. One Japanese soldier confided to his diary, which was suppressed by Japan until 1983,[11] the details of his own murderous actions: "The Chinese are too many for a platoon to kill with our rifles, so we borrowed two heavy machine guns with six light machine guns from the Army company."[12] Armed with this arsenal, the Japanese soldiers slaughtered five hundred additional Chinese, lined up against a wall, because they "might be" soldiers.

Historian Edwin Holt wrote, "So Nanking was looted, the women were raped and murdered, the children were shot or bayonetted if they annoyed the Japanese."[13] And on December 23, George Fitch, an American Protestant missionary in Nanking, wrote of the horror. "Nanking is a city laid waste, ravaged, completely looted. . . . It is hell on earth. Hundreds of innocent civilians are shot before your eyes or used as bayonet practice. . . . A thousand women kneel before you crying hysterically, begging you to save them from the beasts who are preying on them. This is a hell I never before envisaged."[14]

With such crushing military defeats, the Japanese expected China to surrender, thus enabling them to dictate surrender terms that would confiscate as much of China's land as they wanted. The Chinese armies had indeed been steadily pushed back in every battle, each time with tremendous losses. While

Map of Burma and Rangoon
(U.S. Army map)

they had inflicted significant losses on the Japanese forces with ferocious, human-wave counterattacks, they had been defeated at every turn. But Chiang Kai-shek refused to surrender, and his forces once again dug in on a new defensive line. He moved his capital farther west, up the Yangtze River to Chungking. From there he addressed the people and issued his "Message to the People Upon Our Withdrawal from Nanking."[15] It was a message of defiance and pledged to fight on to the end.

The new capital was now far inland, away from the reach of the Japanese except by bombing strikes. But the move also severely decreased China's routes for resupply. The Japanese controlled the entire eastern coastline, along with the waters along French Indo-China. That control included the inland waterways and roadways that had previously delivered vital materiel of war and staples for survival. There was only one route left for resupply, and it was tortuous. To the west was the Burmese port of Rangoon, off the Bay of Bengal, where supplies could still be landed and reloaded onto rail cars and transported 436 miles to Lashio. From there they would be transferred onto trucks for a 726-mile rough ride on the gravel-surfaced Burma Road to Kunming.

The primitive road was an engineering wonder of the world, carved out through impossible country by a swarm of two hundred thousand Chinese laborers. They worked, nonstop, to crush rock into gravel with primitive hand tools. Men, women, and children from each hamlet along the way brought their own food for a day of pulverizing rock. At the day's end, they hauled away the mud in baskets. That gravel was flattened with rollers powered not by machinery or beasts but by hundreds of harnessed Chinese laborers pulling the heavy rollers and graders. Construction had begun in 1937, after the

Ledo Road
(Army photo #199666-S)

Japanese blockaded all Chinese ports, and by 1938 the road was opened. One American engineer, upon seeing the road, exclaimed, "My God, they scratched these roads out of the mountain with their fingernails."[16]

The end of the road was at Kunming, but the resupply journey was still not over. The trucks then had to negotiate the final leg to Chungking: the Ledo Road. It included a twisting, short, two-mile, death-defying stretch of road that seemed almost vertical. It rose nine hundred feet over those two miles and included twenty-four hairpin turns that snaked up the sixty-degree ridgeline. Sheer drop-offs plunged down on either side. Most truck drivers had to make the turns in two stages: first begin the turn, and then back, and forward again to complete the turn. Not all were successful.

In the spring of 1937, as China and Japan squared off to fight the Second Sino-Japanese War, an obscure Army Air Corps captain contemplated his retirement after twenty years' service; not in good health and confined to a bed at the Army-Navy Hospital in Hot Springs, Arkansas. By all accounts, he was old before his time. His face was weather-beaten, and he suffered from deafness, chronic bronchitis, low blood pressure, and general exhaustion.[17] He had been grounded from flying because of his ailments, and was in a state of depression after years of fighting what he called "Blind opposition to the significance of airpower by the foggy-brained brass of the Army and Navy."[18] He had seen the Father of the Air Service, General Billy Mitchell, challenge similar foggy-brained brass during his day; and when he had dared to proclaim that air power could defeat massed armies and big battleships, they attacked him. In 1923, he had proclaimed that one day Japan would attack Pearl Harbor, at 7:30 in the morning, and the first bombs would fall on Ford Island! The foggy brains screamed for his head. This establishment had championed flawed policies, and designed flawed aircraft, and taught flawed aerial tactics that caused the deaths and injury of many airmen. When the dirigible *Shenandoah* crashed while on a navy publicity flight, killing fourteen of its forty-two-man crew, Mitchell could take no more. He hit the roof and publicly called out these senior officers and labeled their actions as "criminal and incompetent, and treasonable."[19]

His enemies hated him and used his fiery spirit to charge him with unbecoming conduct. They court-martialed him, convicted him, and removed him from the service. Eighteen years later, when Japan attacked Pearl Harbor and the first bombs fell on Ford Island at almost the exact time that he had predicted, Mitchell was long dead, and the foggy brains showed no remorse.

If Mitchell had been the advocate for bombers, then Captain Claire Lee Chennault was now the advocate for fighters. "With the development of General Billy Mitchell's concepts of strategic bombardment, popularity shifted away from the fighter boys, who dominated World War I in the air,"[20] he said. Bombers and the bomber boys became the new darlings of the air. Even the senior Air Corps general, Hap Arnold, a subaltern of Mitchell, was forced to admit, "Fighters have been allowed to drift in the doldrums."[21] And the austerity of the Depression years further exacerbated the development of fighter aircraft. The struggle for funding dollars led to the oft-quoted lament, "Fighters are obsolete."[22]

And the Chief of Staff of the Army, Major General Charles P. Summerall, was an avowed enemy of the Air Service, and continued to be a thorn in the side of aviation development. In 1923, he had been publicly humiliated at Billy Mitchell's court-martial. He had sat as the presiding officer on this court composed of the highest-ranking army generals. To aghast onlookers, Mitchell brazenly challenged his fitness to sit on the court and produced a damning report detailing Summerall's unfitness when he had commanded in Hawaii, and his more recent prejudice against Mitchell for writing his Pearl Harbor prediction.

The challenge was upheld and Summerall was removed as unfit "on the grounds of his prejudice, hostility, bias, and animosity against the accused."[23] In humiliation, Summerall had to march out of the packed courtroom in front the smirking press and onlookers. He never forgot that. The animosity had not changed by 1928. Now, General Summerall was the highest-ranking general in the army, and he attended a demonstration that Chennault proudly presented to show the ability of the Air Service to jump fully equipped paratroopers into combat. It was a revolutionary idea.

"We polished this technique until the paratroopers were opening fire with machine guns in less than a minute after they landed,"[24] beamed Chennault. But Summerall was unimpressed and in fact didn't even stay for the jump. Why he even came to witness this presentation, only to leave early, remains unknown; but his animosity for the Air Service was on full display. "Summerall waited until the chutes blossomed," said Captain Chennault, "and then turned his back on the exhibition with the comment, 'Some more of this damned aviation nonsense.'"[25]

Chennault became the senior instructor in fighter tactics in an air service dominated by the bombers and his hated foggy-brained brass. "I was a fighter pilot for as long as I could fly," he said. "My experiences in Hawaii convinced me that an air force could never get along without fighters, and that in any future war they would play as vital a role as bombers."[26] But his was a voice crying in the wilderness. He studied the air documents of World War I and made a remarkable discovery that seemed to have been known only to the German aviators. Two planes could be maneuvered to fly together as a team. Dogfighting was a vainglorious tactic.

That meant applying the ancient axiom concerning firepower, recognized from the time of Napoleon, that the difference in firepower of two opposing units was not the difference in the number of guns, but in the square of that difference. In other words, a two-plane element attacking a lone enemy enjoyed odds not of 2 to 1, but of 4 to 1. Individual dogfights, while romantic, created a distinct disadvantage.

In the winter of 1916–1917, Baron von Richthofen molded his Flying Circus around this firepower advantage, and the Allied airmen were overwhelmed

General Claire Lee Chennault
(U.S. Army photo)

and never defeated the Red Baron's group. When he was killed, Herman Goering took over and immediately dismissed Richthofen's proven tactics and reengaged in individual dogfighting. The result was predictable: Goering's revised air force was crushed by Billy Mitchell's combined Allied air force.

Chennault spent his time as the senior flight instructor at the Air Corps Tactical School at Maxwell Field in Alabama confronting and arguing with those who knew little about flying. He had to fight armament engineers who scoffed at the suggestion that four machine guns could be synchronized to fire through a propeller. They were wrong. He then fought design engineers who pushed for "multiseater fighters that provided interesting engineering problems, but were useless for combat." Chennault scoffed at the sophomoric mindset: "These young engineering officers had no ideas on its tactical employment, but were fascinated with the intricacies of its construction."[27]

Despite all opposition, Chennault stressed teamwork and formation flying, not individual tactics. He had demonstrated super precision flying at numerous barnstorming air shows with his select team that he called *Three Men on a Flying Trapeze.* At times they flew only three feet apart. "From the ground it looked as though our planes collided many times," quipped Chennault. "Actually we never came closer than three feet; and the many thousands who turned out to see us crash were regularly disappointed."[28]

But no matter how many air shows he flew, leaving his audiences gasping, and *oohing* and *aahing,* or how many irrefutable arguments he presented justifying the necessity for fighters, it all meant nothing. "The issue," he lamented, "was not how many, or what kind of fighters we should have, but simply whether there should be any fighters at all. Pilots who merely contended that a well-balanced air force needed fighters were bitterly scorned by the bomber boys."[29]

In the end, all this pressure and angst took its toll on Chennault's health. In the fall of 1936 "the flight surgeons grounded me completely and packed me off to the hospital."[30] Chennault's stint in the hospital was a long one. It began in the winter and stretched into the spring of 1937. He had long hours to reflect on his life and what he had and hadn't done during his forty-four years. He had always wanted to be a soldier, but not just any soldier. "Until flying came along, I never found a segment of military life entirely to my liking."[31]

His grandfather, Dr. William Wallace Lee, had been a surgeon in the Army of Northern Virginia and was related to General Robert E. Lee, and perhaps his love for soldiering had been inherited. "While still in grade school," he remembered, "I pored over history books in my grandfather Lee's library, reading about the Peloponnesian and Punic wars. Although I had no idea where Greece, Carthage, and Rome were, I was enthralled by the charging elephants, armored warriors, and burning ships."[32] But that euphoria gave way to a certain melancholy that gripped him. "It was obvious that I was going round and round, and getting nowhere. Ever since my boyhood in Louisiana I have watched the eddies of the muddy Southern rivers as they boiled to life in the spring. Particularly on the Mississippi, where I have seen them smash a steamboat into kindling wood, and drag a full-grown cypress down into their whirling vortex, these eddies have become symbolic."[33] He reflected on "boyhood friends caught in the eddies of life and sucked under or whirled aimlessly about in ever narrowing circles—some as river gamblers, dead from gunshot wounds; others struck by yellow fever . . . and others still caught in the drudgery of trying to eke a living from cotton farming."[34]

He did not want to follow this destructive path. To prevent this descent into a black hole, he was determined to somehow get back into flying, despite his present ill-health. "I knew that I would have to find something that would give me a chance to keep flying, to fight, and to prove my theories on tactics. When the Army suggested retirement, I accepted the offer. Several aircraft-manufacturing firms made flattering overtures, but I had no urge to fly a desk."[35]

He fondly remembered his first exposure to the discipline of the military life. "My first taste of military life came while a freshman cadet (1906) at Louisiana State University, where I enrolled in the agricultural course [having first lied about his age]. I had no desire to be a farmer . . . I entered L.S.U. when I was fourteen,[36] and like every other student, settled down in the red brick pentagon barracks that enclosed the old campus where military discipline governed the student corps of cadets."[37]

A few days after he had enrolled at LSU, in a class of 146 men and eight women, upper-class cadets practiced the art of hazing young Chennault. One decided to pose as an "*Officer of the Day*," and he rousted Chennault from his room and gave him a rifle with a bayonet fixed to the barrel. As Chennault stood ramrod straight, this senior cadet posted him at the entrance to the barracks with stern orders to allow no one to pass without a special pass. The young cadet acknowledged his orders. "As I paced the post before the entrance, upperclassmen bombarded me with pitchers of water from the second-story windows. I continued to walk my post, drenched to the skin, with my temper so hot I could almost hear the water sizzle as each new deluge hit me."[38] This abuse continued, much to the glee of the culprits hurling water from the upper floor. Then suddenly mess call sounded, and those same hazing upperclassmen now raced downstairs and out of the main entrance.

To their utter surprise, the water-bomb throwers were now confronted by the fierce-looking, bayonet-wielding Cadet Chennault, still under orders to let no one pass unless cleared by the Officer of the Day. His combative stance, with legs akimbo and rifle tipped with its flashing bayonet at the ready, convinced all that he meant to carry out his special orders. His stoney stare was most convincing. No one stepped forward to challenge him. "The joke ceased to be funny as the afternoon wore on without lunch," laughed Chennault, "and nobody was quite willing to see if I was bluffing. It took a long time to find the real Officer of the Day, who officially relieved me of my post and released the hungry cadets."[39]

Nor was fourteen-year-old Chennault squeamish about working the system to his best advantage. He was an avid fisherman, and the best fishing along the Tensas River occurred in the early summer prior to the end of the school year. To take advantage of this prime-time fishing, Chennault devised a scheme to get expelled so he could go fishing early. "I discovered that a carefully timed accumulation of demerits enabled me to be expelled annually just after I passed all final examinations and received academic credits. Expulsion then sent me home a week early and netted seven precious days on the Tensas."[40]

And he always managed to get reinstated in the fall, since his grades were good; and if that was not enough, his Uncle Nelson was well-known as a state educator. But it all finally caught up to him, and once, when Chennault was short a few demerits that would ensure his early expulsion for early fishing, he was called out by the Regular Army officer in charge of the cadets. He was marched to the front and center of the cadet formation. He looked every bit the screw-up that deserved demerits, with his trousers rolled up to his calves to earn the needed demerits for fishing. In full view of the Cadet Corps, Chennault was proclaimed to be a misfit. "Chennault," bellowed the officer, "you will never make a soldier."[41] And so his final college year was spent at Louisiana State Normal School to attempt to qualify for a teaching position.

But those funny, happy times were but a distant memory for Claire Chennault, now confined to his hospital bed in 1937. His only connection with those happy days were the occasional letters sent to him by old friends whom he had taught and were now serving in China. Whenever his mail arrived, he first searched for anything bearing a Chinese postage stamp.

Then, as if by Divine Providence, a special letter with Chinese postage arrived. It was from his friend Roy Holbrook, a former student who had been an Air Corps pilot and gave it up to go to China and serve as an air instructor. Holbrook had previously written Chennault to recommend twelve American pilots to go to the Chinese flying school. But this letter was not about recruiting. Holbrook was now a confidential advisor to the Central Trust Company of China, and in that position was forwarding an offer from Madame Chiang Kai-shek, who had taken over leadership of the aeronautical commission. Madam Chiang wanted to know: "Would I consider a three-month mission to make a survey of the Chinese Air Force?"[42] The terms were generous: "$1,000 a month plus expenses, a car, chauffer, and interpreter, and the right to fly any plane in the Chinese Air Force."[43] On April 30, 1937, Captain Claire Chennault retired from the army and was China bound.

⚜ ⚜ ⚜

In May 1937, the liner *President Garfield* docked in Kobe, Japan. Claire Chennault secretly slipped ashore but left his baggage on board while he set off for a clandestine trip to Shanghai. His mission was to tour Japan's key industrial and military facilities and secretly photograph all that he saw. "Our intentions in Japan were to reverse the role of the hordes of Japanese 'tourists' who wandered over the United States with imitation German cameras, binoculars, and an unhealthy interest in harbors and airfields."[44]

Meeting him on the Kobe dock was his friend and former aviation associate Billy MacDonald, now an instructor in the Chinese Air Force. An application for MacDonald to visit Japan would have been immediately rejected, so Billy decided to hide in plain view. His cover was that he was the assistant manager of an acrobatic troupe touring Japan. When he had to enter the country, his "acrobatic" passport was usually sandwiched in between the stack of the passports of the real acrobats—and routinely stamped with the others. Chennault and MacDonald acted as holiday tourists and even hired an open tour car to take them around. Their prying cameras were concealed under topcoats and came out only at safe moments to click away at what they saw. They filled notebooks with data to correspond with the photographs, and when the touring was over they turned in the car and crossed the Yellow Sea to Shanghai.[45]

Several days later, Roy Holbrook whisked him off to the high-walled compound of the French Concession to meet his new employer, Madame Chaing. They were shown to a sitting room to await her arrival. "Suddenly a vivacious young girl clad in a modish Paris frock tripped into the room, bubbling with energy and enthusiasm," said Chennault. "I assumed it was some young friend of Roy's and remained seated."[46] But he jumped to his feet when Holbrook introduced the young woman to be Madame Chiang herself. She was friendly and vivacious and was all business, especially when it came to the state of readiness of the Chinese air force. She wanted a complete inspection of the air force and an unvarnished assessment of its combat readiness. Chennault promised a report in three months.

He flew to Nanking, his first flying since his grounding in the United States, and quickly discovered that Chinese unrest over Japanese excesses was rampant. Students demonstrated in the streets. Japan poured smuggled goods into China, and customs officials who tried to stop the flow were kidnapped or beaten into submission.

And the Italian government was in complete control of Chinese aviation since Mussolini had seized the opportunity to "send some of my boys over to help you in your troubles."[47] Those boys turned out to be forty Italian pilots and a hundred ground personnel, and the Italians lost no time demonstrating their ostentatiousness. "Italian military pilots swaggered around Nanking in full uniform," observed Chennault. "General Scaroni roared through the streets in a big black limousine, his uniform dripping medals and gold braid."[48]

None of this enhanced the readiness of the air force. The Italians graduated every Chinese cadet regardless of ability. Previously, the American school had weeded out the incompetents, only graduating the best. To influential Chinese families, this had been unacceptable, and the "washed out" cadets soon saved face by graduating from the Italian school.

Other evidence pointed to Italian perfidy against Chinese interests in accordance with Mussolini's political favoring of Japan and Germany. Everything that could weaken the Chinese air forces was put into play. The Italians instituted a novel accounting system for the Chinese air force. Once an airplane was entered into the inventory, it was never removed for any reason. It could be a total wreck, or a plane that had been cannibalized for parts or obsolete and unflyable, but it was always shown as an asset ready for combat. The result was an air force only on paper. Chennault noted, "The Aero Commission roster listed five hundred planes but only ninety-one were fit for combat."[49] But this damning report on the corruption and incompetence of the Chinese air force was never submitted. Just weeks after meeting with Madame Chiang, Japan attacked at the Marco Polo Bridge. "I was inspecting the unique Italian flying school at Loyang, with the 100-percent graduation record, when word reached me of the Japanese attack on the Marco Polo Bridge on July 7."[50]

Chennault immediately volunteered his service to the Generalissimo, and was ordered to take over the final combat training of the Chinese fighter groups. He was called to meet with Chiang and Madame Chaing at their headquarters in Kuling to report on the status of the air force, and he confirmed that it was indeed a force only on paper, with ninety-one aircraft fit for combat, not five hundred.

On August 6, he listened to Chiang Kai-shek's declaration of war. "The Generalissimo spoke briefly, announcing that China would yield no more, even if it meant fighting inadequately prepared, and to the death."[51] And retired Captain Claire Chennault stayed with him. He had no official status in China, and had been ordered by the State Department to leave Nanking and return home. "Consul-General Clarence Gauss . . . threatened to send the United States sheriff of Shanghai to arrest and deport me under armed guard.

Gauss also intimated a court-martial and loss of citizenship were in store for me."[52] But Chennault stayed on and defied all attempts to forcibly remove him from Nanking. "I was civilian adviser to the Secretary of the Commission for Aeronautical Affairs—first Madame Chiang and later T. V. Soong (Madame Chiang's brother). I had no legal status as a belligerent and held no rank other than retired captain. . . . My official job was adviser to the Central Bank of China, and my passport listed my occupation as *farmer*."[53]

For the next two months, China's puny air force faced off against the overwhelmingly superior Japanese fighters and bombers. They fought like demons, knowing that the very survival of their country was at stake, but the numbers finally caught up to them. Whenever they hurled themselves into the Japanese attacking formations, the Japanese fighters brushed them off like annoying flies. "The Chinese Air Force came to the end of its rope in October," lamented Chennault. "Of the eighty fighters that began the fighting, [fewer] than a dozen were left. Many of the best pilots were dead. The remainder were an endless chain of moving ducks in a shooting gallery. It was just a question of time."[54]

As the Chinese retreated up the Yangzi River, the Generalissimo sent out an appeal for help to all the major powers, and amazingly, only the Russians responded. Although they had no love for Chiang Kai-Shek, who had defeated their allies, the Chinese Communists, the Russians were willing to support anyone confronting Japan, which for twenty-years had planned to attack Russia.[55]

Arriving in the fall of 1937, the Russians provided China with more than 200 aircraft, 450 pilots and ground crew, and supporting antiaircraft artillery. That amounted to four fighter squadrons and two bomber squadrons. They would stay for eighteen months and become a formidable foe to the air-assaulting Japanese. And the Russians opened a vital resupply route from Turkestan into northwest China. The pilots were regular Red Air Force, not volunteers, and commanded by General Asanov.

"They were considerably older and more mature than American pilots . . . and wore civilian clothes but retained their Red Air Force rank; and they received automatic promotions when they return to Russia," said Chennault. "Their combat behavior was unpredictable. Some days they just decided not to fight and scuttled off full throttle over the treetops. When they decided to fight they did so with the teamwork and tenacity of ants, swarming over the Japs and overwhelming them with sheer determination."[56] They flew two fighter aircraft: the highly maneuverable I-15 biplane, which brandished four machine guns, and, the I-16 monoplane for high-speed work. With those aircraft they confronted the Japanese, always giving as good as they got.

But on April 29, 1938, the Russians, operating under a battle plan created by Claire Chennault, would deliver the most smashing victory ever over the Japanese air force. After the Japanese army overran and raped Nanking, their next target was the provisional capital of Hankow. Knowing the Japanese penchant for exalting the Emperor in elaborate ceremonies and daring actions, and announcing stunning military victories, Chennault and his Russian and Chinese cohorts became convinced that a smashing blow against Hankow would fit the bill as a suitable gift for Emperor Hirohito. The problem was that recent raids against Hankow had not met with success, and the Japanese seemed squeamish to return.

The plan was to lure them in on April 29, the Emperor's birthday, so "we ostentatiously stripped Hankow of its fighter defense," said Chennault. Without fighter aircraft, the city would seem to the Japanese to be ripe for the picking. In the late afternoon of April 28, all Chinese and Russian fighters took off, making sure their departure was seen and heard by all. Engines roared upon takeoff, and the aircraft then circled at low altitude so that everyone could see the massed formation. When that was done, the entire force roared off to the west. "Japanese agents, who were as thick in Hankow as bedbugs in a country inn, reported the exodus before the planes were out of sight,"[57] wrote Chennault. But the departing flight was only for an hour, and the force turned back toward Hankow and slipped down to treetop level altitude to avoid detection. Chennault and Billy MacDonald then climbed to the top of the tallest building in the city, and from there, in the evening twilight, saw and heard nothing. The trap was set.

The next morning, air raid sirens reported the approach of fifteen bombers and twenty fighters hustling to join them. Twenty Chinese pilots flying Russian planes patrolled to intercept the enemy fighters. Their plan was to either shoot them down or force them to burn up their fuel and not join the bomber force. The main Chinese force was forty Russian fighters, flying high like falcons from above and poised to swoop down upon the unsuspecting bomber force.

"Their first onslaught separated the fighters from the bombers," said Chennault. "While one group slaughtered the bombers, the other chased the fighters and hacked them down. . . . Three bombers escaped but not a single fighter. The Sino-Russian score for the day stood at thirty-six out of thirty-nine. Four

Chinese pilots and nine planes were lost. The Russians lost two planes, but no pilots."[58]

The traditional Japanese mindset that had always denied and would continue to deny reality now sought to save face. Chennault listened over the radio. "That night the Japanese radio cheerily announced that fifty-two Chinese planes had been shot down over Hankow to celebrate Hirohito's birthday."[59]

By the end of 1938, the constant fighting over Hankow had severely depleted the pool of good Chinese pilots, and the Russians began providing materiel support only. Slowly they withdrew their pilots as German saber rattling became louder. Madame Chiang suggested a Foreign Legion-type air force using airmen from around the world. But a brief experiment to do that proved to be a disaster.

Madam Chiang pressed Chennault to give up the combat operations and take on teaching and training Chinese pilots. He refused and requested to return to active duty in the U.S. Army Air Corps. The answer from the army was short: "no funds available for the return of retired officers to active duty!"[60] So, Chennault settled in Kunming with the Chinese Training School and embarked upon a totally thankless job. He would call it "self-imposed exile in the Chinese hinterland."[61] He flew his last combat mission in October 1938, just months before Japan began an aerial onslaught designed to break China's back. The bombing campaign targeted every major population center in Free China.

In 1939, Chennault again requested to return to active duty, and the answer from the army was a second "No!" During that year, Chinese cities erupted in flames from relentless Japanese bombing. Chennault was in Chungking when the bombers came to him. "Twenty-seven bombers came in a perfect V formation. Approaching the bomb run, they swung into line abreast. . . . Open bomb bays sprinkled the city with hundreds of silvery incendiaries that burned the heart out of the capital and left raging fires for three days."[62] He went into the burning capital to try to help suppress the blaze and found himself manning a hand pump amid lines of bucket brigades. "It was like trying to quench a forest fire with a garden hose."[63] The relentless bombing continued for more than a year, and city after city burned to the ground. In the summer of 1940, the bombers again attacked Chungking, this time with almost a hundred bombers a day.

In October 1940, Chennault received an urgent summons from the Generalissimo. Chiang was terribly worried that, although the people had absorbed this enormous pounding, continued exposure to defenseless bombings might force him to succumb. Despite this momentary gloom, he quickly perked up, and told Chennault that "he had a plan for ending the Japanese bombings. He wanted to buy the latest American fighters and hire American pilots to fly them. What did I think?"[64]

Two days later, Chennault was on his way to Washington with instructions to report to T. V. Soong, Madame Chiang's brother, who directed China's efforts to solicit aid from the United States. On his first night, Soong took Chennault to a special dinner with two distinguished newspapermen from the *Chicago Daily News* and the *New York Herald Tribune*. He explained the dire circumstances facing China, especially since Japan had now introduced the *Zero* fighter into their order of battle; and that meant the Chinese would have no chance in aerial combat.

Chennault explained to the dismayed reporters that the only solution was to provide the Chinese with the best available American fighters, flown by American pilots. But the journalists could only frown at the proposal and told him that the desperation of the British, fighting for their lives against Germany, demanded every aircraft coming off an assembly line; and "my volunteer project had little hope for success."[65] His reception wherever he went was the same. "Virtually everybody, to whom I broached the subject, told me, with varying degrees of courtesy, that I was insane."[66]

But the members of the Chinese lobby worked long hours to come up with a list of exactly what China needed to extend its resistance to the Japanese onslaught. The list was daunting: 350 fighters, 150 bombers, 10 transports, 190 training planes, and the necessary allotments of parts, fuel, and material to build airstrips. On November 25, 1940, T. V. Soong presented this laundry list to President Roosevelt's liaison committee.[67]

A high-priced lobbyist, Tommy Corcoran, recognized in Washington as a "fixer," interviewed Chennault on the details of his plan. At the end of an hour, he sent a message to the President that Chennault's plan would be a useful weapon to deter Japan. It was the very model of simplistic logic, and one that the British well understood. He said: "Japan, like England, floated her lifeblood on the sea, and could be defeated more easily by slashing her salty arteries than by stabbing for her heart." An American air base, or bases, in Free China would be in a position to slash at Japanese supply lines and advanced staging areas. "An air offensive from China could have smashed the Japanese southern offensive before it left its home ports."[68] Apparently FDR thought Chennault's plan was good, because shortly afterward Corcoran would write, "Roosevelt sent back orders for me to take Chennault around town and introduce him to . . . influential men who could keep their mouths shut."[69]

Thus was born the American Volunteer Group (AVG). The haggle over the laundry list would dramatically revise that list downward. No bombers made the final cut, and the 350 fighters suddenly became three hundred P-40B *Tomahawks* produced by Curtis-Wright from its remaining inventory. Curtis-Wright had switched to the production of its newer model, P-40E *Kittyhawk*. The final haggling then determined that two hundred of the three hundred planes should go to the RAF. That left only one hundred for the newly formed American Volunteer Group—a far cry from the hoped-for numbers. Still, on December 23, President Roosevelt approved the plan.[70] It was now left for Claire Chennault to take care of the daunting details.

After getting the air frames from Curtiss-Wright, Chennault had to equip and arm them, and find crews to fly and maintain them. The China Defense Supply Agency, one of Corcoran's companies—would pay the bill. Frederick Delano, uncle to the President, signed on as honorary counselor, adding his significant name to the legitimacy of the CDSA. Corcoran promised the army and navy that Chennault would not "pillage" any one service for pilots.[71] But all U.S. military commanders were adamantly opposed to the AVG.

"Orders went out to all military airfields," wrote Chennault, "signed by Secretary Knox and reluctantly by General Arnold, authorizing bearers of certain letters freedom of the post, including permission to talk with all personnel. Field commanders were . . . enraged when they discovered the purpose of the visit was to lure men out of their services. Several spluttering commanders called Washington, long distance, for confirmation of their orders."[72]

The recruiting program had all the trappings of a clandestine operation. It lacked the cloaks and daggers, but in all other aspects it was shrouded in mystery, and the key players were tight-lipped. Consider for a moment a memorandum from the U.S. Navy's Bureau of Navigation, sent to Admiral Nimitz on August 7, 1941. The subject was the "Releases of naval personnel to accept employment in China."[73] Paragraph five stated that none of these requests for personnel would be handled by mail "in the usual way," and that designated parties would act a couriers between the person recruited and the Navy Department.[74] The author of the memorandum, Commander J. B. Lynch, went on to say, "I have at no time seen anything in writing regarding this program."[75]

China's Central Aircraft Manufacturing Company (CAMCO) handled all the details of recruiting. A statement from one of its executives declared, "for obvious reasons individuals engaged in operations of this nature, contemplated in China, must have no connection with the U.S. Government services while so engaged. They must act as individuals and on their own responsibility. . . . All arrangements have been handled orally with no file record of any nature."[76]

And so the recruiting began, but it wasn't as easy as CAMCO thought. "The recruiters were forced to lower their sights a good deal," wrote pilot R. T. Smith, "and by mid-summer of 1941 had signed up a hundred pilots whose experience, with some exceptions, was a far cry from what was originally planned."[77] Smith's own case was an example. "I was a Basic Flight Instructor. . . . I had never flown a fighter plane, had received no gunnery training of any kind, and indeed had never even seen a P-40."[78] The Navy provided about fifty pilots, thirty-five came from the Army, and the final fifteen came from the Marines. Very few had any experience as fighter pilots. "The rest were the damndest assortment you can imagine," said Smith, "from the Navy, former pilots of dive-bombers, torpedo-bombers and flying boats. And from the Air Corps there were some who'd been flying bombers, or ferrying aircraft . . . and a handful of instructors like myself."[79]

Ken Jernstedt was actually a Marine fighter pilot stationed at Quantico. He and fellow pilots Tom Haywood and Chuck Older heard about the AVG and made an appointment to meet with a recruiter somewhere in Manhattan. They listened to what he had to say and signed a one-year contract with CAMCO. "Here you have three guys looking for adventure," Jernstedt said. "I don't think any one of us would have gone on our own. We were offered three times the money we were making . . . [and] I had sympathy for the Chinese at the time. *Life* Magazine had run a series of pictures of the devastation that was being wrought upon China by the Japanese."[80] And the "official" letters of resignation were hardly elaborate. Greg Boyington's was especially terse—just two sentences:

> I hereby tender my resignation from the . . . Marine Corps.
>
> . . . I will accept a position with the Central Aircraft Manufacturing Company.[81]

Chuck Older said, "In late July, 1941, we were separated from the Marine Corps and instructed to proceed to Los Angeles where we would meet with others and receive instructions for our departure to the Orient. On August 26,

1941, we sailed from Los Angeles on the Dutch passenger cargo ship, *Zaandam.*"[82] Jernstedt, Haywood, and Older were preceded in sailing for China by a group that had sailed on July 8, on another Dutch ship, *Jaegersfontaine.*

And if anyone thought that all this sleight-of-hand recruiting, and meeting in hotel rooms, and skulking around had escaped the watchful eyes of the Japanese, they were badly mistaken. "Japanese intelligence was not fooled by passports claiming occupations of musician, student, clerk, banker, etc., with a leader who was a farmer," said Chennault. "The Japanese radio announced that the first group of American volunteer pilots, planning to fight in China, had left San Francisco by ship.

"'That ship will never reach China,' the Japanese radio voice chortled. 'It will be sunk.'"[83]

It took a month to steam from San Francisco to Burma, and along the way it was, for these young adventurers, a chance to party; and party they did. Card games, deck tennis, boxing matches, shuffleboard, rollicking, drinking, and swimming was the order of the day, and they drained the ship of all of its booze. Emma Jane "Red" Foster, one of the two female AVG nurses embarked with this ship full of out-of-control rolickers, said, "They looked like a bunch of boys who had barely survived adolescence! They were good-looking boys, but these are the people who are going to defend the Burma Road?"[84] "All of us were to become familiar with what it meant to be on a slow boat to China,"[85] said R. T. Smith.

The aircraft had a three-month journey. They were loaded into Norwegian freighters in Weehawken, New Jersey, for the long, around the Horn, journey to Rangoon. Each aircraft was packed into two crates: the fuselage and engine in one crate, the wing, tail, and propeller in the second.[86] When the freighters arrived at Rangoon Port, unloading and assembly operations were not mechanized. For the P-40s, it was all about back-breaking manual labor. The four-ton fuselage crate was dragged and muscled by straining arms onto a low trailer and hauled to an assembly area. Then forty to fifty other Indian laborers stripped the crating away and, using four-inch pipes as rollers, rolled it to a hoist that lifted it for further assembly.[87] The 2,500-pound wing assembly was lifted by a sea of hands and "walked" to the fuselage, passed delicately beneath the hoisted fuselage, and placed on two heavily padded wooden horses. The fuselage was then lowered into place on the wing, and bolted, with forty-four bolts.[88] Trucks towed the partially assembled *Tomahawks* to the nearby airfield at Mingaladon for completion and test flights. On June 12, 1941, the first P-40 was test flown.

P-40 *Tomahawks* at Kyedaw Airfield (U.S. Army photo)

Toungoo and the Kyedaw Airfield were 175 miles north of Rangoon and the Mingaladon assembly facility. After a brief inspection, Chennault selected it as his training center. "Toungoo was a shocking contrast to a peacetime Army or Navy post in the United States," said Chennault. "The runway was surrounded by quagmire and pestilential jungle. Matted masses of rotting vegetation carpeted the jungle and filled the air with a sour, sickening smell."[89]

Flight school at Toungoo began with teaching bomber pilots to fly fighters. For some it was an exercise in futility. "Many multi-engined pilots had trouble getting used to the hundred-mile-per-hour landing speed and violent maneuvers of the P-40," complained Chennault. "One morning I watched them crack up six P-40's in landing—heavier losses than the A.V.G. ever suffered in a day's combat."[90]

Chennault began his *school* day at 6 a.m. The lectures were supported with a profusion of maps, mimeographed textbooks, and a blackboard. The curriculum included seventy-two hours of lectures and sixty hours of specialized flying. There was no shortage of lessons on Asian geography (which was desperately needed) and most importantly how the primitive Chinese air-raid network worked, similar to jungle drums rolling out a warning.

From captured Japanese flying manuals, Chennault taught his pilots about Japanese flying tactics. "From these manuals these American pilots learned more about Japanese tactics than any single Japanese pilot ever knew,"[91] beamed Chennault. "The object of our tactics is to break up their formations," he drilled into his students, "and make them fight according to our style. Once the Japanese are forced to deviate from their plan, they are in trouble."[92]

On the blackboard, he drew the outline of each plane and concentrated on where all the vital spots were located and where to aim their fire. He emphasized pounding the vital areas; bullets passing harmlessly through the fuselage were useless, whereas a bullet aimed at an engine might miss, but would hit a fuel tank. He also taught the pilots to focus on the strong features of the P-40 in combat. It was a heavy airplane and had a big 1,040-horsepower engine. It was a true gun platform—armed with four .30-caliber guns and two .50-calibers. That meant higher speed, faster dives, and superior firepower.

Pilot Robert Neale said, "So a pilot had to learn and play it smart—had to know when to dive, how fast, pick out a target, and when to pull the trigger to engage those six teethrattling machine guns."[93] Never dogfight with the Japanese fighters. They can out-maneuver, out-turn, and fly rings around anything currently in the air. Engaging them in a dogfight was a sure death wish. "Use your speed and diving power to make a pass; shoot and break away," Chennault hammered at them. "You have the edge in that kind of combat. All your advantages are brought to bear on their deficiencies. Close your range, fire, and dive away. . . . Accurate fire saves ammunition. Your plane carries a limited number of bullets. There is nothing worse than finding yourself in a fight with empty guns."[94]

And so, day after day, Chennault pounded tactics into their heads. "Fight in pairs. Make every bullet count. Never try to get all the Japanese in one pass. Hit hard, break clean, and get in position for another pass."[95]

He divided his pilots into three squadrons of thirty-three each. The 1st Squadron designed their logo with a bluish-green apple and a serpent wrapped around it, and superimposed upon the apple was a nude Eve pursuing a clothed Adam in the Garden of Eden. Fighters were historically referred to as "pursuit" aircraft, and to the 1st Squadron, Adam and Eve represented the first pursuit. The squadron was called the "Adam and Eves." The 2nd Squadron chose the Panda Bear logo to honor China, and was properly called the "Panda Bears." The 3rd Squadron chose Hell's Angels for their art and moniker. A scarlet nude adorned with a halo and wings graced their fuselages.

Pilot Erik Shilling thought that the lines of the P-40's fuselage, starting from the front of the cockpit to the nose of the propeller, resembled the torpedo shape of a shark. "I went out in the morning and drew the teeth in chalk and went to Chennault and asked him if we could use it as a squadron insignia; and he came out to my airplane, and looked at it and said, 'No. I'd rather have it for all the airplanes.'"[96] For the next week, all AVG hands transformed the P-40s into the menacing image of fierce-eyed attacking sharks flashing their terrible teeth. R. T. Smith would say, "Looks mean as hell."[97]

In the summer of 1941, the United States had embargoed Japanese oil. Chennault felt that, based upon past Japanese actions, the embargo would force them to strike soon "or crawl back into their shell."[98] The Japanese were never known to crawl or back down. In fact, face-saving was the top priority. Predictably, in November, in neighboring Indo-China, the number of Japanese aircraft suddenly increased from forty-five to 275, and there was a great increase of unidentified aircraft that began to fly over Malaya. In mid-November, an anxious Claire Chennault began spending the hours from dusk to dawn in Kyedaw's bamboo control tower. On December 8, 1941, he exited the tower at 11 a.m. and walked across the field, where one of the radio operators ran toward him waving a message. It was December 7 in Hawaii, and the message contained a brief report: an attack at Pearl Harbor.

⚜ ⚜ ⚜

"After Pearl Harbor I considered a Japanese attack on Toungoo a certainty," said Chennault. "My only thought was to meet it with my planes in the air. . . . If I had been caught with my planes on the ground, as were the Air Corps commanders in the Philippines and in Hawaii, I could never again have looked my fellow officers squarely in the eye."[99] In one fell swoop, all the complicated U.S. planning that had produced an array of tactical plans over the years for an eventual war with Japan were swept aside on December 7. And in reality, the only force remaining to confront the Japanese juggernaut was the tiny group of planes and aviators of the AVG, squirreled away in the remote reaches of Burma.

"It was immediately evident that both ends of the Burma Road would have to be defended from heavy air assaults," said Chennault. Rangoon was the only crack in the wall through which supplies could come to China, and Kunming was the only distribution point. Both were on opposite ends of the Burma Road. The Hell's Angels were placed to defend Rangoon; the other two squadrons were stationed at Kunming to the north. To make it worse, Chennault still had twenty-five pilots not sufficiently trained and a dozen aircraft under repair at Toungoo.[100] On December 19, only thirty-four P-40s of the original one hundred stood ready to fight at Kunming.

DECEMBER 20, 1941

At 0945 on December 20, thirteen days after the attack on Pearl Harbor, the Japanese were ready to strike Chennault's AVG. Chennault's special phone to the Chinese commander of the 5th Air Force rang with the terse report that "Ten Japanese bombers crossed the Yunnan [Chinese] border . . . heading northwest."[101] On the ground, the Chinese listening stations began to sequentially report the progress of the bomber flight. Station X-10 reported: "Heavy engine noise"; Station P-8: "Unknowns overhead"; Station C-23: "Noise of many above clouds."[102] All of those fixed-position reports were plotted on a control board and formed a straight line, revealing that the Japanese flight would enter AVG air space fifty miles east of Kunming. Chennault scrambled the Panda Bears to make the interception.

Two four-plane elements took off, one to search for the bombers and the second to fly defensive cover over Kunming. An additional sixteen-plane force of the Adam and Eves stood west of Kunming ready to join the fight. Chennault fired a red flare into the sky signaling all forces into action. He listened for details over the radio, but except for some crackling static, there was nothing. There was a brief Chinese report that the bomber force had reversed course and there was the sound of bombs detonating in the distance, but that was it. Planes of the AVG finally began to land. Most had seen nothing; the Panda Bears had seen the bombers but, except for Ed Rector, had not engaged. They saw the bombers reverse course, jettison their bombs, and hightail it back toward Indo-China. As they retreated, they had lowered their "dustbin" doors, revealing their tail gunners, ready to confront pursuers.

Finally the Adam and Eves began to land, and from the way they sauntered across the field it was obvious that they had been in a gunfight. From their position thirty miles east of Kunming, the squadron had sighted the Japanese formation as it turned to retreat. And the P-40s pounced on it like sharks in a feeding frenzy. But all the discipline taught by Chennault quickly went out the window and it became a wild melee and a running gunfight that stretched for 130 miles across the Burma skies.

Ed Rector, the sole Panda Bear who had raced off to attack the retreating bombers, described his part in the battle: "We caught up with the formation of ten bombers, and I just kept shooting and shooting." He closed rapidly on his targeted bomber and "at the last minute, I shoved the stick forward, and barely missed running into him . . . I don't know how I missed, and I can see in my mind's eye, the camouflage pattern, the rivets, and that rear gunner in the tail. I had shot his lower jaw away, and he was slumped over the gun."[103]

In the end, three bombers crashed in flames and the remainder were streaming smoke in various degrees as they departed.[104] Later, a fourth bomber was reported to have exploded in midair. The pilots speculated that other bombers certainly did not return home. Bob Neale reported that as he pulled out of the fight, he thought that eight Japanese were still aloft, but also noted that he had seen that all of the tail gunners were dead, sprawled and slumped over their guns on their lowered dustbin platforms.[105]

Chennault debriefed them all, and in detail went over all that had happened and berated them for their failure to fly in the disciplined formations he had taught. But his chastising words had little sting on the beaming pilots, who were now battle-tested warriors. They all knew they had bloodied the vaunted Japanese bombers in their very first encounter. Finally the debrief was over, and the wild celebration continued. Chennault left their celebration with the satisfaction of the final word on the subject: "Well, boys, it was a good job, but not good enough. Next time get them all!"[106]

Years later they learned that only one of the bombers had ever returned to the Japanese base. Lewis Bishop was shot down five months after this battle, and while in prison, met the leader of that bomber force, who confirmed that he was its sole survivor. That was not surprising. All the bombers were badly shot up, as evidenced by the 100 percent casualties of the tail gunners, and they were all streaming smoke. The Adam and Eves and one Panda Bear had almost gotten them all. The Japanese had little to say about this battle, not even to claim concocted numbers of American planes shot down as they had after their 1938 lopsided loss to celebrate the Emperor's birthday. What is irrefutable is that after the December 20, 1941, defeat at the hands of the AVG, the Japanese never again tried to bomb Kunming.

DECEMBER 23, 1941

The 3rd Squadron, Hell's Angels, defended the skies over Rangoon. Only twenty to twenty-five aircraft were available for duty on any given day. On December 23, the Japanese Army Air Force (JAAF) made a strong effort against Rangoon with fifty-four bombers and twenty fighters. "The Third Squadron was casually ordered to clear the field," said Chennault, "and while still climbing they were informed . . . 'Enemy approaching from the east.'"[107] Japanese fighters busied themselves diving on the city and strafing gawking citizens foolish enough to come out and watch the raid. Bombers strung their bombs along the vital docks before the AVG fighters interrupted them. The fight was brief: six Japanese planes down at the cost of two P-40s.

On Christmas day the Japanese were back, this time with sixty bombers and thirty fighters. Twelve Hell's Angels prowled the skies above the approaching force. An observer on the ground described the Hell's Angels attack: "Like rowboats attacking the Spanish Armada." Squadron leader Arvid Olson excit-

Iconic photo of Hell's Angels, 3rd Squadron, over the Salween River Gorge, China, 1942. (Air Force Museum)

edly radioed Chennault at his Kunming headquarters: "It was like shooting ducks," he said. "We got 15 bombers and 9 fighters. Could put entire Jap force out of commission with whole group here."[108]

The AVG went on the offensive and roamed the skies over Thailand searching for enemy targets. They found and strafed numerous JAAF airfields, with planes neatly parked, side by side, anticipating an assault on Rangoon. By the end of January, intelligence indicated that the Japanese were again ready to hurl their force against the smaller combined air forces of the AVG and the RAF. On January 28, grim numbers spelled out dismal Japanese failure. Fifty JAAF planes had been destroyed against the loss of two AVG pilots and ten RAF pilots killed.[109] But despite that rock-ribbed defense of Rangoon, the reality was that the 2nd Squadron was down to ten P-40s. Chennault sent the rest of the 1st Squadron to reinforce it.

Then a strange phenomenon took place. At AVG Headquarters in Kunming, Chennault was suddenly swamped with mail containing news clippings from Stateside newspapers. "My men were astonished to find themselves world famous as the *Flying Tigers*. The insignia we made famous [shark's teeth] was by no means original with the A.V.G."[110] But his fighting and tactics against the rampaging Japanese war machine was.

Before he left Louisiana, Chennault had asked friends to send clippings about the AVG, if there ever were any; and now the mail bags poured in. Perhaps that was because the ever-intuitive American magazine magnate Henry Luce decided to convert the first story of the AVG's incredible victory on December 20 into a sensational story to be featured in *Time* magazine. In those terrible days immediately following Pearl Harbor, good news was hard to find. The world was staggered and on its heels. Certainly the newspapers regarded that early AVG victory as good news, but not for the front page. The *New York Times* mentioned it on page 27, and rightfully so. Monstrous other stories like the life-and-death struggle unfolding on the Russian front, the tragic fall of Wake Island after an extended holdout, MacArthur's retreat in the Philippines to Bataan, and the brutal Japanese occupation of Hong Kong deserved the front pages.

Still, on December 29, 1941, eight days after the AVG had dealt a stinging blow to the heretofore undefeated JAAF, Luce rolled out his own story, and Americans were desperate for it. They ate it up: "Battle of China: Blood for the Tigers," was the headline. "For three years the Japanese had been bombing China from the coast. . . . They had laughed at the ineffectual popping of Chungking's worn anti-aircraft guns, had shot down fledglings of the Chinese

Air Force like wounded ducks." But now there was Colonel Claire Chennault, "lean, hardbitten, taciturn," and he had "rounded up U.S. volunteers to fly 100 new P-40s purchased from the U.S. If U.S. aid were to flow in over the Burma Road, U.S. flyers would have to protect it."

During the summer, Colonel Chennault "whipped his volunteers into shape," and then "Last week ten Japanese bombers came winging their carefree way up into Yunnan, heading directly for Kunming, the terminus of the Burma Road. Thirty miles south of Kunming, the Flying Tigers swooped, let the Japanese have it. Of the ten bombers, said Chungking reports, four plummeted to earth in flames. The rest turned tail and fled. Tiger casualties: none."[111]

It was magic. The Flying Tigers were suddenly a household name. "After the initial battles over Rangoon," said Charles Mott, "the Japs were just plain stopped; they'd lost; they quit. They called in reinforcements from the Philippines, and from Singapore, and Malaya, and Indochina."[112]

But despite the heroics and daring-do, Chennault and his AVG were on their last fighting leg. There had never been any spare parts, and the Japanese aerial assaults had been relentless. In January, Chennault was hospitalized by what he called his "annual attack of chronic bronchitis."[113] On February 4, he listened from his bed over an installed radio as the Japanese continuous attack on Toungoo unfolded. It was devastating. "The operations building and a hangar were destroyed by direct hits; three P-40's still under repair were wrecked; and half a dozen R.A.F. Blenheims burned,"[114] he reported.

In mid-February, Singapore fell, and with that victory the Japanese moved massive reinforcements for an aerial assault on Rangoon. "These reinforcements boosted enemy plane strength available to attack Rangoon to four hundred planes," said Chennault. "Before the month's end, they were hammering at the city with two hundred planes a day."[115]

On February 26, the AVG could field only fifteen fighters against this onslaught. The next day they were down to six as a force of two hundred Japanese planes flew in. But against those odds, they still managed to bag eighteen before withdrawing.[116] On February 30 the Japanese army overran Rangoon, and the battle for Southern Burma was over. Defeat was at hand. The Japanese poured their infantry through Rangoon and advanced to the north, toward the Burma Road, for further attacks into China. The Chinese formed up and manned the left of an allied defensive line while the British Army manned the right.

If Japanese intention to now overwhelm China in a massive ground offensive was still unclear to some British generals, it was perfectly clear to Prime Minister Winston Churchill. Unlike his military leadership in China, he did not think that the Japanese had taken all that they wanted in the capture of Rangoon and Singapore. He addressed his assessment to Parliament. "Their best plan would be to push northwards from Burma into China and try to finish Chinese resistance and the great Chinese leader Chiang Kai-shek. . . . Certainly by driving China out of the war . . . Japan would be furthering her own interests. China is the only place where Japan can obtain a major decision in 1942."[117]

So, after a pause in what Chennault called "the slugging match between the Japanese Air Force and the A.V.G.-R.A.F.," the air battle continued. If ever there was a mismatch, this was it. Japanese air strength in southern Burma was now five hundred planes—fourteen air regiments. But the upcoming aerial fight to be waged would not be air-to-air combat. "The character of battle shifted to continual attempts to catch opposition on the ground and shooting sitting ducks," said Chennault.[118]

The advanced units of both the AVG and the RAF were located at Magwi, an unfinished airstrip 250 miles from Rangoon. On March 19, the AVG struck first—but not from Magwi. A two-man flight, flown by Bill Reed and Ken Jernstedt out of the previously bombed Toungoo airstrip, discovered a new Japanese airstrip ten miles from their main field at Moulmein. To their unbelieving eyes, Reed and Jernstedt saw twenty fighters lined up and parked in a row. "We started strafing," said Jernstedt, "Bill down one side while I went up the other and we kept running that circle until everything was destroyed. It was like shooting ducks in a pond. I saw a truck drive up full of pilots and all they could do was sit there and watch."[119]

Twelve strafing runs later, without a shot being fired against them, fifteen fires raged from the burning hulks of the parked aircraft. While their fuel lasted, and still with ammunition, they flew the ten miles to attack the main field at Moulmein. Reed and Jernstedt struck again. When they completed their unopposed strafing runs, three more enemy bombers and a military transport were on fire. Those combined nineteen enemy kills marked the highest individual numbers ever scored by any AVG pilots. "Bill Reed and I had the most productive strafing mission. I received a radiogram from Chennault that said, 'Congratulations, you have set a new record for destruction.'"[120]

The next day, completing a devastating one-two punch, the RAF struck the Japanese at Mingdalon. Another twenty-eight enemy planes went up in flames. The Japanese command was stunned and furious with the two stinging defeats and now planned an all-out attack against Magwi. Their objective was to crush, once and for all, any and all Allied presence in Burma. On March 21, just after noon, waves of Japanese planes attacked. For twenty-five hours, one hundred fighters escorted 166 medium and heavy Japanese bombers in a nonstop bombardment of Magwi. "When the fires burned out," reported Chennault, "there were only three flyable P-40's, and only four out of twenty-five Hurricanes [were] able to get into the air."[121] The survivors fell back to their base in Yunnan Province just across the Burma border. A pilot debriefing revealed that the attacking Japanese had surprised them, flying in from northern Thailand airfields. Since the fall of Rangoon, the AVG had not kept Thailand under routine surveillance. "We paid heavily for this relaxed vigilance," lamented Chennault.[122]

With his ever-dwindling air force, Chennault was still not beaten. He planned a "revenge mission." Six P-40s from the 1st Squadron (Adam & Eves), led by Bob Neale and Greg Boyington (future leader Pappy Boyington of the famous Black Sheep Squadron), and four aircraft from the 2nd Squadron (Panda Bears), led by Jack Newkirk, would strike Japanese bases in Thailand. To conceal their movements and preserve the element of surprise, the ten aircraft took off from Kunming and, after refueling, waited until dusk to slip into RAF airfields along the Thai border. They bunked down for some well-needed rest, but at 0400 took off in the dark on runways illuminated by truck headlights and kerosene lanterns. It was March 29.

The force reached Chiengmai just after dawn. Neale's flight was the first to pounce on the Japanese preparing for yet another strike against Magwi. "More than forty fighters and bombers were lined up in neat parking rows," reported Chennault, "with engines warming up and ground crews doing final work preparatory to take-off. Many pilots were near their planes."[123]

Ed Rector and William "Mac" McGarry stayed aloft as high cover while Charlie Bond led the *Tomahawks* in the attack. Neale, Boyington, and Bill Bartley swooped in after him. Their gun ammunition was generously laced with incendiary rounds. "Not a Jap plane got off the ground," wrote Chennault. "Enemy planes, with a full gas load, burned like phosphorus matches. Pilots and ground crews were cut down as they scampered across the bare airdrome for cover. . . . At least twenty burning planes were counted on the field with ten more riddled beyond repair."[124]

"I made my first strafing run firing everything into the [fighters]," said Charlie Bond. "After turning 270-degrees, I was in position to strafe another line of parked aircraft. They were sitting practically wingtip to wingtip. Hell,

I hadn't seen this many aircraft in years. Seems like the whole Japanese Air Force had tried to crowd into this one little field."[125] Bond made four runs. He reported that on his last run he zeroed in on a larger plane in the parking area. His .30- and .50-caliber guns pounded the aircraft so that it "seemed to shake itself to pieces."[126]

Greg Boyington made two devastating passes on the "aircraft . . . parked mainly in two long lines. . . . All enemy planes were turning up, and the pilots and crews were running about."[127] Three transports were consumed in a single fire, and after his second pass he counted ten fires on the ground. Neale reported, after his third strafing run, that he estimated forty planes on the field, and that they had destroyed half of them. As he streaked away in the face of heavy antiaircraft fire, he counted eight or nine fires on the field, and some were very large.

But the success of the six-plane attack was tempered by the loss of the Panda Bear's leader, Jack Newkirk. He had flown over Chiengmai to attack a second field, which proved to be empty, and while strafing the road, was shot down and crashed in a ball of flame. Mac McGarry was also lost to the AVG. He bailed out of his stricken *Tomahawk* and was taken by the Thais and interned, and sat out the rest of the war.

The Japanese, continuing their proclivity to forever deny anything hinting of military defeat, claimed that the Chiengmai attack was really nothing. Historian Daniel Ford, citing Japanese sources written twenty-five years after the war, wrote the Japanese claimed that those documents showed that on March 29 the Japanese only lost three aircraft, with ten others damaged, but only four seriously. He wrote: "The Japanese mechanics were as industrious as their counterparts in the AVG and by noon, they had eleven planes fit to fly."[128] No mention of the horrific losses.

This continuing penchant for self-deception was already well known by the world, and would reach its zenith of absurdity in the summer of 1944. In the Mariana Islands, as the Japanese were defeated and crushed at every turn, Imperial Headquarters still managed to publish official, face-saving reports. It reported that, in defeat, Japanese forces had sunk or damaged thirteen aircraft carriers, five battleships, seven cruisers, three destroyers, one submarine, and seven transports, and shot down more than 863 aircraft.[129] None of that happened.

The reality of the 1942 AVG attack on Chiengmai was that the Japanese did not return to attack Magwi. If they indeed had lost only three aircraft and had repaired most of the damaged ones by noon, they would have had most of their original strike force intact, not just eleven planes out of forty. They would have launched their planned attack against Magwi. They did not attack, and the British were able to continue their frantic evacuation of British and Indian nationals from the Magwi field. Britain's Air Vice Marshal Leigh Forbes Stevenson signaled Chennault: "Many thanks for the breathing spell furnished us by your magnificent attack at Chiengmai."[130]

Chennault moved the AVG headquarters to a new field at Loi-Wing that he described as "a touch of America" because of its neat white houses and a country club featuring an enormous picture window framing the scenic countryside. It also featured an enormous jukebox. The Japanese made their first attack on April 8 without favorable results, losing ten of the twenty planes they started with. They were back on the night of the 10th and lost eight more.[131]

Chennault was most aware that April 29 was the Emperor's birthday. He expected some sort of repeat performance by the JAAF. Since the previous year's attack had been a disaster, he guessed that they might try surprise and come a day earlier. His patrols discovered "signs of great activity along the entire air front. This confirmed my hunch, that the Japanese were timing a knockout blow against Loi-Wing, to announce the end of the A.V.G. on the Emperor's birthday."[132]

Chennault had his fifteen P-40s in the air, watching the approach to Lashio, and was not disappointed. Bombers with fighter escorts were attacking from the south, following the trace of the railroad tracks leading toward the Burma Road. It became a one-sided contest as the Flying Tigers intercepted the enemy fighters. Chennault noted the results of the Japanese aerial defeat. "An honest Emperor's birthday communique would have announced a 22-to-0 score for the A.V.G."[133]

But it was the last hurrah for the Flying Tigers in Loi-Wing. The Japanese army, pushing up from Rangoon, was advancing steadily on Lashio, and the AVG position became untenable. The next day they hurriedly moved out, setting fire to twenty-two P-40s that were in various states of repair. On May 1, the Japanese army overran the Loi-Wing airfield.

THE FINAL ACT

The Salween River Gorge played host to the final heroics of Chennault's AVG. It was truly the Grand Canyon of the Burma/China theater. The Burma Road crawled over the rocky western rim of the enormous gorge and slowly descended the sheer slope to the Salween River, a mile deep into the gorge.

To reach the river, vehicles had to negotiate thirty-five hairpin turns over a twenty-mile stretch that had, like the Burma Road, been hewn out of a wall of solid rock by thousands of human hands wielding primitive tools. At the river, a destroyed suspension bridge had spanned the churning water. Any crossing to the east bank presented a similar tortuous, serpentine ascent to the heights of the eastern rim of the gorge.

On May 6, the steadily advancing Japanese army had penetrated deep into China and was actually on the western road, twisting and descending into the gorge. The retreating Chinese army had blown the bridge even though some of their forces had not crossed to the eastern side; and that twenty-mile, thirty-five-hairpin-turn road was now jammed with thousands of refugees intermingled with the stopped, bumper-to-bumper Japanese column. They now waited for engineers to throw a pontoon bridge across the Salween River.

Chennault described the desperation of the Chinese military. "There were no obstacles between the Japanese and Kunming but a broken bridge and the A.V.G. . . . If the Japanese got to Kunming, it meant the end of the war for China."[134] But the AVG was powerless to attack the column with anything but machine gun fire. The P-40 was not a bomber, and it had no bomb racks.

"For months . . . we had experimented with homemade bomb racks on the P-40B's without success," Chennault wrote. "We tried everything, including dropping whiskey bottles filled with gasoline as incendiaries.

"Apparently the only people who realized our acute need for bombers were my fellow citizens of Louisiana who raised $15,109.86 by popular subscription in a 'Buy a Bomber for Chennault' campaign."[135]

Governor James Noe was informed by the War Department that even the smallest bomber cost more than what they had raised, but the governor still sent the Louisianans' money to aid Chinese war orphans.

And then, it seemed, Divine Providence intervened. Four new P-40E *Warhawk* fighter bombers bought by the Chinese government had been held up by an Air Corps general for his own use in India, and only the most strenuous protest by Chiang Kai-shek to President Roosevelt was able to pry them from his grasp.[136] The new *Warhawks* (aka *Kittyhawks*) bristled with six .50 caliber Brownings in its wings and a bomb rack to carry a 500-pound Russian demolition bomb under the fuselage. Two wing racks could also carry fragmentation bombs. Chennault now had the means to attack the Japanese column strung out in the Salween Gorge. But if he attacked that column, many refugees and civilians would also perish. He wrote: "I was faced with one of those grim decisions that come to every military man so often in battle—issuing orders that meant the sacrifice of a few to save the many. We had small stomach for bombing and machine-gunning those refugees on the west bank of the Salween, but if we were going to stop the Japs, we would have to slaughter some innocents along the road."[137] He signaled his dilemma to the Generalissimo, who immediately gave the order to attack.

On May 7, 1942, Japan stood on the brink of delivering the knock-out blow to China, and to her British and American allies. With China out of the war, Japan would have sealed off her western flank from any outside interference, cut all supply lines to China, and guaranteed total freedom of maneuver in the Pacific Ocean Area. With the fall of China, Japan's armies would be free to turn to other military tasks. Once across the Salween, her army would make the quick, unopposed march to Chungking and dispose of Chaing and whatever was left of his army. The great victory of 1942 for Japan, predicted by Churchill, would become a reality.

Just after dawn on that fateful day, David Lee "Tex" Hill led the flight of the four new P-40E *Warhawk* fighter-bombers. With him on this bombing mission were Tom Jones, Ed Rector, and Frank Lawlor, all former navy dive-bomber pilots from the carrier *Ranger*. Flying top-cover for the bomber force, in the older P-40B *Tomahawks*, was flight leader and former Air Corps pilot Arvid Olson and R. T. Smith and Erik Shilling, also from the army. The fourth *Tomahawk* was flown by Tom Haywood, a former Marine.

The eight-plane strike force streaked toward its target stalled in the gorge. Armor, artillery, vehicles, and soldiers of the Japanese army were strung out on the twenty miles of the great twisting western wall road like a giant sunbathing serpent. But massive black and purple thunderheads stood like sentinels, barring the way to the gorge. In other circumstances, a pilot would have aborted the mission and turned back from the storm front, but turning back was now not an option. The bridging equipment and the Japanese engineers were already unloaded at the Salween and ready to span the river. Chennault wrote: "'Tex' Hill circled for a few minutes studying the weather ahead and then resolutely plunged to the south, threading the flights between the towering, ice-crested cumulus clouds. A quarter of an hour later [they] emerged from bouncing turbulence into clear cloudless sky, with the muddy Salween ahead."[138] And below them was the target. Chennault wrote, "On the Salween escarpment they were trapped in the open like flies on flypaper—a sheer precipice on one side of the narrow road and a rock wall on the other."[139]

Hill led his strike force in a classic dive-bombing formation and unleashed his Russian demolition bomb, targeting the straight parts of the road at the top of the gorge to collapse the ledge and begin a landslide that would take out lower sections. The other three *Warhawks* followed, and their 500-pounders shook the slope and produced more slides, effectively cutting off any escape from the gorge. "Hill [then] led his flight in a frag bombing attack on the trucks . . . and returned to shuttle back and forth, strafing along the straight stretches of road, lacing the Japanese with bullets from twenty-four .50-caliber machine guns."[140]

The entire escarpment was now ablaze with the burning fuel from destroyed and exploding trucks. On the ground, pandemonium was the order of the day. Finally Hill's strike force was out of bombs and ammunition and flew out of the Salween canyon. But there was no relief for the Japanese trapped in the gorge. Arvid Olsen now went into action, and his top-cover assassins plunged into the gorge and emptied their machine guns, laced with incendiary rounds, into the pontoons and bridging equipment. In short order flames engulfed it all.

"For four more days I threw everything we had against the Salween gorge and the Burma Road," said Claire Chennault. "The Japs sent another group of light tanks up from Burma. . . . A flight led by Frank Schiel . . . scattered them with bombs and machine-gun fire. Every town and village along the road that could serve the Japanese as shelter or a supply depot was bombed and burned . . . a truck column loaded with gas left behind at least fifty billowing gasoline fires."[141]

Finally, on May 12, Chennault was able to signal Madame Chiang: "Reconnaissance this morning along west bank Salween discovered no sign of Japs."[142] It was over. The AVG had written the final chapter of its remarkable combat record. Its year-long contract was declared to be over on July 4, and the regular Army Air Corps was moving in to take over operations. But that takeover was done without honor and became a disgraceful chapter in the Air Corps' history. Except for a few of its general officers, most of the hierarchy had little use for the AVG and held its pilots in great disdain.

"Hap Arnold didn't like us and did everything he could possibly do to sink us out there. He refused us spare parts from the beginning when we could have really used them," said pilot Robert "Catfish" Rainer.[143] It had taken direct personal intervention from FDR to pry the pilots away from the top brass, and it had taken FDR's intervention to release the P-40Es from the hoarding Air Corps general in India. An air of resentment infected the thinking of the senior officers, and jealousy and envy reared its ugly head among them. Early on, Chennault had labeled these obstructionist officers as "foggy-brained brass."

As the July 4 takeover date approached, the army wanted the soon-to-be released pilots of the AVG to rejoin the Air Corps. "They sent a man over to recruit us and he was so obnoxious that most were just anxious to get out," said Dick Rossi, a flight leader and a six-kill ace. "They just told him to go jump in the lake. It was this man . . . Colonel Bissell, and they had made him a general one day before Chennault so he would outrank him."[144] Erik Shilling added: "Bissell said that if we didn't join up, no one was to help us return to the states, and that when we arrived . . . he would see to it that the draft board was there to meet us and we would all be put into the army as buck privates!"[145] "The boys sat in stony silence," said Chennault. "Later they came to me individually or in small groups. Their stories were all the same. 'If that's any sample of how the Army is going to treat us, we want no part of it.'"[146] Here was the most famous group of aviators, personally approved by President Roosevelt, having met and defeated the vaunted Japanese air force in its all-out assault against China, being lectured to as if they were draft dodgers by a stuffed shirt, envious colonel. But the Flying Tigers decided on a little insult of their own for the despised Clayton Bissell.

From the time of their arrival in China, the AVG all learned at least two words of Chinese, and that was the expression *Ding Hao!* As they greeted the Chinese people, especially the children, they would give the thumbs-up sign to them, and the children responded with the Chinese translation, "Ding Hao." These crusty AVG pilots spread the word, especially among the children, that whenever anyone gave them a thumbs-up, they were to respond loudly with a new translation. "So," Dick Rossi explained, "we taught all these little kids, instead of saying *Ding Hao,* they were now to say, 'Piss on Bissell.' So everywhere he went he heard 'Piss on Bissell' and he wasn't too happy about it."[147]

The result of Bissell's buffoonery was that only five pilots and twenty-two ground personnel opted to join the Army Air Corps and stay in China. To cover up Bissell's total failure, the Air Corps resorted to its own face-saving lying and deception. "The War Department never stopped pretending that all went well with the A.V.G. induction," wrote Chennault. "As late as January 1945 the War Department wrote the House Military Affairs Committee of Congress that 220 of 250 A.V.G.'s had been inducted into the Army in China on July 4, 1942."[148]

The pilots and ground crews, other than those who stayed to fly for CNAC,[149] worked their way back to the States as best they could with little help from anyone, as promised by Bissell. Ken Jernstedt, a five-kill ace, hitched rides with Pan American until he got to Florida and took a train home. Twenty pilots had achieved ace status, and seven more would attain ace status with further kills made during the rest of the war.[150] The most famous of those who added to their total was Greg Boyington, who returned to the Marine Corps in 1942 and eventually commanded the legendary "Black Sheep Squadron." During the Solomon Islands aerial campaign, he shot down an additional twenty-two Japanese aircraft, and was himself shot down and taken prisoner. He was released at war's end and awarded the Medal of Honor and the Navy Cross.[151]

At midnight on July 4, 1942, the AVG passed into history. It was a melancholy evening for all who knew what the American volunteers had done, and how they had now been dismissed out of hand. Chennault would write: "The group that the military experts predicted would not last three weeks in combat had fought for seven months over Burma, China, Thailand, and French Indo-China." The Flying Tigers had destroyed almost three hundred Japanese aircraft in the air and on the ground. They had lost twelve P-40s in combat, and another sixty-one on the ground, and that included the twenty-two that they burned at Loi-Wing to prevent falling into Japanese hands.

Four pilots had been killed in aerial combat, and another ten by antiaircraft fire. Three had died as a result of Japanese bombing. Ten more died from flying accidents, and three were taken prisoner. But the AVG had prevailed in fifty encounters with the JAAF and never lost a battle. Their victory in their final battle at the Salween River was nothing short of astounding. With fewer than one-third of its aircraft, the AVG had saved China from a final collapse.[152]

Claire Chennault and his AVG had managed this incredible military achievement with absolutely no support from the military braintrust of the United States. In fact, the senior generals threw roadblocks up along every step of the way and deceived President Franklin Roosevelt concerning the real situation in Burma and China, and would continue to do so. Chennault bitterly regretted "that circumstances had prevented me from ever throwing the entire group at the Japanese in a single battle, and that the Army had forced the group to disband."[153]

Franklin Roosevelt wrote: "The outstanding gallantry and conspicuous daring that the American Volunteer Group combined with their unbelievable efficiency is a source of tremendous pride throughout the whole of America."[154] Winston Churchill compared the AVG performance to the RAF's defeat of the

German Luftwaffe in the Battle of Britain.[155] General Hap Arnold penned a perfunctory congratulatory letter to Chennault saying that he was "personally directing an intense effort to enroll in the Army Air Forces all of your ex-American Volunteer Group combat personnel who are now in the States." But if that were so, he failed to explain why his Air Transport Command had issued instructions that in their return to the States "absolutely no priority would be given to former A.V.G. members."[156]

CONCLUSION

The AVG was succeeded by the China Air Task Force (CATF), which was an impressive name for an outfit that was part of the newly formed 23rd Fighter Group. But the Group existed mostly on paper, with only twenty-nine flyable aircraft.

Later, at a peak strength of forty fighters and seven bombers, the CATF was expected to face off against a Japanese force of four hundred aircraft over a two-thousand-mile front. And there was still no resupply and few tools of the trade. Chennault would lament: "The China Air Task Force, unbroken in combat, was facing death from acute starvation."[157] Chennault had stayed with the CATF, but was really in charge of nothing. Until April 7, 1942, he had been a civilian farmer running the AVG. On that date, he was promoted from his old rank of captain to temporary colonel, and nine days later to brigadier general.

The overall China operation was run by Lieutenant General Joseph Stilwell, a sour man who richly deserved his nickname "Vinegar Joe." He was a darling of General George Marshall and had the support of the entire War Department. But he carried with him "a strong prejudice against air power,"[158] hardly an asset since it was minimum air power that had prevented the Japanese takeover of China. To make matters worse, he had a dark side demonstrated by his venomous tongue that lashed out at everyone. He was a pompous charlatan. He referred to President Roosevelt as "Old Rubberlegs" and called British allies "bastardly hypocrites"; and called Chiang Kai-Shek a "peanut dictator . . . the little dummy . . . an ungrateful little rattlesnake."[159]

The air operations were controlled by General Clayton Bissell—made one day senior to Chennault even though Chennault had been senior to Bissell during his army career. Chennault called the ploy "an old and effective Army routine."[160] But that one-day difference was the handiwork of Stilwell, who had no use for Chennault and his air tactics and said, "I spoke for Bissell, and insisted that he rank Chennault."[161] "My status was simply that of deputy commander," said Chennault, "subject at all times to Bissell's orders. . . . I obeyed his orders to the letter, but I never developed much respect for his ability, either as a military administrator or tactician."[162]

The Stilwell-Bissell team failed miserably and never met the supply quotas for the CATF. Its records documented receipt of only 30 to 50 percent of the supplies promised by Stilwell and Bissell.[163] Stilwell's answer to Claire Chennault's supply complaints was, "Chennault, you must realize that the air force can't have everything. You've got to learn to do without things."[164]

At one point in January 1943, the CATF remained grounded for thirty-three days with no fuel. Then, in March, as if it were manna from heaven, Washington revealed that a new 14th Air Force would be formed in China under Chennault's command. And just as the AVG had passed into history, so did the CATF.

"Despite its grim beginning and starvation diet the China Air Task Force left a proud combat record," wrote Chennault.[165] In nine months it had destroyed 149 Japanese aircraft with the loss of only sixteen P-40s. It also conducted sixty-five bombing missions that the Japanese air defenses were unable to stop. Only one bomber was lost. But there was a rising tide of criticism against Stilwell for his lack of performance. To attempt to quiet his critics, he threw up a wall of secrecy around his command. No information could be released unless by him, and all correspondence from subordinates was pigeonholed. "The War Department, from General Marshall down, naturally received no information except from Stilwell. Thus Stilwell's distinctly distorted version of the Chinese situation became the official version. . . . All references to the C.A.T.F.'s pitiful size were carefully censored."[166] said Chennault.

But suddenly, without warning, Stilwell's stone wall was smashed. Wendell Wilkie, President Roosevelt's personal envoy, stopped in China as part of his around-the-world fact-finding mission. He wanted to confidentially meet with Claire Chennault, and on October 11 the two men met for two hours while Stilwell sat in an outer office. When Chennault told him that the CATF was making bombing raids with fewer than a dozen aircraft and that the whole of China was being defended by fifty fighters, Wilkie was astonished. He was even more amazed to learn that despite these small numbers Chennault's force had racked up an 8-to-1 kill ratio against the Japanese. "Wilkie asked me to state my case in a detailed letter that he could present directly to President Roosevelt."[167] It took Chennault all night to compose his lengthy, fourteen-paragraph, detailed letter, and it was in Wilkie's pocket as he departed China the next day. His opening sentences were eye-catching: "Japan can be defeated in China. . . . It can he defeated by an Air Force so small that in other theaters it would be called ridiculous . . . given real authority in command of such an Air Force, I can cause the collapse of Japan."[168]

On March 3, 1943, Chennault was promoted to Major General and finally separated from Clayton Bissell. A month later, Stilwell announced he was coming to Kunming. Chennault met him, only to be asked where were his bags, and wasn't he ready to go? A bewildered Chennault then followed Stilwell to a secret meeting behind his airplane, where Stilwell announced that he had been summoned to Washington, "the result of my finagling behind his back."[169]

In Washington, the battle lines were drawn: Stilwell against Chennault. Stilwell presented his side, forcing Chennault to constantly refute his commanding officer. This was the Trident Conference: the meeting of the heads of state of the United Kingdom and the United States. Through it all, Stilwell was miffed having to defend his positions against a subaltern. He was "curt, surly, and short tempered."[170]

In the end, Chennault prevailed and Trident endorsed his plan. Chennault would write: "The Trident decision put Stilwell in the embarrassing position of being charged with execution of a subordinate's plan with which he violently disagreed."[171] In a private meeting with Chennault, Roosevelt candidly asked if a China-based air force could sink a million tons of Japanese shipping. When he replied that with 10,000 tons of supplies a month, he could sink more than a million tons, the President exploded with delight. "If you can sink a million tons, we'll break their back."[172] With considerably less than the 10,000 tons per month supply, the 14th Air Force sank, over a two-year period, almost 3 million tons of shipping and thirty-two naval vessels.

On October 19, 1944, Stilwell was relieved from command and replaced by Lieutenant General Albert Wedemeyer.

In mid-June 1945, a shameful deception unfolded in China. The war was almost over and the Japanese were on their last leg when a concocted plan to reorganize the China air forces was floated by Generals Marshall and Arnold. It made no sense to anyone, and a brief examination of its folly proved it would be militarily counterproductive. General Wedemeyer dismissed it out of hand. But its thinly veiled purpose had nothing to do with efficiency. The proposed reorganization would get rid of that thorn in everyone's side, Claire Chennault. Stilwell had constantly whined and blamed all his troubles on him, and his concoctions, no matter how ridiculous, always found sympathetic ears among his legion of "old boy" cronies that populated Washington and the army high command, including Marshall and Arnold.

Seeing that Wedemeyer had sniffed out their skullduggery, Marshall and Arnold were undaunted and pressed on to achieve their desired result. Incredibly, on June 20, 1945, a secret meeting was called for all general officers—no stenographers, no aides, no persons of lesser rank—to an "ears only" briefing. Chennault reported the meeting: "Wedemeyer read Arnold's special-delivery letter stating that it was the wish of Marshall and Arnold that reorganization of the China air forces be carried out as planned *regardless of the consequences*. . . . On July 8, eight years to the day that I first offered my services to China, I wrote Wedemeyer requesting relief from active duty and retirement from the Army."[173]

Just two months later, on September 2, Japan's formal surrender took place in Tokyo Harbor on Battleship *Missouri*. The elite army ruling class ensured that Chennault was not invited. Many thought that Claire Chennault deserved the Medal of Honor. He certainly was recommended for it. Every man and officer of the CATF signed a petition that it be awarded to him. That petition was promptly dismissed by General Stilwell.[174]

On July 18, 1958, Claire Chennault was promoted to the three-star rank of lieutenant general. Nine days later he died from cancer in New Orleans. In 1972 he was inducted into the Aviation Hall of Fame, and in 1980 the U.S. Postal Service honored him with a 40-cent stamp in its Great Americans series. His statue is prominently displayed on the grounds of the Louisiana State Capitol in Baton Rouge.

In 1991, the U.S. government finally recognized the AVG veterans as having been a part of the U.S. military service during its months in combat against Japan. In 1992, the AVG was awarded the Presidential Unit Citation for "professionalism, dedication to duty, and extraordinary heroism." Five years later, in 1996, each of the pilots received the Distinguished Flying Cross, while the members of the ground crews each received the Bronze Star Medal.

In 1991, coinciding with the unit's fiftieth anniversary, historian Daniel Ford published his book *Flying Tigers*. He sought to challenge the validity of the official 296 kill record of the AVG and suggested that the actual record of destroyed Japanese planes had been inflated by almost 300 percent. In Ford's mind, the Japanese Army Air Force (JAAF) lost "115 aircraft to the AVG in Burma, Thailand, and China."[175] In his revised edition published in 2007, Ford revealed that critics had labeled his work "Japanese-sponsored revisionism" and "the surviving Tigers . . . roundly condemned the book, as a defamation of Claire Chennault and the fighter group he readied for combat in the fall of 1941."[176]

Ford's reasoning was flawed. By his own admission, there were no Japanese Central Records and "all we have are reconstructions . . . as time went on, JAAF veterans wrote their memoirs."[177] Ford seemingly ignored the Japanese penchant for self-deception when it came to the realities of war: denial of lost battles, lost men, and lost equipment. He wrote, "As for using Japanese sources after more than half a century, a writer shouldn't have to apologize."[178] He complained that "No such skepticism is shown toward German reports of their losses in the Battle of Britain."[179] But Germany never tried to suggest that they had won the Battle of Britain or had shot down most of the RAF, or had prevailed in the Battle of the Bulge. Quite the opposite is the recorded history of Japanese false reporting.

"The Japanese accounts are solid," wrote Ford, "they're convincing; they have the kind of unambiguous detail that can't be faked."[180] These same Japanese accounts, written forty to fifty years after the Japanese defeat, conclude that the JAAF only lost 115 aircraft to the AVG. Ford agreed to that exact number.

But Daniel Ford also maintained a blog, Warbird Forum, covering every aspect of the Flying Tigers and their war in Burma and China, and in it, he seemed to contradict himself. One of the featured pages is entitled "About Those AVG Victory Credits."[181] He wrote, "AVG pilots were paid $500 for each Japanese plane they destroyed. . . . Unlike the practice in most air forces, a plane burned on the ground was given the same weight as one shot down in aerial combat."[182] He further wrote: "In 1986, aviation enthusiast Dr Frank Olynyk worked through the AVG records, tossing out claims against aircraft on the ground and restoring air-to-air credits to the pilots who actually scored the kills, as shown by their combat reports and other documents. His work looked good to me, and I used [it] in my 2007 and 2016 revisions of Flying Tigers."[183] But as a matter of fact, Ford did no such thing. Olynyk's work concluded that the air-to-air kills of the Flying Tigers were 230, not 115. And adding back in the numbers he dismissed as "killed" on the ground, his final figure would have been very close to the official 296.

Frank Olynyk was a master on the subject. "His goal was to establish those who were indeed 'Aces,' pilots credited with destroying five or more enemy aircraft in air-to-air combat. . . . [His] 700-page book *Stars and Bars: A Tribute to the American Fighter Ace, 1920–1973* was published in 1995 . . . a more concise list, of the 1,301 U.S. Aces credited with five victories."[184] American Fighter Aces Association officers called Olynyk "the world's most knowledgeable historian regarding the fighter aces of the world."[185] His research should end the subject.

CHAPTER 12

Robert Hilliard Barrow

FROM THE CADET CORPS TO WAR BELOW ZERO

It was dark in the hallway outside the third-floor dorm room in the Pentagon Barracks at LSU, and "Dog" Barrow was ready to strike. His name wasn't unusual: all the freshman in LSU's ROTC were insultingly named "Dog." And the harassment and hazing expected from upperclassmen could be pretty severe. Paddling was the order of the day; humiliation and denigration were expected; freshman, like obedient dogs, were required to do any and all biddings from the dominating upperclassmen and be at their beck and call.

"The freshmen all had their hair cut absolutely short," said Barrow, "like you were going to Parris Island—almost skinheads. And you wore a little beanie cap with the bill turned up. For any fellow who showed a little spirit, a little resistance to any of this—that was an invitation to experience more of the same. So, I was fair game to the upperclassmen; and I became a retaliator, without fear."[1]

Barrow's favorite technique for retaliation was to give his selected target what he called a "barber pie." It was his signature tactic, and once administered there was no doubt among the upperclassmen that it was Barrow who was the offending culprit. Tonight he was working alone. Once he had had an accomplice, but he preferred solitary missions. He was not a hit-and-run assassin. He liked to live dangerously and hang around to poke his target and wait for the sleeping lion to awaken to his unfolding fate—the barber pie.

From the hallway, Barrow approached the door and quietly let himself into the bedroom. The doors were never locked. He tiptoed into the common bathroom, serving the three occupants, and fashioned a giant cone out of stiff paper. "I whipped up this concoction that was only limited by one's imagination," he said, "but most of it was thick soap suds and maybe some shaving cream, and if you really wanted to get with it, you might put a little ammonia in it."[2] Armed with his breath-taking concoction, Barrow approached his deep-breathing target and held the cone with its wide end just inches above his nose. The thick, soapy mixture quivered like foam in the upside-down cone, and faint whiffs of ammonia escaped. Barrow gently shook his sleeping adversary and lowered the cone so that the foam barely touched his nose. His target stirred slightly but drifted back to his deep sleep. And there was a second poke, and the unsuspecting target slowly emerged to semiconsciousness. His tongue flicked out to make contact with the hovering, frothy billows, and his nose sucked in a breath in anticipation of stretching.

At that exact moment, as his nostrils recoiled from the ammonia, the full barber pie was administered, and the cone smothered the senior cadet's face. It was Barrow's signal to run like hell while enjoying the sounds of gasping and choking and spitting behind him. When asked if he ever won any of these retaliatory contests of daring, he admitted that he never did. "No. And as a consequence, I did a lot of Saturday afternoons marching off demerits. . . . And I missed most of the afternoon LSU football games"[3]

This adventurous streak was something previously unknown to anyone familiar with Robert H. Barrow. He had grown up with loving parents in what was certainly the most isolated place in Louisiana, if not the entire South: West Feliciana Parish. Sharing a northern border with Mississippi, the parish was sparsely populated, and there was no electricity. Robert had two brothers, but they were much older and had their own interests. They had little to do with him. He was mostly a loner.

The family was a by-product of the privations of Reconstruction after the Civil War. They had enjoyed some early prominence during the run-up years to the conflict, but at war's end Reconstruction scattered most of the family. The Barrows of West Feliciana decided to stick it out and stay at home; and they sought to survive by scratching out a living as sharecroppers in a post-Reconstruction world that offered mostly tough living. "I didn't, at the time, think that there was anything unusual about reading and studying with a kerosene lamp," said Barrow, "and not having an automobile or a radio, and all these other things."[4]

Perhaps this could have been a lonely atmosphere for a youngster growing up, but for Robert Barrow it forced him to become one with nature. "It was a happy environment. . . . I did have some black playmates, and some black men who [taught] me the night hunt. I was quite a night hunter at the age of ten, eleven, twelve, thirteen . . . the one thing that probably shaped my life more than anything else was a lot of isolation. I turned to books and reading, and I had a ferocious appetite for the written word."[5]

But there was more than just an insatiable reading habit that motivated young Barrow. "I had a desire, I would say, starting at age fourteen or fifteen, to go to West Point. . . . I read a lot of things related to the military; and I wanted to go, but you had to be right politically. In those days, politics was much more important than it is today. There were not that many appointments, and they were highly sought after."[6]

If politics was the path to secure an appointment to West Point, the Barrows of West Feliciana had two strikes against them. Politics in Louisiana meant Huey Long, and unfortunately the Barrows and most others in the parish were anti-Long. "Out of sixty-four parishes in Louisiana," said Barrow, "this was one of only two parishes that voted against him in one of his big elections."[7]

So Barrow pursued his next, most viable option—enrolling at LSU. Although the tuition and fees were very cheap, and attracted many northern students because of the reasonable price, it was still a lot more than he had. He took on every possible job that he could find and saved everything that he earned like a miser. Still, at enrollment time he was $150 short of the cash he needed. Providentially, a loan from a sympathetic Episcopal minister put him over the top.[8]

When he entered his new world on the campus of LSU for the fall semester of 1939, he crossed the line into a world he had never seen and was as far removed from West Feliciana as the desert was from the North Pole. "There was a lot of . . . of happy times," he said, "A socializing kind of life. LSU had fraternity dances, big bands every Saturday night, sororities had tea dances on Sunday afternoon, and there were fraternity and sorority parties of this kind and that. And it was a good-time school. And I availed myself of every opportunity."[9]

Also, the world news had the campus buzzing. On September 1, Germany invaded Poland, and amidst the fun and gaiety of dances, the big bands, and football, the anticipation of the United States possibly becoming involved in the war dominated conversations. And the military world was very visible on the campus. The elaborate Friday drill and parade attracted throngs to see the cadets strut their stuff. It was a heady time.

"LSU ROTC in those days, people today cannot possibly understand fully what it represented. It was a super institution." said Barrow. "It was called the

Old War Skule, spelled S-K-U-L-E. And it had a tradition, and it produced more Army officers, albeit reserved, than the military academy did."[10] Part of the curriculum of LSU as a land-grant college was compulsory ROTC for at least two years. "So, the parade on Friday afternoon was as fine a parade as I have ever seen. Particularly in the spring months. You had, white, gray, white: the white color for your cap, and you had your gray tunic and then white trousers and black shoes . . . in many ways it looked like a West Point uniform; and it was a large cadet corp. It was really quite, quite something."[11]

But his fascination with the military at LSU, and dances, and social activities, and not going to class left him with less-than-stellar grades: a few A's and B's, and a few more F's. "I just was not a good student. I was capable of being a good student, but I just went for too many other things. I didn't have my priorities straight. And I stayed out one semester as a consequence."[12]

He worked back in St. Francisville, in West Feliciana, for the rest of 1940 and felt more in control to return to campus life in the spring of 1941. Once back, his grades improved and he did not make the same mistakes that he had committed as a wide-eyed freshman. That spring semester turned into the fall, and in those final days Japan attacked at Pearl Harbor.

The American population answered the call to arms in numbers that few had ever predicted. LSU's campus became a hotbed of patriotism, and the young cadets itched to go and get it on with this dastardly enemy. It was a similar emotional reaction that had occurred eighty years earlier, when cadets of the Louisiana State Seminary of Learning and Military Academy strained at the leash to join General Beauregard at Manassas Junction—before the war would end.

"I thought about, wanting to go as soon as possible," said Barrow, "and agonized with . . . trying to go as a commissioned officer, which, had I stayed in school like I should have, would have led to a commission in the Army."[13] Barrow's agonizing ended when in January 1942 a red-haired, poster-type, impeccably uniformed Marine, Major Williamson, showed up on the campus to recruit. Patriotic excitement was sky-high, as the cadets of LSU and the rest of the world had just followed the day-by-day news of the fifteen-day drama that had unfolded on the tiny Pacific atoll of Wake Island.

Simultaneously with the attack at Pearl Harbor, forces of Imperial Japan launched an attack on Wake Island, hardly a blip on the map, but defended by 450 Marines of the 1st Marine Defense Battalion. Hopelessly outnumbered by an overwhelmingly superior Japanese amphibious force that included three light cruisers, six destroyers, three submarines, two patrol boats, and two troop transports, Major J. P. Devereux and his Marines thwarted the Japanese attack at every turn. On December 11, the Japanese launched their powerful amphibious landing force to seize the island and add it, as a trophy, to their long list of victories. But that landing was hurled back into the sea, and expert Marine gunners added to the defeat by sinking two destroyers and damaging a cruiser.

Springing to life, the navy ordered Admiral Jack Fletcher's Task Force 14 to mount a relief of Wake Island. On December 9, President Roosevelt's fireside chat had been one of doom and gloom. After he painfully outlined the numerous defeats at the hands of the Japanese, he concluded: "The reports from Guam and Wake and Midway Islands are still confused, but we must be prepared for the announcement that all these three outposts have been seized."[14] Two days later, just after Devereux's men had given the Japanese attacking force a bloody nose, American newspapers picked up the story and announced a "victory" at Wake Island. Major General Thomas Holcomb, the Marine Corps Commandant, said, "A cheery note comes from Wake, and the news is particularly pleasing at a time like this."[15] On December 12, President Roosevelt, anxious for anything that was not bad news, triumphantly told millions of Americans that the Wake Island garrison was "doing a perfectly magnificent job."[16]

And so, the next two weeks were weeks of high drama. Devereux and Wake Island became famous names bantered around the kitchen tables of America. Where was Task Force 14? Would they get to Devereux in time? This was a cavalry to the rescue story. But on Christmas Eve, December 24, 1941, all hope was dashed when the bad news was broadcast to the American people. Wake Island had fallen, and its garrison had surrendered to the Japanese. But in that defeat, Wake Island became a pillar of hope. A strange phenomenon gripped the American people. Wake Island had suddenly risen to a new level. Its magnificent defense became the lore of a second Alamo. It became a rallying point, a new battle cry.

General Holcomb sensed all this and wrote: "Out of such actions as this, a people's strength and ultimate victory must come. America remembers Wake Island and is proud. The enemy remembers Wake Island and is uneasy."[17] And at LSU, Cadet Robert Barrow was as moved as any other American over Wake Island. "I had never thought about the Marine Corps in my life. . . . Wake Island and this marvelous stand-off of the Marines, created a special awareness, and when Major Williamson showed up on campus to recruit, that was it, and I signed up for the Platoon Leaders' Class."[18]

Barrow was enlisting as a PFC, with the hope of a future commission when he graduated—if he maintained a C average. He knew little about the Marine Corps. All he knew was that there had been a band of heroic Marines fighting an overwhelmingly superior enemy force on a tiny atoll called Wake Island, and now this impressive red-haired Major Williamson stood before him on LSU's campus, inviting him to become one of that breed.

But by the fall of 1942, that future Marine Corps commission seemed a long way off—maybe two years—and Barrow was anxious for action. In two years the war might be over, and he wanted in now! He had seen a double-page advertisement spread in the *Advocate* newspaper that he could not resist. It was a recruiting ad for the Marine Corps, and there was this World War I Marine emerging out of a trench, in his tin helmet and canvas leggings, and he was prominently carrying his Springfield .03 rifle. And he spoke to the Robert Barrows of America. It was not the "Uncle Sam Wants You" appeal. It wasn't an appeal at all, it was a challenge: "Join the Marine Corps, and we will have a rifle in your hand and a man to show you how to use it within forty-eight hours."[19]

Barrow immediately posted a letter saying he wanted out of the PLC program and wanted to enlist as a PFC, and the Marine Corps welcomed him in. The train from New Orleans to Marine Corps Recruit Depot, San Diego, California, took a little longer than forty-eight hours to deliver Barrow, but as soon as it did, as promised, he had a rifle in his hand—not a Springfield .03, but an M1 Garrand.[20]

Six weeks of recruit training led to an extension as an assistant drill instructor, and while he was serving as such, a team of officers came out from Headquarters, Marine Corps, to interview prospects for Officer Candidates School at Quantico; and they selected young Barrow to join 235 other selectees for the first class.[21] The program was eight weeks long, followed immediately by another ten weeks of Reserve Officer's Class. "Regular commissions were given to the first five graduates of the Reserve Officer's Class, and I was one of those,"[22] said Barrow. Three months later, he was in Camp Lejeune, North Carolina, at a place called Tent City—a place where replacements were trained before movement overseas.

In the very early spring of 1944, after weeks of training, the rumor around Tent City was that an overseas order was just around the corner. Barrow and four others took a 1934 Plymouth automobile, with enough ration cards for fuel, to have a last fling in New York. After an overnight drive, they registered at the Belmont Hotel, but as they were signing in the manager handed Barrow

Robert H. Barrow, USMC
(U.S. Marine Corps photo)

a telegram. He was to immediately call his battalion commander, Colonel Trabbot, at Camp Lejeune's Tent City.

Expecting bad news, he nervously dialed the number, and Trabbot answered. "Barrow, do you want to go to China?" were his first demanding words. "Well, I was so relieved that it wasn't bad news," Barrow said. "I leaped to my feet and said, 'Yes, sir, I want to go to China.'"[23]

He was briefed in Washington, and it was indeed a strange assignment. It all had to do with forecasting the weather and disrupting Japanese communications. The navy wanted better weather forecasting for future operations in the Western Pacific, and the Chinese agreed to let the navy build weather stations wherever they wanted in China. In return, China wanted Americans to train and equip their guerrilla forces, sometimes operating in Japanese-occupied territory; and the navy would provide that.[24]

After six weeks of waiting, Barrow's orders took him to Norfolk, and to the dock where the large transport ship *A. E. Anderson* prepared for a long trans-Pacific movement. Upward of five thousand personnel boarded, the vast majority were army, many headed for duty with Claire Chennault's 14th Air Force. There were only two other Marine officers with Barrow, a few Red Cross nurses, and a very few female WAC officers. Everyone was headed for the BCI (Burma, China, India) theater of war.

The ship weighed anchor out of Norfolk and began a long, forty-five-day voyage—first south through the Panama Canal, and then even further south, zigzagging all the way because of the Japanese submarine threat. The ship passed Tahiti, Pitcairn Island; went between the North and South Island; and then to New Zealand and refueled in Melbourne. It proceeded on, well out into the Tasmanian Sea and around through the Indian Ocean to Bombay.[25]

When Barrow disembarked from his very slow boat, rather than speeding along to his final destination, he transferred to a very slow train across India. "I had a weeklong train ride across the breadth of India, in a wooden rail car, British-style; that opened to the outside. And the train took so long because it went slowly and stopped for refueling, and watering, and cattle on the tracks, sacred cows that you had to be very careful about moving." And then it was six more weeks in Calcutta before he proceeded on to China by flying the famous "Hump" over the twenty-eight thousand-foot peaks of the Himalayan range.[26] He had been assigned to something called the Sino-American Cooperative Organization (SACO) and found himself in a primitive base camp area called Yuangling, with a Chinese guerrilla column commanded by a Colonel Wong, who was called "Yellow Tiger."

In 1944, the navy's submarines and Claire Chennault's 14th Air Force had so interdicted Japanese supply lines that little was getting through to their forces operating in China and Indo-China. To solve this resupply problem, the Japanese opened a corridor throughout Occupied China to move supplies and personnel. It was a combined rail, land, and water trail, connecting all their enclaves—all mutually supporting to keep the lines moving.[27]

"So, we were ordered to go behind all of that," Barrow said, "inside what was called 'Japanese occupied territory,' and operate against the railroad line that went from Hankow to Canton."[28] His five-man team included two navy chief petty officers who were demolitions experts, a radio operator without a radio, but who was adept at fixing everything that might need fixing (he also served as the armorer). The fifth man was a navy corpsman. "Everything we had was transported by man-pack. We had 'coolie trains' as they would be called: peasants recruited for a day's work."[29] And these "coolie trains" were the very essence of ingenuity, using upward of two hundred peasants for fast transportation.

Each porter had his own yo-yo pole—very flexible with the center flattened to fit comfortably on a person's shoulders, and the entire length was designed to flex like a bow with the rhythm of each step. On each end of the pole, thirty-inch cubed, woven baskets were secured to carry cargo. The typical load was about one hundred pounds: fifty in each basket. "They assume a kind of fast walk, moving at a pace that causes the yo-yo pole to flex up and down. . . . The center seems to go up a little bit, because the ends go up, and it's not a constant drag weight."[30]

"It was amazing to see. Once the baskets had been loaded and secured to the ends of the poles, the coolies picked up their individual loads, positioned them on the right spot on their shoulders, flexed their knees to feel the flexing of the pole, and got into their unique shuffle. There was never a gap in the line, and the train took off like a well-oiled machine." A two-hundred-man coolie train could move upward of ten tons of supplies. "They were single file on a path, rarely anything that resembled a road—uphill, downhill, wherever it went, and go as many as twenty-five to thirty miles a day. The only thing they got out of this was a meal at the end of the day, and then they turned around and went back the thirty miles to where they came from."[31]

The whole purpose of this remarkable supply train was to keep Chinese columns, scattered throughout China, on the offensive against the Japanese. The train provided materiel, medical supplies, and communications—all brought about by Barrow's five-man team. The Americans brought encouragement wherever they went and stimulated the scattered guerrilla forces to continue to attack and hammer away at Japanese communications, especially the railroad.

And that fast-moving coolie train was an invaluable source for collecting intelligence as it snaked its way through the scattered civilian hamlets, picking up every detail from the villagers. "Our mission was open-ended," said Barrow. "Go into this area and do the most you can to interdict this line of communication, and when the war ends, your job is over. . . . With no radio communication, I had no follow-on instructions, and no way to report what it was we were doing."[32] It was as simple as that.

An obscure citation detailing Barrow's actions in February 1945 reveals dramatic action of this American-led 4th Column of the Chinese Commando Army. In an attack against the Japanese garrisons at Ningshan and Singtan in Hunan Province, "it was necessary to walk a distance of 125 miles in heavy snowfall, severe cold, and under road conditions which had brought civilian activity to a standstill. . . . It resulted in loss by the enemy of 100 officers and men killed in action and a considerable number of wounded . . . without loss to your own forces."[33] Barrow was also awarded the Bronze Star for heroism.

Finally, the war against Japan ended in August 1945, and Barrow learned about it strictly by word of mouth. He visited a Japanese-held garrison and talked to the commander, who was very cordial and waited for the Chinese 94th Division force to enter the town. Meanwhile, the Japanese forces remained armed and walked around with weapons, including swords.[34] Barrow knew nothing of the atomic bombs or anything about what had brought about the end. "It wasn't until after the war that I learned there had been something called Iwo Jima and Okinawa, and the things that happened in Europe that brought about the surrender."[35]

✤ ✤ ✤

Barrow's next four years after the war were hardly typical. He stayed in China for another year and then spent twenty-eight months as a general's aide before attending the Amphibious Warfare School in Quantico. That led to an assignment with the 2nd Marine Division as commander of Company A, 1st Battalion, 2nd Marines (Regiment).

Barrow was ecstatic, saying that "having a Marine rifle company, may be the ultimate assignment for a Marine of any rank."[36] But the Marine Corps' rifle companies of 1949 were not the rifle companies of World War II. At the end of the war, the Marine Corps was a force of six divisions, with a Corps strength of over five hundred thousand. Postwar cutbacks reduced that strength down to barely seventy-two thousand.[37] And all units were at half strength. A rifle company that rated over two hundred men was barely fleshed out at one hundred. And six divisions had shrunk to only two—one on the East Coast and one on the West Coast.

But, to the surprise of everyone, on June 25, 1950, the North Korean People's Army (NKPA) struck, suddenly and swiftly, across the 38th Parallel that divided North and South Korea and drove the South Korean army into a narrow pocket at the southern tip of the Korean Peninsula and threatened it with annihilation. The United Nations, powerless to stop this military advance, was reduced to fecklessly condemning the invasion and issuing an authorization for a United Nations Command to dispatch forces to confront the NKPA. But proclamations cannot create armies from military cupboards that are bare. And those cupboards were bare because they had been stripped at the end of World War II. And intelligence units that might have suspected this June 25 attack were as bad as any in the history of warfare. As for "dispatching forces to confront the NKPA," the only forces available were the four American divisions assigned to occupation duty in Japan; and they were seriously understrength, and hardly trained or equipped for combat.[38]

On June 29, Major General Lyman Lemnitzer, the director of the Office of Military Assistance, sent a memorandum to the Secretary of Defense acknowledging that Washington knew "for many months that the North Korean forces possessed the capability of attacking South Korea . . . but had [not] centered attention on Korea as a point of imminent attack."[39] To some Washington observers, "The surprise in Washington on Sunday, 25 June 1950 resembled that of another, earlier Sunday—Pearl Harbor, 7 December 1941."[40] By the end of the first week, the Republic of Korea (ROK) forces, overwhelmed in the onslaught, were reduced from an original ninety-eight thousand to fifty-four thousand. The rest had been killed, wounded, captured, or missing.

Back home, reaction to the attack was frantic in the Marine Corps. A brigade was formed from the 1st Marine Division in Camp Pendleton, California, and sent to the Pusan Perimeter to try and stem the tide. It departed California on July 14. The denuded remnants of the 1st Division were to be filled with units from the 2nd Division, based at Camp Lejeune, North Carolina. "We were ordered almost at once to ready ourselves and then execute a move to the West Coast,"[41] said Captain Robert Barrow. Suddenly the Marines at Lejeune witnessed a strange phenomenon. Trains began pouring into the rail depot. Trains of every type, equipped with the rail cars necessary to move personnel and hardware, including tanks and artillery.

Barrow's train pulled out of the station on July 31 to the blare of a band sending them off, as so many others had been sent in other wars. It was a six-day trip across the southern United States to Camp Pendleton. When he arrived, his battalion was immediately placed in the 1st Division, and his Company A became part of the 1st Marines (Regiment). Fifty reservists quickly joined them. They had been waiting for the battalion's arrival and came from units from Oklahoma City, Tulsa, San Antonio, and a few from Arizona and California. Another fifty came from post and station Marines—those serving at Marine Barracks on naval stations.

"So, from about a hundred and ten," said Barrow, "we found ourselves up to table of organization strength, well over two hundred, and double staffed in some cases. I went out with two first sergeants, two company gunnery sergeants."[42]

Among Barrow's newly fleshed-out company were some reservists from Tucson. There were nineteen Mexican Americans, full of the spirit of the Marine Corps, who had joined the reserve but had not yet been to boot camp.

"They had signed up and were to go to boot camp when it was time," said Barrow, "and meanwhile they were on the rolls, so when their unit got called up, they got called up. They probably thought it was a big lark, so they came along; no one raised a question, until the eleventh hour, that they had not even had boot camp to say nothing of any active training."[43]

So, Barrow grouped them together for camaraderie, and they formed half of the force of Lieutenant Jack Swords's 3rd Platoon. With no active training for combat other than what they had seen actors do in the movies, they pressed on, blissfully unaware that they could face hard fighting against a surrounding enemy on a frozen battlefield. "They got some very valuable OJT [on job training] in combat from Inchon to Seoul," Barrow laughed, "and after that, you'd have a hard time distinguishing that they hadn't even been to boot camp. . . . They were good Marines, and they did the job."[44]

In California, General Oliver P. Smith had been assigned to command the 1st Marine Division, which only had a strength of thirty-five hundred after most of its Marines were pulled out to form the brigade. "We've only got 3,500 short-timers," he complained. After the biblical equivalent of searching the highways and byways for anybody in the Marine Corps, he was sent Force Troops, and elements of the 2nd Marine Division, and men from posts and stations, and from the reserve. Meanwhile, Congress extended all enlistments for a year.[45]

"Our initial problem," said Smith, "was to assemble, equip, and load out, 15,300 men of the main body of the Division, in 15–20 days."[46] This extremely short and frantic timetable came about quite unexpectedly because of a chance meeting between Lieutenant General Lemuel Shepherd, the Commanding General of the Fleet Marine Force Pacific (FMFPAC), and General of the Army Douglas MacArthur, Commander in Chief. It happened at the Dai Ichi Building in Tokyo.

Marine Corps generals had routinely been excluded from any and all high-level military meetings. They had no voice, and the Commandant was not included in the Joint Chiefs of Staff (JCS). General Clifton Cates, the Commandant, decided enough was enough and planned to simply barge in on a high-level meeting being held at the Pentagon to see what was going on. That high-level meeting proved to be a strategy session concerning getting troops over to South Korea to stop the ever-advancing enemy.

Cates boldly announced that the Marine Corps could provide a brigade of infantry, including air support for deployment, as quickly as the navy could provide ships. That seemed to catch everybody's attention, and by July 6 the

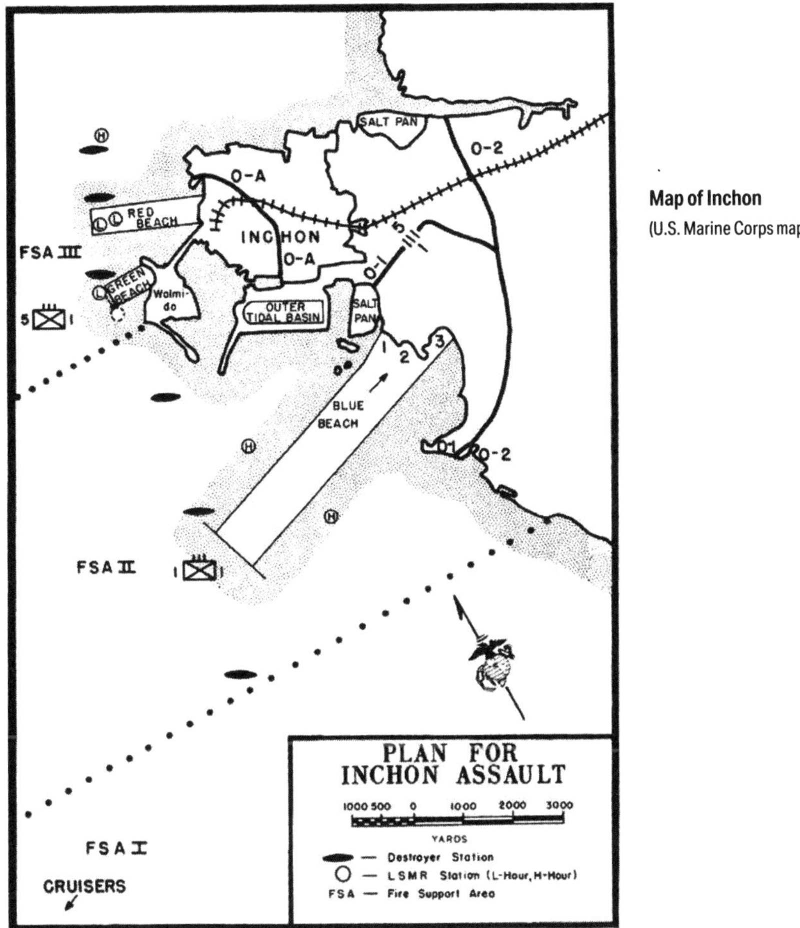

Map of Inchon
(U.S. Marine Corps map)

Marine Brigade was loading out of San Diego.[47] A few days later, the Marines learned that MacArthur had held a high-level meeting on July 4 at his headquarters in Tokyo and announced his intentions to land behind the lines of the North Korean army at the western port of Inchon. His plan was to use the Marine Brigade, scheduled to depart the States on July 12, along with an army division, provided the brigade could arrive in time. He wanted to attack on July 22, but the plan was too ambitious and a renewed North Korean massive offensive further threatened the beleaguered South Korean army.

So, with this frantic army planning to round up and deploy a reaction force being discussed behind every closed door, the navy's Commander in Chief, Pacific Fleet, Admiral Arthur Radford, ordered General Shepherd "to see

MacArthur and find out what all this thing is about."[48] Shepherd met with MacArthur on July 10, and the general took him to stand in front of his wall map. MacArthur fondly remembered that he had commanded Marines during World War II and was well pleased. Then he stabbed his pipe on the map at the port of Inchon. "If I only had the 1st Marine Division under my command again, I would land them here and cut off the North Korean armies from their logistic support and cause their withdrawal and annihilation."[49]

Watching MacArthur tapping the stem of his pipe on the map at Inchon, General Shepherd saw his opportunity. Not knowing whether the Marine Corps could even do it, he told MacArthur that he could have the division ready by mid-September! MacArthur was ecstatic. The die was cast, and Marines were committed. That translated to the initiation of a great roundup, as the Corps began to move with great speed to find the men to flesh out this division. The Marines were filled with the spirit of adventure and eager to join in the fray, even though none of them, including their division commander, had the slightest idea of where they were going.

MacArthur seemed to be a majority of one regarding his plan to land at Inchon. His opposition was just about everybody else. General J. Lawton Collins, the Chief of Staff of the Army, questioned the feasibility of the Inchon landing.[50] General Omar Bradley, Chairman of the Joint Chiefs, mocked the whole plan as "a blue-sky scheme." And called it "the riskiest military proposal I ever heard of."[51]

Bradley was no friend of the Marine Corps and was known for his vitriol. He had fought mightily to remove the Corps from existence during the 1948 congressional hearings on an amendment to the Navy Reorganization Act. He infamously proclaimed that amphibious operations, like those conducted in World War II, would never be needed again.[52] Now he spoke for the rest of the chiefs about this newly planned amphibious operation that he had emphatically deemed would never be needed. "We determined to keep a close eye on the Inchon plan and, if we felt so compelled, finally cancel it."[53]

If Bradley was trying to save face and flex a little authoritative muscle, it didn't work. MacArthur's answer to this seeming parental scolding was blunt. He stated that he was absolutely convinced that this Inchon operation would end the NKPA's ability to win. He swept aside Bradley's and the Joint Chiefs' squeamishness and wrote: "There is no question in my mind as to the feasibility of the operation and I regard its chance of success as excellent. I go further and believe that it represents the only hope of wresting the initiative from the enemy and thereby presenting an opportunity for a decisive blow. To do otherwise is to commit us to a war of indefinite duration, of gradual attrition, and of doubtful results."[54]

Indefinite duration, gradual attrition, doubtful results: those were poisonous words to make any commander cringe. They were anathema to the principles of war. Could the uncertainty that the Joint Chiefs had in MacArthur's plan have been caused by their own lack of study of the terrain and the tides? That hardly seems possible.

All the chiefs and senior military officers had either studied—or participated in—the great Normandy Invasion in 1944: an invasion that confronted twenty-foot tides and difficult terrain as part of the formula to achieve surprise and success. In fact, General Omar Bradley had commanded the First Army across the beaches of Normandy, and General J. Lawton Collins, now the Army's Chief of Staff, commanded VII Corps at Utah Beach.

But what MacArthur saw at Inchon—and what seemed to have eluded the eyes of his doubters—was that Seoul, the capital city of South Korea and MacArthur's military objective, was only eighteen miles from the landing area. No other possible landing area was as close. Alternative landing sites at Wonson on the east coast and Kusan on the west coast were uncomfortably distant from Seoul.[55]

Even more obscure was a detail that MacArthur considered critical, and seemingly had been overlooked by both enemy and friendly strategists. There was only one day in all of the month of September that provided a high tide giving maximum water depth over the Inchon mud flats that would permit an amphibious landing: September 15, the date MacArthur had selected for D-Day.

The high tide on that solitary date in September would measure 31.2 feet (minus 5 feet at low tide). No other date came close to this 36-foot differential. There were a few other dates that approached 27 feet. While that would be sufficient for most of the landing craft, the LSTs (Landing Ship Tank) needed 29 feet. It would not be until the second week in October, as the dreaded Korean winter approached, that the tides would again crest at 30 feet—and that would still be a foot lower than on September 15.[56]

The Marines, arriving from the East Coast at Camp Pendleton, had little time for anything but administrative details. There was no training except for a little familiarization firing and zeroing of weapons. But it did include familiarization firing of a new weapon—the 3.5" rocket launcher. "Suddenly we're given the 3.5" weapon," said Barrow, "which was much more capable, and this was at a time when the news coming out of Korea was that the T-34 Russian-made tank was invincible . . . this was the most modern anti-tank weapon that was available, and everyone fired one round."[57]

The division pulled out of San Diego on August 14 and sailed on board the Attack Transport *Nobel* to Kobe, Japan, where it disembarked. It had been administratively loaded to make maximum use of cargo space, but now needed to be reloaded tactically: whatever the division would need first in combat had to be loaded "on top." However, it seemed that even Mother Nature was against the Marines; that reloading took place in the midst of a raging typhoon.[58]

Sometime during this unloading and reloading operation, Captain Barrow was briefed that the objective was Inchon, and as soon as the reloading operation finished his Company A sailed on USS *Nobel*. It was one of nineteen ships that brought the 1st Marine Division to the Inchon invasion area. There was no time to rehearse for the September 15 attack. MacArthur formed X Corps as a separate command from Eighth Army for the invasion. He designated the Army's 7th Infantry Division and the Marines' 1st Marine Division as the assault regiments.

Any student of the battlefield would conclude that Inchon was a terrible place to land. At low tide, the inner harbor was a vast quagmire of mud and swamp. Cutting through this mud was a single, dredged, twelve-foot channel. When the massive high tide rushed in, it would allow an invasion force a chance to reach the inner harbor, but that chance would only last for three hours. Then it would be gone—drained off into the sea as if someone pulled a giant plug.

The planners had designated three invasion beaches: Red, Green, and Blue. But "beaches" was just a word. There were no beaches—only high, stone floodwalls that would have to be surmounted from the landing craft using makeshift wooden ladders. Protecting Inchon from any attempted invasion across Red Beach was the rocky, wooded island of Wolmi-do, situated to deliver devastating flanking fire into invading boats. Therefore, the 3rd Battalion of the 5th Marines would make the initial landing on Wolmi-do's Green Beach to neutralize any defending force. If successful, the rest of the 5th Marines would land on Red Beach, but only after an agonizing twelve-hour wait for the return of the flooding afternoon high tide. This sentinel island of Wolmi-do was connected to the main harbor by a six-hundred-yard causeway. The 1st Marines were to simultaneously land on Blue Beach and would be completely isolated from the rest of the landing force. There was nothing to like about landing at Inchon. It had no redeeming qualities except one: the element of surprise!

During World War II, an amphibious operation similar to Inchon would have had at least three months of planning and rehearsals. None of that was possible at Inchon due to the need for speed. Adding to the angst of the high command, Typhoon Kezia now appeared, bearing down on Japan and threatened the invasion area. The navy advised the invasion fleet to sail early, further compressing the timetable. "There was not even time for landing exercises by the LVTs," lamented an officer on the 1st Marine Division's staff. "Some of the LVT crews had not even had the opportunity to try their engines out in the water and paddle around."[59]

On board USS *Nobel,* Captain Robert Barrow went over all the details for the landing: the plans, the boat tables, and boat team assignments. He tried to address his A Company Marines in the confined spaces of the transport. "I addressed my company twice in the hours before Inchon, once on the main deck . . . and down on the mess deck, where I stood on a bench and talked to them about what they were expected to do."[60]

All his Marines' eyes were focused on him and the charts and maps that he referenced, describing the area where they were to land and what to expect. Their total invasion preparation had been limited to filling out forms, zeroing their weapons, and firing one round of the 3.5" rocket launcher—in case they were to confront the dreaded Russian-made T-34 tank. In Barrow's cramped, below-deck classroom was Lieutenant Jack Swords's 3rd Platoon, and his wide-eyed Marines, including his nineteen Tucson Marines who had never been to boot camp.

Just hours after midnight on September 15, General MacArthur stood by the rail of his flagship, *Mount McKinley,* and pondered the enormity of this most daring operation and its uncertain outcome. He'd personally guaranteed its success in the face of utmost skepticism by the Joint Chiefs of Staff and other high-ranking officers. L-hour for the assault on Wolmi-do was close at hand. The landing craft would hit the beach at 0630, and there was no turning back now. He would write: "Within five hours 40,000 men would act boldly, in the hope that 100,000 others manning the thin defense lines in South Korea would not die. I alone was responsible for tomorrow, and if I failed, the dreadful results would rest on judgment day against my soul."[61] A young ensign aboard remembered his own apprehensions. "As the morning of September 15 approached, we realized we had all the ingredients for a disaster on our hands."[62]

At first light, elements of the 5th Marines went over the sides and down the cargo nets into bobbing landing craft. Three of the navy's rocket ships shattered the gray and silence of the early morning as they launched thousands of 5-inch rockets against Wolmi-do. The island disappeared under the smoke of this thunderous bombardment. The first landing craft touched down on the fifty-yard beach at 0633 and moved to engage the few defenders. At 0800, Wolmi-do was secured. Marine casualties were very light: seventeen wounded. The enemy lost over three hundred, most of them killed.[63] Now began the great wait for the evening tide. Marine and naval air prowled over the highways approaching Inchon, ready to attack any intruding North Korean forces, and naval gunfire covered the closer approaches.[64]

Finally, it was time. At 1530, the assault companies of the 5th and 1st Marines went over the sides of their transports and down the cargo nets. The landing craft began their circling in the rendezvous area until all the boats were loaded and formed an assault line. Their engines then roared, and the boats crossed the line of departure for the run to the beach. One officer from the 5th Marines lamented, "Two things scared me to death. One, we were not landing on a beach; we were landing against a seawall. Each LCVP had two ladders, which would be used to climb up and over the wall. Two: the landing was scheduled for 5:30 p.m. . . . only two hours of daylight to clear the city and set up for the night."[65] A naval bombardment commenced, and rocket ships moved in close to Red and Blue Beaches and launched two thousand rockets. Landing craft crossed the line of departure at 1645, and the first wave of the 5th Marines breasted the seawall on Red Beach at 1733.

Three miles to the right of Red Beach, Colonel Chesty Puller's 1st Marines ran in on Blue Beach. Robert Barrow's Company A was the battalion reserve company. Barrow observed: "It was not a very clear, distinguishable beach: seawalls and a lot of salt flats, and geometric trapezoids and squares and rectangles of things sticking out of the water. We had a Navy wave commander who was not in my boat who was hell and determined to take us to the wrong spot." Barrow's entire wave headed for the wrong landing area and he could not signal the error to the wave commander. The boats landed in the wrong place. "Everybody piled off. Some of the ramps were down, some were half down; it was not a beach, it was a salt flat." Barrow ran quickly, up and down the landing area, and shouted to "get back aboard." His sergeants quickly rounded up the men.

Barrow found the navy's wave commander and yelled, "You've put us in the wrong place. Back off of this and then we are going this way."[66] He pointed to

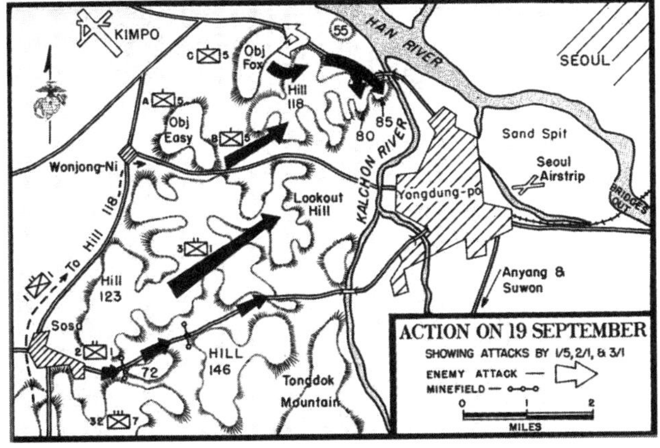

Action on September 19 to Hill 118
(U.S. Marine Corps map)

the distant landing area where the other boats had gone ashore. By the time they landed in the right place, it was dark. There was smoke and fire, but Barrow didn't know exactly where other units were. Luckily, there was no action that night, or for the next several days. On September 19, trucks rolled up to take Company A into battle. The route was a twisting ride through a maze of intersecting roads completely across the landing area to the village of Wongjong-ni, just south of the Han River. From there the final miles were on foot to relieve the 1st Battalion, 5th Marines, now occupying three prominences: Hills 118, 80, and 85. Those hills faced the western approach to the town of Yongdung-po, guarding the final approach to the city of Seoul. "If Yongdungpo is lost, Seoul also will fall," warned the boisterous Communist Korean commander defending the city. He urged his soldiers to fight at all costs.

Barrow hustled his Marines forward at a grueling pace to reach the crest before dark. Sergeants barked at stragglers, pushing and prodding them along as if they were in boot camp. Finally, with the men gasping for air, the column arrived at the top and saw that from this elevated position Hills 118, 80, and 85 looked down upon Yongdung-po and dominated it. The Marine company that had been designated to occupy the forward Hills 80 and 85 had not yet arrived, and the 5th Marine defenders were pulling off. Sensing its importance, Barrow suggested that he move forward and temporarily occupy 80 and 85, but was told no. So, nighttime fell and those prominent positions were left vacant and undefended.[67]

Unknown to Barrow's Marines, the North Korean commander in Yongdung-po had also recognized the importance of those two hills, and now planned to attack and seize them. In the darkness just before dawn, Company

Assault on Yongdung-po, September 21
(U.S. Marine Corps map)

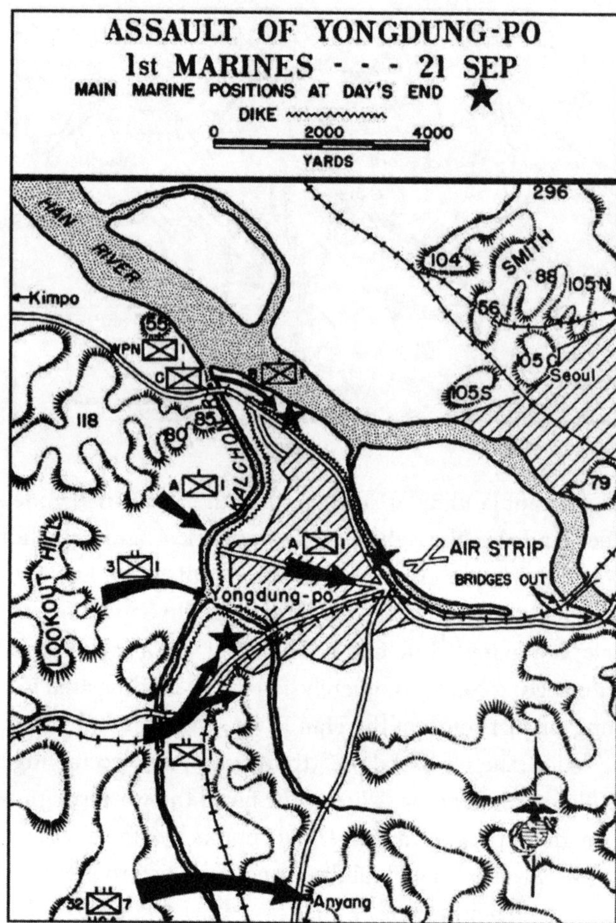

A's attention was drawn to the clatter of small arms and machine gun fire coming from the east. Daylight revealed an attacking North Korean assault line cresting Hills 80 and 85 in a full-blown attack against the vacant positions. The surprised attackers, seeing there were no defenders, ceased fire and began digging in. Then they continued the attack toward Company A, upward toward Hill 118, but Barrow's Marines hurled them back with heavy losses.[68] Companies B and C took most of that day, September 20, to recapture those hills with hard fighting. It was frustrating to know that they had been there for the taking the previous night.

On the 21st, those companies continued the advance down from Hills 80 and 85 to attack the enemy's northwestern defenses of Yongdung-po. Three miles to the south, two battalions of 1st Marines attacked the southwestern

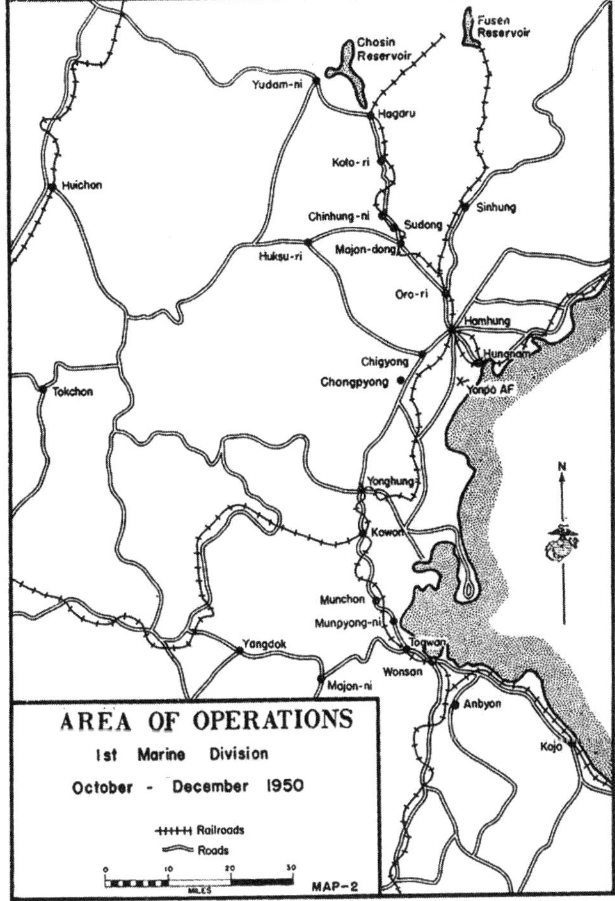

Area of operations, 1st Marine Division

(U.S. Marine Corps map)

defenses. Through the din and confusion of those two attacks, no one noticed that Company A had begun its own advance in the center of the attack zone. Through an undefended approach, they advanced across an open rice paddy that was partially concealed by tall vegetation. To some observers it was as if watching a parade ground training exercise. The company advanced with two platoons up, and one platoon trailing in reserve.

On the left was Lieutenant John Swords's 3rd Platoon, including his non-boot camp Marines. On the right was Lieutenant Don Jones's 2nd Platoon, and following in trace was Lieutenant William McClelland's 1st Platoon. Barrow was between Swords and Jones. In the distance, both to the left and right, Barrow's advancing Marines could see Marine aircraft engaged in dive-bombing runs. Still, they pressed on through the high grain stalks. The advance seemed reminiscent of another long-ago Marine advance, through the wheat fields at Belleau Wood in World War I.

The battle line emerged from the concealing field and stepped into a shallow stream (Kal-chon Creek) and waded across. They were now in full view of anyone who cared to look. But no one did, and Company A entered the outskirts of Yongdung-po without a shot being fired.[69] They marched into the city and moved, unopposed, down the main street. Halfway through town, the main Inchon/Seoul highway dovetailed to form the route of their advance. Suddenly, a Marine from the 1st Platoon held up his hand and pointed to the front. All eyes saw an enemy column advancing toward them, but completely unaware of their presence. They were high-spirited and sang a martial song as they trudged along; and they were still singing when the shattering fire from McClelland's platoon cut their formation to pieces.

Swords, seeing the action, raced his platoon forward through the rest of the town and broke out of the other side to face a large, thirty-foot-high dike. It was topped with a road that they quickly mounted, and dug a hasty defense line. In front of them stretched a vast sand spit with a small airstrip, and beyond that was the road to Seoul. To the left, Swords observed a large body of enemy soldiers falling back in disarray, most likely retreating from the pressure of Company B's advancing attack along his left flank. He immediately directed fire against them and brought his light machine guns to bear as Barrow rushed his heavy guns into action. The combined firepower was devastating.

The rest of Company A now raced up to the dike to join Swords and Barrow and extend the defensive line farther to the right. Despite its advantageous and dominating defensive position, Company A was now dangerously out in front of the attacking line. It also had no communications because of a shortage of radio batteries and was a mile and a half in front of adjacent units. Barrow's defensive perimeter, on top the long dike, resembled an elongated oval, like a stretched-out sausage, sitting in No-Man's land.

"Our machine guns were at each end and pointed down the macadam roads." Across the road was a five-story abandoned building, and next to it, in a vacant lot, was a large pile of coal. Almost immediately, a small enemy attacking force moved into position to engage and hurl back the intruding Marines. It was quickly defeated by superior Marine firepower. But during the firefight one of the Marines lobbed a grenade toward a soldier trying to take cover behind the coal pile. The grenade bounced and rolled into the pile. When it went off, the entire pile detonated in a devastating explosion that sent out a shock wave that rocked the countryside. It had been a stockpile of ammunition, camouflaged by the coal, and the ensuing mushroom cloud pinpointed Barrow's exposed position for all the world to see.

Now aware of the exposed position, other small enemy attack units moved to eject the Marines from the dike. All failed, but with each successful defense Company A's ammunition supply dwindled. As daylight faded, the ominous sounds of clanking from moving armor crept across the battlefield. All eyes were glued to the left from where the retreating infantry had been engaged. In the fading light, five T-34 tanks approached. They crawled forward along the Inchon-Seoul highway and turned onto the approach to Company A's dike.

The Marines crouched in their holes and prepared to engage this enemy armor. Each man gripped his weapon tightly and sighted on the lead tank, knowing all too well that small arms fire was no match for a tank. But at the base of the dike were gunners armed with the new 3.5" rocket launchers. Each of them had only ever fired one round. The assistant gunners slid the nine-pound high-explosive projectiles into the breeches, made sure that the rounds were securely seated, and pulled up on the arming knob. They gave the gunner a pat on his helmet to signal that he could fire when ready. Whether they knew it or not, this weapon had the massive hitting power of a 105 mm howitzer.

As the T-34s began their pass in review in front of Barrow's position, their main battle guns were slewed to the right to deliver the equivalent of a naval broadside against the Marines. They would unleash, in ripple fashion, machine gun fire and 85 mm cannon fire from their main guns. "They were probably 25 or 30 yards away from the nearest man down at the base of the hill [dike]," said Barrow. "They were in line . . . their guns all to the right and fired as they continued to move. At my direction, the gunners got right up on that macadam road."

The T-34s made their first run and fired armor-piercing ammunition, which was fortunate for the Marine defenders since the rounds penetrated the earth before explosion. Had they been firing point-detonating fuses, the flying shrapnel would have been devastating. The Marine gunners fired back, and their lack of experience showed. "They were firing kind of clumsy; kind of down and low," said Barrow. "But there were hits and some misses. They were fired on every time they made a pass. They went past and there was one that was destroyed and left there. Another one went past and was limping, reversed itself and came back and was fired upon again, and turned back. . . . We knocked some out, we crippled some; they did not succeed in doing us any harm."

No sooner had the tanks retreated into the city than the enemy began infantry attacks against the dike. They fell upon the northwest part of the position, against Lieutenant Swords's platoon. Throughout the night, the attacks probed Swords's position, and all were hurled back with tremendous loss to the attackers. At daybreak, 210 enemy dead were counted in front of the platoon—another tribute to his men and their fire discipline, especially the discipline shown by those untrained Marines who had not been to boot camp.

Courage and esprit de corps had never been in short supply in Swords's 3rd Platoon. Most daring was the action of Billy Webb, a young corporal from Tulsa, Oklahoma. "He was a picture-book Marine," said Barrow, "a handsome fellow, athletically built and bright-eyed, and his eyes flitted back and forth when you'd talk to him. He was a reserve called to active duty. And after we had experienced the third or fourth attack that night; after they failed, they would regroup over in the city, not too far away." After each attack, Corporal Webb would hear, above the din, this Korean voice shouting and seemingly berating the attacking troops and encouraging them to attack again. Shortly after each of these harangues, the next attack would begin.

Webb listened and listened and concluded that this was the case. He crawled out of his fighting position and told his fire team that he was going forward, and to be aware that he would return along the same line and to not get trigger-happy and shoot him when he came back. He told his team, "I'm going to get that son of a bitch."[70]

"Webb worked his way down the alleys and backyards to this sort of town square," said Barrow, "where, in fact, this guy was standing up on the back of a truck. Webb simply got a bead on him, and in the semi-darkness, killed him; and then scurried back. And that probably more than anything else, that single act of courage, discouraged any subsequent attacks."[71]

At 1410 on September 26, General MacArthur announced: "Seoul, the capital of the Republic of Korea, is again in friendly hands. United Nations forces, including the 17th Regiment of the ROK Army and elements of the U.S. 7th and 1st Marine Divisions, have completed the envelopment and seizure of the city."[72]

The general had been slightly premature with his announcement in order to make a political point. It had taken the United Nations forces, under his command, just ninety days from the initial onslaught of the North Korean People's Army to stop its advance, attack its rear, and bring about its complete destruction and defeat. By any measure this was remarkable. Compared to the Normandy operation on June 6, 1944, the landing at Inchon was infinitely more challenging. The Normandy landing had been planned for over two years; the nation had been on a wartime footing for three years; the assault units had rehearsed the landing many times; and intelligence had completed detailed studies of the land and the enemy capabilities and briefed the invading soldiers of these details. And the invading force included many veteran units with previous combat experience, and General Eisenhower had planned to use the twice-daily, twenty-foot tide to his advantage and land at low tide—although landing at high tide was an option.

At Inchon, none of this existed. The nation was not on a wartime footing; the Marine Corps had been reduced to only 14 percent of its previous wartime strength; the invasion force included troops with little or no previous training, and were mostly reservists; there was no time for rehearsal—the invasion force moved directly from its points of gathering to the invasion area; there was little intelligence and few maps; and a thirty-plus-foot tide was necessary for the landing—and that tide was available on only one day in all of September 1950, and that was in the late afternoon.

Despite MacArthur's jumping the gun for his "victory" announcement, he was only off by three days. On September 26, X Corps held only half the streets of Seoul and Captain Robert Barrow's company fought in the battle known as the Battle of the Barricades. Historian Roy Appleman wrote: "In the middle part of Seoul, the barricades stretched across the streets from side to side and were usually placed at intersections. Mostly they were chest-high and made of rice and fiber bags filled with earth. From behind them and at their sides enemy soldiers fired antitank guns and swept the streets with machine gun fire. Other soldiers were posted in adjacent buildings. Antitank mines belted the streets in front of the barricades."[73]

The Marines quickly developed a routine to eliminate these fortified positions. Navy and Marine aircraft streaked in to bomb and strafe. Attacking Marines delivered a steady stream of covering fire to allow the engineers to clear the mines, and then tanks approached to deliver shattering fire directly into the position. On extra tough positions, flame tanks rolled forward to burn the enemy out.

On September 28, enemy resistance in Seoul was over. The remnants of the NKPA retreated to the north. The accolades for General MacArthur poured in, and he posted them for all his soldiers to see. Noteworthy was the message from President Truman: "Few operations in military history can match . . . the brilliant maneuver which has now resulted in the liberation of Seoul."[74] General Bradley and the members of the Joint Chiefs of Staff who had worked mightily to disparage and block the Inchon operation now chimed in: "Your transition from defensive to offensive operations was magnificently planned, timed, and executed. . . . We remain completely confident that the great task entrusted to you by the United Nations will be carried to a successful conclusion."[75]

✶ ✶ ✶

On October 29, 1950, a *New York Times* editorial proclaimed the obvious. "Except for unexpected developments along the frontiers of the peninsula, we can now be easy in our minds as to the military outcome."[76]

Victory was at hand. The North Korean army was routed, in full retreat, and defeated. And the Eighth Army had issued the order for a full pursuit of the retreating enemy: "Commanders will advance where necessary without regard to lateral security."[77] Jubilation was the order of the day among every military command. The army began planning for a redeployment of the Eighth Army and a drawdown of its troops. Six ammunition ships were diverted to Hawaii. 1st Cavalry Division began planning a Thanksgiving Day parade in Tokyo wearing their distinctive yellow scarves.[78]

In Washington, the Joint Chiefs of Staff agonized over what the next stage of the war should be. They had been chastened and embarrassed by their unified opposition to the Inchon landings. By any standard, their combined tactical mindset was far removed from the military thinking of General MacArthur. Still, they sought to control the game and pushed for overall control of decisions better left to battlefield commanders.

On October 27, 1950, they agreed that Eighth Army should cross into North Korea and issued a directive to that effect. They proclaimed the obvious: MacArthur's first objective was to be the destruction of the North Korean army, an impossible objective if he did not cross the 38th Parallel. But they added a strange and confusing caveat: these military operations north of the 38th Parallel were not to be conducted if Chinese or Soviet forces had crossed into that country. Did that mean that their own proclaimed "first" objective—the destruction of the North Korean army—was really not the first order of business at all? They further ordered that under no circumstances were his U.N. forces to cross into Manchuria or over Soviet borders—fair enough—but then added that he was to submit his plan of operations north of the 38th Parallel to them for approval.

The absurdity of this armchair generaling would have been obvious to a first-year cadet, so it was no surprise that the Secretary of Defense, General of the Army George Marshall, chimed in to defend the field commander. Marshall immediately sent MacArthur a classified note, marked for his eyes only, stating "that he [MacArthur] should feel free tactically and strategically to proceed north of the 38th Parallel. President Truman himself had approved this message."[79]

Obviously emboldened by this high-level support, MacArthur shot back to the Joint Chiefs spelling out his orders and intentions, "unless I receive your instructions to the contrary."[80] He wrote:

> Under the provisions of the United Nations Security Council Resolution of 27 June, the field of our military operations is limited only by military exigencies and the international boundaries of Korea. The so-called 38th Parallel, accordingly, is not a factor in the military employment of our forces. . . . [Our] troops may cross the border at any time. . . . If the enemy fails to accept the terms of surrender . . . our forces, in due process of campaign will seek out and destroy the enemy's armed forces in whatever part of Korea they may be located.[81]

The Joint Chiefs' squeamishness to pursue the invading enemy forces wherever they might be was further derided by South Korea's President, Syngman Rhee. "We have to advance as far as the Manchurian border until not a single enemy soldier is left in our country . . . we will not allow ourselves to stop."[82]

Getting the 1st Marine Division into the pursuit phase of the attack was as difficult as originally getting them ashore. The 23,591 Marines of the refurbished division out-loaded as they had come ashore—through the port of Inchon, with all the difficulties of its thirty-one-foot tides. It began on October 9, and the Marines were condemned to a miserable sixteen days afloat while Eighth Army crossed the 38th Parallel in the west and followed the path of the South Korean army already fighting its way northward.[83]

The Marines sailed around to the east side of the peninsula and made an administrative landing. The enemy had abandoned the territory at Wonson, a hub of road and rail traffic, so the entire division put ashore, 110 miles north of the 38th Parallel. But Company A, of the 1st Battalion, was pulled out for a special mission. "My company was pulled out of the battalion by Colonel Chesty Puller," said Captain Barrow, "and assigned a mission to bring relief to the 3rd Battalion, thirty miles west from Wonsan at a place called Majon-ni."[84]

After landing, the 3rd Battalion had pressed forward to the west, and now found itself pinned down by rear elements of the retreating North Koreans.

Their route of advance had taken them along a narrow, twisting, gravel road that snaked its way through mountains that rose almost vertically on the right side of the ever-narrowing path. The left side was a steep plunge into the valley below. It was perfect for ambush, and the 3rd Battalion was soon caught in its web, pinned down, desperate for resupply, and bogged down with prisoners. "This shelf-like road twists precariously through a 3,000-foot pass [and] abounds in hairpin turns and deep gorges which are ideal for setting a tactical trap. . . . Although traversable by tanks, it offered too much danger from roadblocks and landslides to permit their dispatch."[85]

The threatened battalion's call for help was assigned to Barrow's relief force. He formed a substantial convoy for the task, including his own Company A and a platoon of engineers. Additionally, Puller assigned him sections of 81 mm mortars and 75 mm recoilless rifles. Thirty-four vehicles departed Wonson on November 4 and proceeded to the west, but in short order Barrow found himself in the same pickle as the 3rd Battalion. A substantial NKPA roadblock took his relief force under fire. The only way to advance was to continue forward along the narrow ribbon of road, and that was impossible.[86] "We were on a shelf," said Barrow. "Down below was a deep valley, up above was the mountains. There was no opportunity for maneuver. The frontage was the width of a road."[87]

Darkness was approaching, and Barrow had been stopped cold. The enemy's fire now began to move toward the rear of his column as enemy soldiers moved along the steep sides of the cliff to try to trap his column into a killing zone. The obvious decision was agonizing: turn around and go back and try something else the next morning. But the decision to turn around and actually turning around were two different things.

The thirty-four truck drivers had to inch their vehicles, first to the very edge of the road, then cut the steering wheel hard and back up until the tailgate touched the cliff wall. Then it was cut the wheel in the opposite direction and, with one foot on the brake and the other on the accelerator, nudge the truck to the edge of the drop-off. They repeated this agonizing back and forth maneuver until they all had turned 180 degrees and faced the opposite direction. Barrow observed: "Those 6×6 big trucks, loaded with food, ammunition . . . somehow had to be miraculously, turned around on that narrow shelf of a road. By then it was dark, and we passed the word, 'Lights out,' and we're just going to drive in the dim light of night. And that was fine until one of the trucks plunged over the side."[88]

It was loaded with troops, and as it began its slide down the steep cliff it was in an area where there was scrub growth protruding from the mountain face. The men were either thrown out or jumped, and scrambled and crawled to grasp onto bushes and small trees to halt a plunge into the valley. Miraculously no one was lost. "We formed the human chain to get them out, and passed the word, 'Turn the lights on.' And we resumed our march and got back to Wonsan."[89]

A humiliated Captain Robert Barrow reported his failure to Puller. "Colonel, I have failed you."[90] But Puller brushed him off, asked what else he needed for a second try, and assigned a forward air controller to move with him at the head of the column. Early the next day, the convoy moved out and headed back to what was now dubbed Ambush Alley. Barrow would have extra daylight to operate and a forward air controller to call down thunder upon the North Koreans perched above the road.

The road was twenty-eight miles long; the last twenty-four were the torturous and cliff-hanging part. But Barrow also had developed a plan to hopefully deal with NKPA ambushes. "I reasoned that they were so sure of themselves that they probably were not actually occupying positions and would not occupy positions until they heard the sounds of the convoy approaching with all its grinding and rumbling noises. I'm sure it could be heard for at least a couple of miles."[91] Accordingly, he placed a reinforced platoon one thousand yards in front of the rumbling column; and in front of that platoon was a four-man fire team moving tactically and silently. At the first bend in the road, the team poked its nose around the corner. The steep, almost straight up and down cliffs gave way to a gentler, sloping ground; and on that slope were North Korean soldiers.

Just as Barrow had predicted, a one-hundred-man ambushing force was relaxing, eating, and talking. All the soldiers were well away from their fighting positions and their weapons. The fire team immediately opened fire as the trailing platoon hustled forward. "They took immediate action," said Barrow. "They fell down into firing positions and opened up, and that was all that the rest of the platoon needed to rush forward and join them. We just laid them out."[92]

Some of the surprised North Koreans tried to run as the whine and impact of bullets cut into them, but few made it, and many were pinned down. In short order, the main mechanized column rumbled forward and added its firepower to the complete smashing of the enemy force. Barrow then proceeded on to Majon-ni, delivered the resupply, and evacuated more than six hundred prisoners.

It was a great victory and proof that the enemy was in full retreat. Back at Wonson, optimism that the war was almost over was the talk of the day. MacArthur had publicly proclaimed it. "There was the rumor that we would all be home by Christmas, widespread and, in fact, MacArthur himself was the cause of that," said Captain Barrow, "but the 1st Marine Division was asking, 'If that's true, why are we going with such determination so far away from our seaports.'"[93]

The 1st Marine Division was indeed moving away from its seaports to the east and was on a northward track. On November 1, the 7th Marines had boarded trucks at their assembly area at Hamhung for a twenty-six-mile transport up the north/south road leading to Chosin Reservoir. Its orders were to relieve an ROK regiment that had fallen back five miles. A surprising, unexpected attack from Chinese Communist Forces (CCF) had overwhelmed the ROKs and put them to flight.[94]

Now the 1st Marine Division was hopelessly spread out. The 7th Marines were up the northbound track toward the Yalu River, and the 1st and 5th Regiments were at Wonson preparing to move north. General O. P. Smith, the division commander, voiced his objections to X Corps Commander General Edward Almond, but to little avail. Almond had routinely dismissed the severity of the reports of a Chinese presence. "At one point," Smith said, "my southernmost battalion was 200 miles from the northern battalion. I complained to the Corps Commander, but his idea was there was nobody out there! He wanted to get us started, via the Chosin Reservoir, up to the Yalu [River] and he directed me to have the 5th go up and join the 7th."[95]

So, reluctantly, the Marines surged forward, but their northward trek began just as Almond and X Corps received reports that there were indeed CCF operating in North Korea, south of the Yalu River. The CCF had crossed the Yalu and traveled only by night to avoid aerial detection, but late in October they left their calling card in a dramatic way. The 8th U.S. Cavalry Regiment and the 6th ROK Division were ambushed and badly shot up. One regiment of the ROK division was completely shattered. Fewer than nine hundred of the thirty-five-hundred-man force had escaped.[96]

On November 2, Marines of the 7th Regiment confirmed this Chinese presence when they visited the command post of the ROK unit they were relieving at Sudong, twenty-seven miles south of Chosin Reservoir. They were told that the Chinese had crossed into North Korea on October 16.[97]

Still General Almond ordered the Marines to advance. But the right flank of the Eighth Army was not on line with the advancing Marines in X Corps zone. In fact, Eighth Army's lines were sixty miles behind the Marines, who now had no flank security. "Under these circumstances," General Smith complained, "there was no alternative except to continue forward in the hope that the Eighth Army situation would right itself."[98] But the Eighth Army situation did not right itself, nor did it even undertake an effort to right itself. Although the 7th Marines advancing toward Chosin Reservoir had not encountered any resistance on their earlier trek from Wonson to Hamhung, rumors quickly spread that big trouble was brewing.

By November 1, just six days after its initial October 25 surprise attack, the Chinese had crippled the ROK II Corps, bloodied the 8th Regiment of the U.S. Cavalry, and were positioned to roll up the Eighth Army's right flank.[99] And still the Marines were ordered forward, to attack all the way to the Yalu.[100]

And Army Intelligence continued to ignore the obvious. It unwittingly called the significant Chinese presence "some further reinforcement of North Korean units" and there was "no indications of open intervention on the part of Chinese Communist Forces in Korea."[101] All this while there was a dramatic change in the weather. In fact, the severe fighting on October 25 had been in near-freezing temperatures.

On November 2, the Marines had been able to identify enemy soldiers from the 124th CCF Division—the 370th and 372nd Regiments. The next day, the lead elements of Regimental Combat Team 7 (the reinforced 7th Marines) stopped short of the village of Sudong to relieve a ROK regiment. That ROK regiment had been probed by a small unit of Chinese soldiers, and as the soldiers passed the relieving Marine column, they were frantically pointing over their shoulders, repeating one word: "Chinese!"

The Marines were unknowingly almost surrounded by the 124th CCF Division. Its three attack regiments were hidden in the hills to the north, west, and east of the approaching Marine columns. Just after midnight, they struck, mounting a strong attack on the 7th Marines and fought bitterly throughout the night. The Communist attack was only suppressed when Marine aircraft finally came in and pummeled them with bombs and napalm, leaving the corpses of seven hundred burned and frozen Chinese on the killing field.

On November 4, the Marines realized just how vulnerable their position was. That initial Chinese attack had featured flares and blaring bugles, and it had been a knock-down, drag-out, close-quarter affair. A battalion commander remarked: "When daylight came, we found that we were in a dickens of a mess. The rifle companies were well up in the hills, and the Chinese were occupying the terrain between the CP and the companies."[102]

The fight raged on for the next three days until the battlefield suddenly fell silent and the 124th CCF Division departed. The casualties inflicted by RCT-7 had been horrific, and the entire enemy division was reported to be "militarily noneffective."[103] By November 10, the birthday of the Marine Corps, RCT-7 had advanced northward sixteen miles to Koto-ri. The tiny hamlet was situated on an elevated plateau as the road ascended higher and higher into the mountains. Nighttime temperatures dipped below zero.

Forty miles to the south at Wonson, Colonel Chesty Puller, commanding the 1st Marines, cut his ceremonial birthday cake in cool, pleasant weather. And still the Marine column pressed on. The road from Koto-ri led to Hagaru-ri, eleven miles to the north. General O. P. Smith became more alarmed over General Almond's seeming disregard for enemy capabilities. He wrote to the Commandant, General Clifton Cates: "Someone in high authority will have to make up his mind as to what is our goal. . . . We should not push on without regard to Eighth Army. We would simply get further out on a limb. . . . I believe a winter campaign in the mountains of North Korea is too much to ask."

On November 15, RCT-7 occupied Hagaru-ri, at the southern tip of Chosin Reservoir. A rail line ran off to the east side of the reservoir while the road continued to the west. The first night in Hagaru-ri saw the nighttime temperatures dip to 4 degrees below zero. Despite ominous reports from Chinese POW interrogations that a huge force was building up in North Korea, an uneasy pall hung over the countryside. On November 4, RCT-7 finally repulsed the four-day attack, while to the west the Army's 1st Cavalry and the 6th ROKs had been mauled; but then there was nothing—no follow-up. The enemy had vanished.

However, several alarming reports drifted into Marine intelligence from a Marine air fighter squadron that was nightly bombing Sinuiju during the first week of November. Sinuiju was located at a crossing point on the Yalu River. These Marine night fighter pilots did not mince their words when they described what they had seen: convoys of trucks moving south, into North Korea. They described the convoys as, "*heavy . . . very heavy . . . tremendous*" and even "*gigantic.*"[104] Seemingly, those night-fighter reports went unnoticed by the army high command. Army intelligence dutifully modified its original underestimates but was not alarmed.

It now listed Eighth Army forces in the west with 120,000 men, and superior air and artillery, to be facing 100,000 enemy. That was up from its original enemy estimate of only a "few volunteers." But it hardly mattered because X Corps, on the eastern approach to the Yalu, was to continue to attack forward and sweep to the west as part of a giant pincer movement. That would put the enemy to flight.

In reality, the numbers of enemy on the battlefield were vastly different from the army's estimate. It was four times larger. What had actually crossed from Manchuria into North Korea were the entire Chinese IX and XIII Army Groups, with their combined thirty divisions and an excess of 300,000 soldiers. Twelve of these divisions—120,000 men—were on the eastern front facing X Corps, including the 1st Marine Division; and eighteen divisions—180,000 men—were prepared to confront the Eighth Army on the Western Front.[105] The ability to move these massive forces undetected was a tribute to the Chinese discipline of marching only at night:

> A *CCF* army of three divisions marched on foot . . . 286 miles to its assembly area in North Korea, in the combat zone, in a period ranging from 16 to 19 days. One division of this army, marching at night over circuitous mountain roads, averaged 18 miles a day for 18 days. The day's march began after dark at 1900 and ended at 0300 the next morning. Defense measures against aircraft were to be completed before 0530. Every man, animal, and piece of equipment was to be concealed and camouflaged. . . . When CCF units were compelled, for any reason, to march by day, they were under standing orders for every man to stop in his tracks and remain motionless if aircraft appeared overhead. Officers were empowered to shoot down immediately any man who violated this order.[106]

✽ ✽ ✽

On November 17, RCT-7 was ordered to protect the left flank of the 1st Marine Division between Hagaru-ri and Yudam-ni, west of the reservoir. Four days later, in Eighth Army sector, the leading element of the Army's 17th Infantry victoriously reached the Yalu River without encountering a single Chinese soldier. Four days after that, RCT-7 was ordered to move, again to the north, and occupy Yudam-ni. RCT-5 was to continue to advance up the eastern side of Chosin Reservoir, clear any enemy, and then move to the west and join RCT-7 at Yudam-ni. Together, the two Marine regiments would lead the attack of X-Corps to the Yalu. RCT-1 fleshed out the rest of the division and deployed just south of the reservoir, in the vicinity of Hagaru-ri and Koto-ri, to provide security for the Main Supply Line.

On November 24, General MacArthur announced: "This morning the western sector of the pincer moves forward in general assault. . . . If successful, this should for all practical purposes end the war, restore peace and unity to Korea, [and] enable the prompt withdrawal of United Nations military forces. . . . It is that for which we fight."[107]

The attack began. The entire line advanced unopposed. In the east, the supporting X Corps attack was to begin in the morning of November 27. The spearhead was the 1st Marine Division. But on the evening of November 24, disquieting reports from the Western Front filtered into Marine headquarters. The right wing of Eighth Army, manned by the ROK II Corps, had been stopped cold.

Despite this setback, Eighth Army was not alarmed, and chalked it up to an "active defense with local counterattack in strength."[108] And it noted that the ROK II Corps had only one battalion of light artillery and no armor, and had to deal with the most rugged terrain on the entire Eighth Army front.[109] Not to worry, the advance began anew on the 25th.

That advance made no contact, and the enemy seemed to have vanished. As the early pre-winter nightfall settled across the battlefield, Eighth Army began the routine of digging and preparing defensive positions. But well to its front, out of artillery range and unseen in the absolute blackness of the night, the massive XIII Chinese Army Group, like a menacing specter, rose from its hidden assembly areas and began to move. On November 26, tens of thousands of soldiers began a several-hours-long approach, at a rapid dogtrot pace, in order to arrive on the assault line to attack at midnight. No one heard this force coming; there was no warning. It was swift and silent, and eighteen divisions smashed into the Allied defensive line. It struck first on the Eighth Army right, against the ROK II Corps, and shattered it.[110]

The attack was relentless and offered the South Koreans no respite. Wherever they broke through, the Chinese pressed on in a pell-mell dash to the rear to set up barriers to block any attempt to retreat. It was a tactic bent on annihilation, an encircling trap from which there was no escape. The overwhelmed South Koreans "broke and ran, tossing rifles and equipment aside in a frantic attempt at survival."[111] This hammer stroke continued, like the wave of a tsunami—rolling right to left. Next it crashed into the U.S. Army's I Corps and IX Corps, stretched across the width of the battlefield. Eighth Army was cast into a frantic battle for survival and soon began its own headlong retreat. It ceased to be a disciplined fighting force. Roy Appleman described the panic: "By the end of November . . . the Chinese 2nd Phase Offensive had decisively defeated the Eighth Army. . . . The days and nights . . . [were] crowded with a churning, hectic, often bizarre, series of battles, large and small, clear across the Eighth Army front."[112]

The brutal truth was that General MacArthur's offensive had been halted. General Walker, who commanded Eighth Army on the ground, no longer had any illusions about reaching the Yalu River. He was now faced with the task to save the army from complete destruction.[113]

✶ ✶ ✶

Sixty miles to the east, the 1st Marine Division and X Corps were the target of the IX Chinese Army Group and its 120,000 soldiers formed in twelve divisions arrayed in an arc across the advancing road and mountains from Yudam-ni to Chosin Reservoir. On the night of November 27, elements of IX Chinese Army Group attacked on all points along their blocking arc. At midnight, the temperature dipped to -20 degrees. The 79th and 89th CCF Divisions assaulted all along the arc while another division made a wide end run around the frantically defending Marines to attempt to cut the vital fourteen-mile road between Yudam-ni and Hagaru-ri.[114]

The relentlessness of the Chinese attacks was part of the tactics. There was little to no battlefield communications, so once an attack began, the soldiers pressed on until they prevailed or their ranks were shredded to extinction or they ran out of ammunition.[115] At daybreak, the issue was very much in doubt, with both sides clinging to portions of high ground, fighting off encirclement, and repairing broken lines. The Chinese had interdicted the MSR in several places. East of the reservoir, the Army's RCT 31 had been overwhelmed by the attacking Chinese, and small bands from that shattered regiment now fled for their lives.

On December 1, the attempt to break out east of the reservoir began. It was a murderous, ten-mile gauntlet run to Hagaru-ri, with the Chinese holding the high ground on both sides of the road. As the army column struggled to the south, it was attacked and blocked from every direction. Finally, individuals and small groups began to break away to attempt to cross the ice of the frozen reservoir on foot.[116] Casualty estimates of 75 percent of the three-battalion army force were confirmed when only 670 stragglers stumbled into the Marine warming tents in Hagaru-ri.[117]

General O. P. Smith had set his division headquarters at Hagaru-ri. The Chinese were attacking everywhere. Try as he might, he could not get any direction from X Corps headquarters. Every four hours he sent a report on his situation, but X Corps was silent. "Apparently they were stunned," he said sarcastically, "they just couldn't make up their minds that the Chinese had attacked in force. They had to re-orient their thinking. It took them two more days before we were actually told to withdraw to Hagaru-ri and advance to the coast—that took them two days to figure out."[118]

Finally, on November 30, the X Corps commander, General Almond, visited General Smith at the 1st Marine Division command post. He was very concerned about the Chinese attacks on the Marines, and especially the attacks on the Main Supply Route, which threatened the very lifeline of X Corps. He stressed the need for speed to fall back, all the way to Hamhung and the port area. His level of concern was most evident in his authorizing the burning and destroying of all equipment that might delay Smith's withdrawal to the port area. He promised them aerial resupply.[119]

But General Smith tactfully replied that he intended to bring everything out, including his dead and most especially his wounded. There would be no needless destroying of supplies and equipment. To Smith this was not a withdrawal. He was surrounded. He would have to attack his way out, and he issued orders exactly to that effect. He could not afford to discard equipment.[120]

The 5th and 7th Marines first had to extricate themselves from their combat lines facing the Chinese and fight their way back to Yudam-ni. From there it would be a second fight to Hagaru-ri. The two Marine regiments to the west cut themselves free from the front-line embrace of the Chinese army. General Smith then ordered the 5th to capture all the high ground that overlooked the road while the 7th Marines advanced to the south along it to Yudam-ni.[121]

Six new inches of snow fell overnight, and on the morning of December 3 the temperature was near 20 below zero. A single tank would lead the attack toward Hagaru-ri, with the trucks and infantry following. Elements of the 5th Marines deployed again as "Ridgerunners," slogging uphill through the snow and ice to sweep along the ridgelines and attack the sniping Chinese along the high ground. The column on the road would move at the ridgerunners' pace. The 1st Marine Air wing flew 145 sorties to escort the column along the way, and smashed Chinese roadblocks with bombs, rockets, and napalm. Only the truck drivers, the severely wounded, and the dead rode; everyone else walked.[122]

At the rear of the column, Sergeant Robert B. Gault, with Graves Registration, trudged along with his five-man team and his vehicle for the grisly job of picking up any dead along the way. To him, this column was a unit of rugged and ragged Marines. "There was no more 5th or 7th," he said, "you were just one outfit, just fighting to get the hell out of there, if you could."[123]

It was a fight every inch of the way. The Chinese army continued to attack at every turn. Finally, the head of the column stumbled into Hagaru-ri after dark on December 3, and the Marines collapsed into warming tents. The division mess provided pancakes, syrup, and coffee. It had taken fifty-nine hours to run the gauntlet of those fourteen miles. And it took twenty more hours for the rear of the column to arrive.[124] The exhausted Marines were in no shape to continue without rest. General Smith said, "The men were pretty well beaten down. In view of their condition . . . December 6 seems to be the first day that we can start out to Koto-ri."[125]

The breakout that had begun at the Marine front line on December 1 resulted in the regrouping of the column at Hagaru-ri late on December 4. It had cost the 1st Marine Division 1,140 casualties, including 164 killed, and an additional 1,194 non-battle casualties as exhaustion and the extreme cold took its toll.[126] They would now embark on the next leg of the breakout: Hagaru-ri to Koto-ri.

Despite the horrible conditions and the continued attacks by seven surrounding enemy divisions, calm seemed to prevail at Hagaru-ri. Major General William H. Tunner, the Air Force Chief of the Combat Cargo Command, was astonished to see 537 Marine replacements fly into Hagaru-ri rigged for combat, including cold-weather gear. Tunner had flown in to offer his C-47s to evacuate the Marines, and hardly expected to see replacements flown in.[127] A television reporter interviewed General Smith and photographed some of the grim scenes: exhausted Marines with gaunt faces manning the freezing defensive line in the ice and snow, and casualties on stretchers waiting to be evacuated. "These pictures and recordings were later shown in the United States under the title, *Gethsemane*."[128]

At 0630, December 6, Marines of the 7th Regiment, with attached tanks, pushed out of the southern edge of Hagaru-ri. Koto-ri was ten miles to the south, and the 1st Marines were holding it against the Chinese. But the Chinese held the high ground all around the town, and the mountains were massive—some over three thousand feet (one thousand meters).

The long breakout column began arriving in Koto-ri in the late afternoon, and just after midnight of December 6 the tail end had made it in. Hagaru-ri was left a burning, smoldering, exploding wreck.[129] The advance had been a running gunfight for the entire ten miles. It had cost the division another 103 killed and 500 wounded. The division would move out on the next ten-mile run—Koto-ri to Chinhung-ni—at first light, December 8.[130]

Just south of Koto-ri, the enemy had destroyed the key bridge along the entire seventy miles of the MSR. It was a trestle bridge that could not simply be patched, and the success of withdrawing the entire division depended upon its replacement. Nor could the infantry simply descend to the bottom of the cut and crawl up to the other side in the face of the enemy. The walls of the bridge supports were a steep drop. As long as the Chinese held the high ground, no bridge replacement was possible, but without the bridge, tanks, trucks, heavy equipment, and anything else that rolled or was towed would have to be abandoned.

The key mountain dominating the terrain, including the bridge, was Hill 1081, thrusting up almost vertically from the road. Four days earlier, the CCF had attacked and probed the Marines' defensive perimeter. A patrol of Captain Robert Barrow's Company A finally discovered where this Chinese battalion was hiding. During daylight hours, they hid in farmhouses in the valley; at night they came out to attack the bridge site. Barrow's company, reinforced with 81 mm and 4.2-inch mortars, made quick work of the hiding Chinese. Artillery laid down a devastating barrage, and Barrow's infantry attack "ran the Chinese right out of the country."[131] They left behind fifty-six dead.

One week later, Barrow's Company A was called upon again. It was ordered to seize Hill 1081. "My battalion commander had the wisdom beforehand to do some *recce* (recon)," said Barrow. He went up to the most critical point on the entire road going north. It was in a narrow, twisting mountain path where there was a bridge over a deep chasm, and a small number of people could control that bridge. They could blow it out and control the movement of the entire 1st Marine Division.[132]

That bridge was actually the viaduct that crossed over the four giant penstocks or pipelines that carried water from the reservoir down the steep mountainside to the power turbines in the valley below. The original concrete bridge had been blown, leaving a gap of twenty-four feet. There had been two attempts to repair it: the army's engineers had attempted a wooded structure, but the Chinese immediately destroyed it. The second replacement was a steel one, but Chinese explosives took care of that one too.

Now, with the long Marine column inching down from the reservoir, the army planned to airdrop eight sections of a steel Treadway Bridge at Koto-ri. Only four of the twenty-five hundred-pound sections were needed, but eight

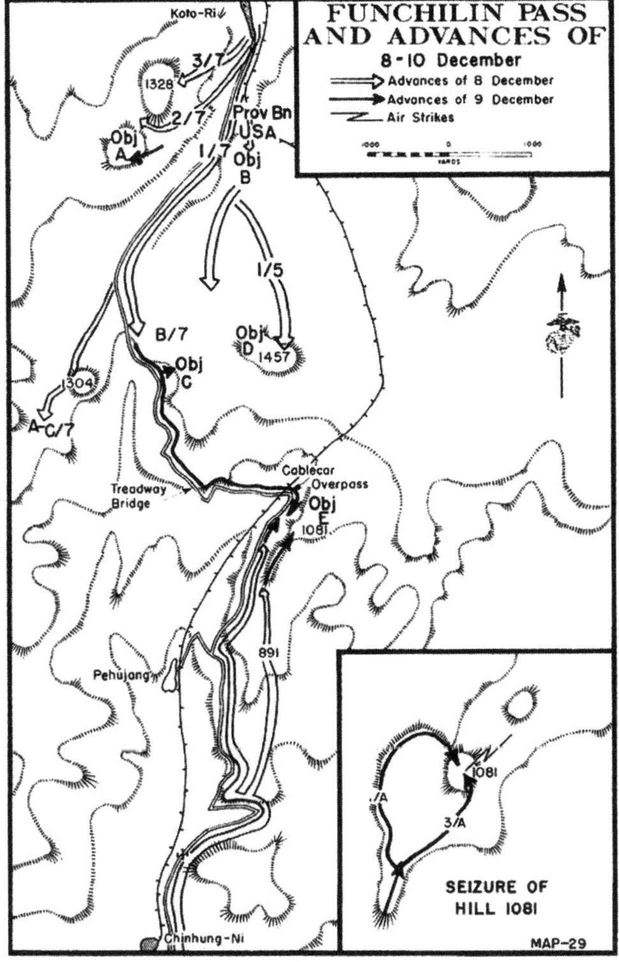

Funchilin Pass, Hill 1081, and the bridge site
(U.S. Marine Corps map)

provided a safety margin. Assembly would be at Koto-ri, and special trucks would transport the sections the final three and a half miles to the blown bridge site. But replacing the bridge was not possible as long as the Chinese dominated Hill 1081.[133]

✻ ✻ ✻

At first light on December 8, the vanguard of 7th Marines began its run to the bridge site. A blinding snowstorm limited visibility to fifty feet, and although it covered their movement from Chinese eyes, it also precluded any covering air support. The 1st Battalion, with Robert Barrow leading Company A, moved to the north to attack at 0800 against the forces occupying Hill 1081. Intelligence had pinpointed that the 60th Division of the CCF occupied the high ground.

Two thousand yards after the jump-off, the road hair-pinned to the left, and a less severe slope touched the road. It was only a small finger of a ridge, but it offered the only access to the otherwise impregnable up-thrusting of Hill 1081. Everywhere else along the narrow road, the vertical sides of the mountain were so severe that ascent was impossible. Barrow broke off his company from the rest of the battalion's advancing column and moved to the base to begin the climb. This back door to Hill 1081 was a very narrow, razorback ridge. It required a single file climb along the very peak ridgetop.

Barrow took the lead, followed by his gunnery sergeant, a mountain of a man nicknamed "Tiny." King D. Thatenhurst was a six-foot-six, 240-pound Marine Corps legend. He had fought in bloody battles in World War II and was exempt from service in Korea, as were all recipients of two Purple Hearts. He had seven but brushed that exemption stuff aside. He was ready for combat on the slopes of Hill 1081.

"We had to negotiate the climb on our knees—climbing, pulling yourself up by rocks and various small trees protruding out of the rocks," said Barrow. "From where we were in the valley, up one thousand, eighty-one meters is a pretty long piece of vertical terrain to negotiate, and that's where the Chinese were, on top of Hill 1081."[134] He started the daunting ascent with his full company of 223 Marines; each man agonizingly pulled himself up, foot by foot, yard by yard, along the long spine of the ridge. If they had been visible, the column would have resembled a long snake inching steadily upward.

Falling snow shrouded the column and blanketed everything. Visibility was only fifty feet, and the only thing to be seen by a man climbing was the rear end of the Marine directly to his front. Finally, Barrow halted the climb as the ridgeline seemed to level off. "We got to where it was the top of something," observed Barrow. "It may not have been the ultimate top, but it was the top. And we hadn't encountered anything." He was actually still 150 meters from the summit of Hill 1081 and was on the plateau of Hill 891.

Sergeant Thatenhurst brought the first part of the column forward, Lieutenant Don Jones's 2nd Platoon, and deployed them in a makeshift assault line. One squad (twelve men) squeezed in across the crest of the narrow ridge, and two other squads deployed off each sloping flank. The rest of the column formed up in the rear of the attack area, and Barrow prepared to give his attack order. Everyone stared to the front, as if they could see what lay ahead, but the snowstorm that had concealed the Marines' ascent up the mountain also obscured the Chinese defenders and their positions. But, as if fate decided to lend a helping hand, the obscuring veil of snow suddenly lifted.

"The snow momentarily lifted," said Barrow, "and peering to the north, we could see lots of Chinese. They had expected no one, they had heard nothing, they had seen nothing, and were out of their dug-in positions, stumbling around. . . . And we got it, just a snapshot; it happened for seconds." And then it was gone. The snow was back, as thick as ever, and obscured the mountain. But in those brief seconds Barrow and his men had glimpsed all they needed to see, and the Chinese had not seen them lying prone and motionless in the snow. The Marines had also seen that just behind the Chinese defensive positions to their front was the rise of another small ridgeline that led to the summit of 1081—and the top bristled with more Chinese soldiers.

Barrow's plan was now clear. He would use artillery and mortars positioned in the valley to hammer the dug-in enemy to his front. But since he could no longer see to adjust the artillery fire, he would attempt to "walk" the fire onto the enemy. Calling for a few spotting rounds to be fired onto a ridgeline that was shown on the map to be east of his position, he gave the guns corrections by ear. Listening for the sound of the explosions, he steadily moved the impact area until it sounded due north of his position, and he could hear and feel that the explosions were directly to his front.[135]

And then Jones led his 2nd Platoon forward. The firefight was fierce but brief, as artillery and mortar barrages did their jobs. Although the Marines had some casualties, the Chinese positions were smashed and their infantry overwhelmed. There was no way to continue the attack to the top of Hill 1081, so Barrow ordered his men to dig in, not only to protect them from a possible counterattack by the surprised enemy but also from the stabbing and paralyzing, 20 below zero cold winds. "The wind out of Siberia, upon the top of that mountain was just unbelievable," said Barrow, "and my concern was that we would freeze and be put out of action. So, I spent the entire night going from one group to another, making sure that they were awake."[136]

During the night, the Chinese made a feeble attempt against Company A's lines but were easily repulsed. In the valley 3,000 feet below, the 1st Battalion headquarters recorded a low of 25 below zero.[137] On top of the mountain, the temperature was unknown. For the Marines of Company A, there could be no warming fires. Those would attract enemy mortar fire. And as difficult as it was, each man pulled off damp socks for dry ones, and tucked the damp ones inside their shirts to dry them with body heat. Rations were frozen solid, so there was no eating, and canteens were frozen too. Some of the men were suffering from frostbite.

When the sun rose on December 9, the snowstorm was gone. The sun was brilliant in the clear, blue sky and the mountaintop was dazzling white in the blinding sun. "We looked north," said Barrow, "and there they were. We had overrun one of their external positions . . . and there they were on the top. We, again, employed supporting weapons, including, air strikes."[138]

Supporting artillery and mortars pounded the mountaintop. Shattering explosions and flying shrapnel blanketed the entire summit. Lieutenant Bill McClelland braced to lead the 1st Platoon in the final assault to the top when the artillery lifted. But as soon as he stepped off, the Chinese opened up with blistering small arms and machine gun fire. Barrow moved to the front with his Forward Air Controller to direct an incoming flight of four Marine Corsairs. They came swooping in, over his lines, and were guided to their targets by convenient "aiming stakes" silhouetted on the white landscape. Two high-wire electrical power poles pointed to the sky and acted as reference points to guide the Corsairs to deliver their deadly loads on top of the Chinese defenders.[139]

The 1st Platoon assault was joined by the 2nd and 3rd Platoons, which made a difficult double envelopment of the round-top summit. Included with the 3rd Platoon's maneuver were the veterans from Tucson, who had never been to boot camp. As the attack progressed, the men broke out in cheers and redoubled their efforts to hurl the enemy from the top of the mountain. From their elevated vantage point, they saw the Treadway Bridge maneuvered into position over the chasm. On the far side, waiting to cross, was the long, marching column from the north. Nothing now barred the way for the 1st Marine Division to complete its breakout to Hungnam, and out of the encircling trap of the CCF.[140]

Captain Robert Barrow's Company A had been reduced to half of its original strength. Seventy-five were casualties of the freezing cold. Only 111 of the original 223 were able to participate in the evacuation of the wounded and dead. It was a most agonizing task, with each able-bodied man carrying the weapons and gear of a fallen comrade. A patrol from the battalion came up to help in the harrowing descent.

The Chinese army had paid a terrific price trying to hold onto the hill. It left 530 of its dead on the mountain top.[141] "There were no prisoners," said Barrow. "Supporting arms had crushed their bunkers. The whole hill took on a different appearance. It was blackened. The snow had been turned into dirt and blasted into powder. It was tranquil the first day when we saw it, but after we were finished with it, it was scarred.

"The next day we watched all day long, the patches of the 1st Marine Division, down below us, steadily going south to the sea. Late in the day, we took ourselves off of the hill. . . . We left nothing on it worth having."[142]

Marine losses during the Chosin campaign, from October 26 to December 15, 1950, were 700 dead, 192 missing, and 3,508 wounded. There were more than 7,000 non-battle casualties. The enemy paid a terrible price in its attempt to annihilate the 1st Marine Division—37,000 casualties, of which 25,000 had been killed. Chinese documents added great credence to these estimates.[143]

CONCLUSION

Captain Robert Hilliard Barrow received the Navy Cross for his heroic actions on Hill 1081. He went on to a forty-year career in the Marine Corp. In 1968–1969 he commanded the Ninth Marine Regiment in combat in Vietnam, and eventually rose to its highest rank its 27th Corps Commandant, joining General John Archer Lejeune as the second Commandant of Marines to come from LSU's Cadet Corps.

Along the way, from 1957 to 1960, he served in New Orleans as the Marine Officer Instructor at Tulane University's Naval ROTC.[144] In 1959, he recruited Midshipman Ronald J. Drez into the Marine Corps option program. Captain Drez went on to be commissioned and served in Vietnam at the same time as Barrow. He was the Commanding Officer of Company H, 2nd Battalion, 5th Marines—and is the author of this book. In 2002, at a small reunion with some of his former "recruits" from the Naval ROTC, General Barrow commented to Drez that he was pleased "with what you have done with your pen." Hopefully that pen has now honored some of the history, heritage, and military excellence of LSU, and its Cadets of the Ole War Skule.

CHAPTER 13

Conspicuous Gallantry and Heroism of the Ole War Skule

Rene T. Beauregard, Army of Tennessee: In 1860, he was one of the original cadets of the old Louisiana State Seminary of Learning and Military Academy at Pineville. This nineteen-year-old interrupted his schooling in 1861 with the outbreak of the Civil War to enlist with Slocomb's Battery, 5th Company, of the Washington Artillery. Later he became a first lieutenant in the South Carolina Regulars and fought in the West in the Army of Tennessee. His bravery and fortitude were especially noted in the battles of Franklin, Missionary Ridge, and Lookout Mountain.

Charles MacDonald (LSU Archives)

Charles H. MacDonald, USAAC: MacDonald finished LSU in 1938 and later was commissioned in the U.S. Army Air Corps. He became the third-ranking ace in the Pacific Theater during World War II. During the Japanese attack on Pearl Harbor, he managed to get a damaged fighter airborne to confront the Japanese attack, but the Japanese aircraft had returned to their carriers. In October 1943, MacDonald commanded the 475th Fighter Group, flying the twin-boomed P-38 Lightning out of Doba Dura, New Guinea. He and his fighter group became the scourge of the skies to Japanese aviators, and during one seven-week period between November 10, 1944, and January 1, 1945, MacDonald and his P-38, *Putt-Putt-Maru,* shot down thirteen Japanese aircraft. He finished the war with a total of twenty-seven kills. His awards included the Distinguished Service Cross (2), Silver Star (2), Legion of Merit, Distinguished Flying Cross (6), and Air Medal (11).

Arthur DeLaHoussaye
(LSU Archives)

Captain Arthur J. DeLaHoussaye Jr., USMCR: After graduating from Baton Rouge High School in 1960, DeLaHoussaye attended LSU, and was later commissioned in the U.S. Marine Corps and became a Marine aviator. Flying the F-4 Phantom, he deployed to Chu Lai, Vietnam, in November 1967 as part of squadron VMA-323 and flew eighty-nine missions.

On November 20, 1967. Captain DeLaHoussaye launched from the Chu Lai Airfield for an attack mission, but was diverted before he reached his target. The airborne Forward Air Controller directed him to provide close air support for a Marine platoon desperately fighting against a numerically superior enemy. The controller flew in and marked the targets, but heavy ground fire greeted him, shattered his canopy, and wounded his co-pilot. But his white phosphorus rocket was on target and pinpointed the enemy position.

Captain DeLaHoussaye rolled in for the attack through the increasingly intense antiaircraft fire. He dropped his bombs with perfect accuracy, pulled out of his shallow run, regained altitude, and swooped in for a second run. He made eleven runs on the enemy position, continuing until he had expended all his ordinance. He would be awarded the Distinguished Flying Cross.[1]

On January 30, the combined North Vietnamese Army and its Viet Cong allies launched the surprise Tet Offensive across all South Vietnam. The next day, enemy gunners pummeled the Chu Lau Airfield with 122 mm rockets. During the attack, one of the rockets struck the bunker in which Captain DeLaHoussaye had taken cover. The blast killed him instantly. He was additionally awarded the Purple Heart and four Air Medals.[2]

Jill Prattini Klooster
(Courtesy of the Klooster Family)

Jill Prattini Klooster, USAF: As a cadet in the Air Force ROTC, she graduated from LSU in 2002 and was commissioned as a second lieutenant in the U.S. Air Force. Her specialty was intelligence, and she became a first lieutenant only eight months after completing the Intelligence Schoolhouse.

In 2004, as part of Joint Task Force 7 at Victory Base in Bagdad, Iraq, she flew drone intelligence gathering missions to target enemy positions and personnel. From 2007 to 2009, she deployed four times to Al Udeid Air Base in Qatar, the forward base for U.S. Central Command. She flew over eleven hundred combat hours in the E-8C Joint Surveillance Target Attack Radar System (STARS) to provide ground surveillance to support attack strikes. She was awarded five Air Medals (four for operations over Iraq, and one for operations over Afghanistan).

Campbell B. Hodges
(U.S. Army photo)

Campbell B. Hodges, U.S. Army: Hodges served as a Professor of Military Science and Tactics, Commandant of the Cadets, and a professor of Spanish at Louisiana State University from August 22, 1910, to December 15, 1912. He served with the 31st Division in France on the Meuse-Argonne front in 1918 and was awarded the Victory Medal and two Bronze Stars. After serving with honors in World War I, he went on to serve as Commandant of the Cadets at the U.S. Military Academy at West Point. He was then appointed President of LSU in 1929, but that was cancelled when he was called to serve as the top military adviser to President Herbert Hoover. He finally accepted the appointed as President of LSU on July 1, 1941, and served until 1944, when he suffered a slight stroke and retired. He had attained the rank of major general. Hodges Hall on the LSU campus bears his name.

Simeon A. (Alex) Box
(LSU Archives)

Simeon A. (Alex) Box, U.S. Army (First Lieutenant): He was an outstanding athlete in all sports while in high school, and in 1938 turned down a baseball offer from the Cincinnati Reds and multiple scholarship offers from other universities to accept a football scholarship to LSU. He graduated in 1941 in petroleum engineering and was commissioned second lieutenant in the army from the ROTC.

On August 2, 1942, Box sailed for England for further infantry training and was based at Tidworth Barracks with the 1st Engineer Combat Battalion of the 1st Infantry Division. In November 1942, the division sailed aboard twenty-two ships to land on the beaches of North Africa on November 8 as part of Operation

Torch. On November 9, 1942, at Arcole, Algeria, Lieutenant Box mounted a half-track and attacked and destroyed enemy machine gun emplacements blocking the 1st Division's advance. For his extraordinary heroism he was awarded the army's highest award, the Distinguished Service Cross.

In February 1943, after dogged fighting, Field Marshal Erwin Rommel launched an all-out attack against the American forces at Kasserine, and Lieutenant Box was tasked with laying mines and constructing roadblocks to confront the attacking Germans. On February 19, one of those mines accidently detonated, killing him instantly, along with four other soldiers. He was posthumously awarded the Purple Heart. In May of 1943, the LSU Board of Supervisors voted unanimously to name the university's baseball stadium in honor of Alex Box.

Ralph Thompson Brown (LSU Archives)

Ralph Thompson Brown, Lieutenant Colonel, U.S. Army: He attended LSU on a working scholarship. He lived and worked in the cattle barn and earned a bachelor's degree in 1937 and a master's degree in horticulture in 1938. After Pearl Harbor, he was commissioned a second lieutenant in the 94th Infantry Division and fought in northern France and Germany. In 1945, while commanding Company K, 376th Regiment, 94th Infantry Division, he led his company in the Allied spearhead into Germany, crossing the Saar River and hurling back the enemy from the Siegfried Line. In eighty-one days, his division captured the Rhine-Moselle Triangle along with many enemy prisoners. As the Reich began to collapse, Brown and his company attacked with the division to liberate more than two hundred towns and capture 13,400 German prisoners.

On March 16, 1945, Brown led his company in an all-out attack on a strongly held German position at Sitzerath, Germany. A well-camouflaged German 77 mm artillery piece dominated the route of advance and was protected from American tank fire by well-positioned German rocket launchers. Captain Brown armed himself with a grenade launcher and courageously crawled toward the German gun. The German defenders saw him and seriously wounded him with shrapnel with point-blank artillery fire, but he managed to force the enemy soldiers manning the rocket launchers to retreat.

Without their protection, friendly tanks silenced the artillery piece and led the infantry units in a successful assault upon the town.

Of the forty thousand men of the 94th Division, Captain Ralph Brown was one of only twenty to receive the Distinguished Service Cross for gallantry in action.[3] After the war, Brown returned to LSU and worked with the Citrus Research Station. His research helped develop new varieties of satsuma.

Germaine Laville
(U.S. Marine Corps photo)

Corporal Germaine Laville, USMC: She lived in Plaquemine, Louisiana, and enrolled at LSU in 1938 and graduated in 1942 with a bachelor's degree in education. She began teaching the fifth grade at Shady Grove Elementary School in Iberville Parish.

She was the oldest of seven children, and since there was no one in the family that was old enough for military service, she decided to stop teaching and patriotically represent her family in World War II. She enlisted in the Marine Corps Women's Reserve as a private in July 1943. After completing Boot Camp at Camp Lejeune, North Carolina, she was then stationed at the newly created 3rd Marine Aircraft Wing at Cherry Point, North Carolina, in December 1943. At the Aerial Gunnery School, she taught the art of aerial gunnery and was promoted to the rank of corporal on February 10, 1944.

On June 3, nine civilian workers were cleaning and waxing the upper floors of the two-story, wooden Synthetic Training Building when they accidentally sprayed a highly flammable liquid floor wax onto a frayed live electrical wire of a flight simulation machine. The highly flammable liquid exploded into flames, and within minutes the entire building was engulfed. Germaine Laville was teaching fifty other Marines in a classroom on the ground floor and escaped. But seeing that many others were trapped inside, she reentered to assist her fellow Marines. By this time the fire was consuming the entire building at over 1,000 degrees. Five Marines died in the blaze, including twenty-two-year-old Germaine Laville, and another thirty-seven were injured. She became one of only two Women Marines killed in World War II.

The Marine Corps recognized her heroism by naming the final obstacle on their grueling Crucible obstacle course at Parris Island after her. That last obstacle is called "Cpl. Laville's Duty." In 1948, East and West Laville Halls were dedicated on LSU's campus. Today, the halls are home to the Laville Honors House.

Jefferson Joseph DeBlanc
(National Archives)

Jefferson Joseph DeBlanc, Col., USMCR (MOH): As a child, DeBlanc was always fascinated with planes and played cowboys and Indians. "An airplane was forced to land a mile from our home, and we all ran there," said young Jeff. "And since I was first, the pilot picked me up and put me in the cockpit, and I looked at all the dials, and I was hooked. There was no doubt in my mind; and I switched from cowboys and Indians to Eddie Rickenbacker's Sam Browne belt and uniform."[4]

He had enrolled in Southwestern Louisiana Institute and studied and excelled in pilot training, but after the outbreak of war in Europe, DeBlanc left college. In July 1941 he enlisted in the Naval Reserve as a Seaman Second Class, and then received flight training at the Naval Reserve Aviation Base in New Orleans. Two weeks later he reported to the Naval Air Station in Corpus Christi, Texas, and on October 15 became an Aviation Cadet.

On May 4, he was commissioned as a second lieutenant in the United States Marine Corps Reserve and reported to the 2nd Marine Aircraft Wing in San Diego for advanced carrier training. For the young pilots during this early stage of war, it became a life of daunting night carrier landings, frequent malfunctions, and outdated dogfighting tactics.

Finally, on August 6, as a qualified fighter pilot, he entered the dreaded new pilot's pool—a limbo-like place where pilots waited to be called. August turned into September, and September into October, until finally in mid-October the call came to report to Marine Squadron VMF-112 to fly the F4F Wildcat, a plane in which he had only ten hours flying time. Two weeks later he was in Guadalcanal.

On November 13, Japanese aircraft attempted to torpedo Allied ships and were intercepted by VMF-112. DeBlanc shot down three of them. But his good luck as a flyer seemed to run out on January 29, 1943, when mechanical failure forced him to ditch his Wildcat into the hostile waters around Guadalcanal. But he crashed in the wake of an American destroyer, itself fleeing from a possible Japanese air attack, and was rescued. Two days later, he was back in the skies leading eight Wildcats escorting a strike force of twelve Douglas *Dauntless* dive-bombers on a mission to attack Japanese shipping.

"Word came down from fighter command that a Japanese invasion fleet was coming down with ships to reinforce Guadalcanal," said DeBlanc. "So, they scramble us, eight fighter pilots, to escort twelve dive-bombers to hit the Japanese destroyers and ships in the Kolombangara area—250 miles out from

Guadalcanal." But that was beyond the range of the Wildcat, so the aircraft had to be outfitted with auxiliary fifty-gallon wing tanks.

"About one hundred miles out, I'd already used my fifty gallons, and the main tank was not full," said DeBlanc. "There was a leak somewhere. I knew I could make it to the target area but could not make it back." At that same time two other Wildcats experienced similar problems and turned back from the mission. That left the mission with only six protecting fighters. "I figured I knew enough about survival that I could survive if I hit the islands, so I decided to keep going." Pulling out would have left only five fighters to protect the bombers.

As they arrived in the target area, they engaged many Japanese Army *Oscar* fighters at fourteen thousand feet that attempted to drive off the Marine bombers. But the Wildcats held them off, and the bombers completed their runs. "Our dive bombers were rendezvousing to go home when they suddenly yelled that they were being hit by float planes." He immediately placed his aircraft into a steep dive from fourteen thousand feet to where the float planes were attacking at one thousand feet and engaged them with his four .50-caliber wing guns. The first plane exploded, and DeBlanc maneuvered to attack the second one, sending a stream of fire into its fuel tanks, setting it ablaze. "I burned two of them," said DeBlanc, "and the rest of them left. About that time, when I was over twenty-five feet of water, that's when the *Zeroes*[5] came. They didn't see me first, so I got the first crack at them and burned the first guy, their leader. Everybody else broke off into dogfighting."

But the day's fighting for DeBlanc was not done. "I shot down five aircraft that day and had one probable. The last two kills were above me coming down; so, I had the better position. I had a mid-wing fighter which is the best platform to shoot from, so I put the pipper [sight] right above his cockpit, and he immediately exploded, and I flew through the pieces."

The remaining Japanese wingman, having watched the destruction of his flight leader, moved to attack in a high-speed steep dive. "When I saw him coming down on me," said DeBlanc, "he was picking up speed. And he was too anxious for the kill. He was coming down so fast that he didn't realize that he had overtaken me. So, I chopped my throttle and popped up my flaps to slow me down and he couldn't stay on my tail, and we went wingtip to wingtip and looked at each other. There was no way to go above me, he couldn't go under me, he had to pull up [or crash] and, when he did, I burned him." It was DeBlanc's fifth aerial victory of the day and would have qualified him as an ace-in-a-day if he had not already attained that designation with his three previous kills. His total was now eight; and his day's action was still not over.

"So, then I stood out fat, dumb, and happy," he said, "and that's when I got hit." A Japanese fighter lurking out of sight in his blind spot closed on him at high speed and fired bursts from its two 20mm guns that ripped into his cockpit. They tore his wristwatch from his arm, smashed his instrument panel, and set fire to his engine. His only choice was to bail out. He released the cockpit canopy and climbed onto his seat. "So, I jumped out, treading for the edge of the wing and pulled the rip cord; and floated through the air, and it was beautiful. I'd never seen that. I was at 3,000 feet. Man, that was the most sensational thing I ever did in my life."

DeBlanc floated down to the Vella Gulf, near the Japanese-held island of Kolombangara. As he splashed into the sea, he became aware of wounds to the backs of his arms and neck. A shoreline was in the distance, and supported only by his life jacket, he painfully paddled toward it. After a six-hour struggle, the waves washed him up on the beach.

Expecting the arrival of a Japanese patrol to capture him, he limped into the dense jungle and hid out for two days, subsisting only on coconuts. On the third day, a party of friendly natives stumbled upon him and escorted him to their own camp. In a vigorous bartering session, they traded him to another group for a sack of rice and then transported him by outrigger canoe to the home of an Anglican missionary. From there, the missionary forwarded him to two coast watchers, who radioed for help. On February 12, two weeks after his heroic aerial combat action, a navy PBY *Catalina* patrol bomber landed in the sea off the Kolombangara. The islanders paddled DeBlanc to it.

⚜ ⚜ ⚜

In November 1944, after recovering from his wounds, Captain DeBlanc returned to the Central Pacific for a second tour of duty and flew in the Marshall Islands campaign. In May 1945, he transferred to VMF-212 for the Okinawa campaign, where he shot down one more Japanese plane, bringing his total to nine.

On December 6, 1946, Jefferson Joseph DeBlanc was presented the Medal of Honor by President Harry S. Truman in the White House "for conspicuous gallantry and intrepidity at the risk of his life above and beyond the call of duty" for his actions in the Solomon Islands on January 31, 1943. His other awards included the Distinguished Flying Cross, the Purple Heart, and five Air Medals.

After the war, DeBlanc became an educator. He earned a bachelor's degree in physics and math from Southwestern Louisiana Institute in 1947, a master's degree in education from LSU in 1951, a second master's degree from LSU in education (mathematics) in 1963, and a doctorate in education from McNeese State University in 1973.

Sanderford Jarman
(LSU Archives)

Sanderford Jarman, Major General, U.S. Army: In 1901, Jarman won a scholarship to enter LSU, and he later became Commandant of LSU's cadets in 1906–1907. At the end of his junior year, he accepted an appointment to West Point and graduated in 1908. When the United States entered World War I, he was ordered to France with the advance party of the 29th Division, and later served with the 2nd Army Railway Artillery.

From France he wrote home, "We have all of our big guns on railroad track and move them along the track and fire them. It is some job getting them loaded at times. There is constant firing and all kinds of troubles to be met. Everything is at least six inches in mud, and it rains practically all the time."[6] Jarman was in the St. Mihiel and Meuse-Argonne offensives and was awarded his first Distinguished Service Medal. During the 1930s, as the world seemed to be on another collision course with war, he was tasked to reorganize the defenses of the strategic Panama Canal.

During World War II, from June 15, 1944, until April 25, 1945, he was the Commanding General of the Marianas, which consisted of the Islands of Tinian, Guam, Saipan, Auguar, Ulithea, Iwo Jima, and Pellilieu. But on June 22, as the battle for the island of Saipan raged, the Army's 27th Infantry Division was inserted into the center of the line between the two Marine attacking divisions. Unfortunately, the 27th's infantrymen were unable to keep up with the fast assault pace of the battle-hardened Marines fighting on their flanks, and it fell behind, and the assault line bent like a horseshoe. The Marines were then exposed to Japanese infiltration and a galling flanking fire. By the end of the day, Marine Corps Lieutenant General Holland M. Smith could no longer hide his concern over the poor showing of the 27th Division and approached General Jarman, the senior army officer on Saipan. He asked Jarman to visit the commander of the 27th to impress upon him the need to keep up on the battle line. Jarman found the 27th Division commander, General Ralph Smith, to be despondent over his division's lack of fight, and he told General Jarman that if his division did not do better, he should be relieved. Unfortunately, the division did not keep up, and the next afternoon General Holland Smith exercised his prerogative and replaced him with General Jarman, who successfully led the division for the critical next four days. This episode would create much bitter interservice rivalry between the Marines and the army.

After the island was secured, General Jarman developed Saipan into a B-29 base for the long-range bombing of Japan, and the first bombing mission originated from Saipan. His awards include two Distinguished Service Medals and the navy's Distinguished Service Medal, recommended by Admiral Chester Nimitz.

Richard Joseph Keller
(LSU Archives)

Richard Joseph Keller, Captain, U.S. Army Reserve: Keller was born and raised in Hahnville and graduated from LSU in 1939 with a bachelor's degree in chemistry. After commissioning through LSU's Army ROTC, he was assigned to the Ohio National Guard's 37th Infantry Division during the Solomon Islands Campaign and participated in the battles of Guadalcanal and New Guinea. He then commanded Company E, 2nd Battalion, 148th Infantry Regiment, for the invasion of Bougainville.

On November 1, 1943, the 3rd Marine Division and the 37th Infantry Division assaulted the beaches on the western side of the 125-mile-long, 48-mile-wide island of Bougainville. The landings were made at Empress Augustus Bay, and the ultimate goal was to provide a closer base for light bombers and escort fighters to join the long-range heavy bombers in the aerial offensive against Rabaul. The massive Japanese base was two hundred miles to the west and was the headquarters to the Japanese Southeastern Army and the Southeastern Fleet.

The entire island of Bougainville was not the objective. The military brain-trust did not want a long, bloody campaign, as had been fought at Guadalcanal and New Guinea. It just wanted a five-mile-wide, two-mile-deep perimeter of flat beachhead where Seabees could carve out an airfield to assist in the Rabaul assault.

The Marine and army divisions landed against moderate opposition, in-

Building the airstrip on Bougainville
(U.S. Marine Corps photo)

cluding Japanese air attacks, and pushed out to the foothills that led to towering volcanic mountains. The Japanese were completely fooled by the landings and had expected the Americans to land on the opposite side of the island. The strong garrisons of the Imperial Japanese Army's 17th Army were in the northern and southern ends of Bougainville, far removed from the central landing beaches. There were no roads on the island, just some primitive trails into the interior and around the perimeter, so the defending Japanese could not easily move to attack. It would take months to hack their way through thick jungles and swamps. Along the way they would have to improve primitive trails to move artillery, bridge fast-running streams and rivers, and carve out assembly areas to be able to attack.

Meanwhile, on November 29, 1943, the Seabees began work on an eight-thousand-foot bomber strip. In record time they bulldozed the land and laid down the steel matting for the bomber runway. After one month of construction, the first bomber went into operation. On December 10, 1943, they began constructing a fighter strip, parallel to the bomber field. The fighter strip was completed on January 3. The remote Japanese forces were powerless to stop any of this. But by February 1944, intelligence detected that the twelve thousand men of the Japanese 6th Division were on the move.

On March 7, Japanese infiltrators began cutting the concertina wire in front of the American positions, and the following morning Japanese artillery unleashed a massive barrage against the defensive lines and the airfields,

Marine Corsairs on Bougainville
(U.S. Marine Corps photo)

forcing the aircraft to scurry off to the safety of the neighboring islands of Munda and New Georgia. The American perimeter was thinly spread along the two-mile front. In the center of the line was Hill 700, towering over the entire American position and giving the Japanese a bird's-eye view of the bomber and fighter strips. Hill 700 also dominated the defensive positions to its right and left, and the collapse of the hill would mean the collapse of the entire defensive line, and the end of airfield operations.

Patrols had routinely been sent out forward of the American lines, and the capture of several prisoners revealed that the importance of Hill 700 was well understood by the attacking Japanese.[7] On March 8, the onslaught began. The attack was unrelenting against the 37th Division's positions, and by dawn on March 9 the enemy had seized and occupied several positions along the north slope of the ridge. More importantly, they took two vital positions atop Hill 700 itself. The hill's reverse slope was filled with fighting holes, ample proof that the Japanese 6th Division had succeeded in moving its entire force, against all odds, through the hostile jungles of Bougainville and into a threatening attack position.

A desperate fight for control of Hill 700 continued for the next two days with high casualties, and at dawn on March 11, the Japanese launched a fanatical banzai charge. "They came on in waves, one whole battalion attacking on a platoon front. Brandishing their prized sabers, screeching '*Chusuto!*' ('Damn them!'), the enemy officers climbed up the slope and rushed forward in an admirable display of blind courage. The men screamed in reply, '*Yaruzo!*' ('Let's do it!') and then '*Harimosu!*' ('We will do it!'). As they closed with the Americans, their leaders cried, '*San nen kire!*' or 'Cut a thousand men!'"[8]

The 37th Division mowed them down with machine guns, ripping holes in their tightly packed formations, and by 0800 the surviving Japanese pulled back, leaving behind hundreds of their dead comrades. They had failed to break the American line, but they had succeeded in fully occupying Hill 700. If the 37th Division was to retake this vital high ground, it would have to be with a frontal, uphill attack against galling firepower.

The 2nd Battalion of the 148th Infantry was called for this seemingly suicidal job. Captain Richard Keller's Company E was ordered to envelop the enemy positions on the top. In the early afternoon, Keller and his men began the slow crawl up the steep slope. The first squad of eleven men gained the crest, but their victory was short-lived. They were ambushed by murderous fire from their front and flanks. Eight were killed instantly, and a ninth later in the afternoon. By 1600, Keller could not gain another inch against the enemy's rock-solid defense and was ordered to cease his attack and consolidate the ground that he held. Throughout the night, the two sides exchanged small arms fire.

At 0800 the next day, Keller's Company E was joined by Company F to double envelop Hill 700 and break the Japanese position. The soldiers again began the slow crawl up the slope around the right and left of the looming crest, but Japanese machine guns still dominated the battlefield from above and swept the approaches with deadly fire. Keller deployed every weapon at his disposal, including grenades, flamethrowers, rocket launchers, and explosives to inch his company up the slope for an assault against the crest. Finally they were in position, and the men deployed for a final attack.

"By noon, Captain Richard J. Keller of Company E and Lieutenant Sidney S. Goodkin of Company F reported by radio to the battalion commander: 'We believe we have got them. We are going over the top together.' They personally led their company assaults, shouting defiance at the Japanese and encouragement to their own men."[9] They charged over the crest and down the reverse slope, driving the Japanese remnants in front of them. Fifteen minutes into the final attack, a Japanese bullet struck down Captain Keller, severely wounding him in the chest as his company continued to rout the enemy. He received the Distinguished Service Medal for his gallantry. His citation read in part:

> For extraordinary heroism . . . while serving as Commanding Officer of Company E, 148th Infantry Regiment, 37th Infantry Division, in action against enemy forces during the Battle on Hill 700 at Bougainville, Solomon Islands, on 11 and 12 March 1944 . . . Captain Keller pressed a vigorous, determined advance until darkness halted the action. Throughout the 5 1/2-hour attack on the following day, he was in the forefront of the lines. . . . After all but one dominant fortification had been destroyed by riflemen, flame throwers, rocket launchers, and grenades he leaped over the crest of the hill into the face of the enemy and led a final, furious charge which ultimately overran the Japanese position and liquidated all enemy opposition.

His other medals also include the Bronze Star, two Purple Hearts, Combat Infantry Badge, and Presidential Unit Citation.

Stanley Martin Maillet
(LSU Archives)

Stanley Martin Maillet, Lieutenant Colonel, USAF: Maillet graduated from LSU in 1958 and was commissioned as a second lieutenant in the United States Air Force. He served in combat in Vietnam with the 17th Special Operations Squadron. He also flew B-52s with the 2nd Bomb Wing, including 280 combat missions. Four of those missions were in the air offensive over Hanoi and North Vietnam. He was awarded three Distinguished Flying Crosses, four Air Force Commendation Medals, and fourteen Air Medals.

Thomas Rhame
(U.S. Army photo)

Thomas Rhame, Major General, U.S. Army: Rhame graduated from LSU in 1963 and commissioned as a Second Lieutenant through the Army ROTC. He served in Vietnam from 1967 to 1971 during a thirty-four-year army career. On August 2, 1990, Iraq invaded Kuwait. The 1st Infantry Division was put on alert for deployment on November 8, and the division deployed over twelve thousand soldiers and seven thousand pieces of equipment to Saudi Arabia during the following two months.

On February 23, 1991, General Colin Powell, the Chairman of the Joint Chiefs of Staff, stoically announced, "Our strategy to go after this army is very simple. First, we're going to cut it off, and then we're going to kill it." The following day, Major General Thomas G. Rhame's 1st Infantry Division, the Big Red One, spearheaded Operation Desert Storm's armored attack, leading the way for the army's massive VII Corps to liberate Kuwait. The division smashed into the Iraqi 26th Infantry Division, taking over twenty-five hundred prisoners.

During the six-month build-up to the invasion, numerous news agencies and armchair generals had warned to expect upward of thirty thousand casualties, but General Rhame thought those estimates were grossly inaccurate. He thought the way to attack this enemy was hard and fast, with overwhelming power. He said:

> I didn't come here to fight fair. I came here to put maximum destruction on this son of a bitch with as few American casualties as possible.
>
> We knew that after the Air Force had taken away Saddam's communications and intelligence gathering capabilities—and damaged his transportation infrastructure, there wasn't much he was going to be able to do to counter us.[10]

Rhame's division was tasked with cutting lanes in the Iraqi defensive line so that the British 1st Armored Division could pass through to continue the attack. Eight lanes were cut over a ten-mile stretch, and then the engineers cut eight more, and widened them all. Through it all were Iraqi fortifications, high berms of sand, oil-filled trenches, minefields, and barbed wire. The engineers used line charges to cut into the minefields, and rakes and blades mounted to tanks. During this operation there were no American casualties although many Iraqis fired from defensive positions only to be destroyed by fire or buried by the clearing plows.[11] "They chose not to surrender," said General Rhame. "They chose to fight. So, we buried them."[12]

At 0800 on February 28, 1991, the war was over when a ceasefire was called. The war was called the "100-hour" war, and the Big Red One had fought through 156 miles of enemy-held territory, destroying 550 enemy tanks, 480 armored personnel carriers, and took 11,400 prisoners. Eighteen of the division's soldiers were killed.

General Rhame's decorations for valor included two Silver Stars, three Bronze Stars with the Combat V, the Legion of Merit, and three Army Distinguished Service Medals.

Robert James Reeves
(U.S. Navy photo)

Robert James Reeves, Senior Chief Petty Officer and Navy SEAL, U.S. Navy: Reeves graduated from Caddo Magnet High School in 1998 and enrolled for one year at LSU before enlisting in the U.S. Navy. After recruit training, he attended the grueling Basic Underwater Demolition/SEAL School, at Coronado, California, and graduated as a SEAL in December 1999. From 2000 to 2004 he served with SEAL Team 5, and deployed with SEAL Team 7 to Iraq in support of Operation Iraqi Freedom.

On April 4, 2010, as an assault team member in support of Operation Enduring Freedom in Afghanistan, he was in an overwatch position on a roof as the main assault force moved to attack a targeted house. He identified an armed enemy sentry patrolling in front of the building, and when the sentry raised his weapon in the direction of the silently approaching SEAL team, Reeves quickly eliminated him without alerting the other enemy soldiers sleeping inside. When the main team had positioned itself at the building and called out anyone inside, the enemy unleashed a tremendous volume of fire. Reeves and the entire attack team returned fire until all of the heavily armed fighters were eliminated.[13]

By 2011, Reeves had become part of the Gold Squadron of the navy's elite SEAL Team 6.[14] On May 2, SEAL Team 6 famously attacked the compound of Osama Bin Laden in Pakistan and killed the mastermind of the 9/11 attacks. On August 6, 2011, SEAL Team 6 was in action again. Army Special Forces were on a mission in the Wardak Province of Afghanistan to engage a high-value target secluded in the mountains. While on that mission, they came under heavy fire, and the success of the operation was in doubt. A team of U.S. Army and Afghanistan military members, along with seventeen members of SEAL Team 6, responded to suppress the opposition forces and assist with the capture and evacuation of the high-value target. But the attempt was unsuccessful, and while returning the team to base, the CH-47 Chinook helicopter, *Extortion 17*, was struck by a rocket-propelled grenade fired from the window of a building. The Chinook exploded and crashed in flames, and Senior Chief Petty Officer Robert Reeves was among the thirty killed in action. There were no survivors.

Reeves was buried at sea off the coast of Virginia, and a cenotaph was placed in Arlington National Cemetery in memory of him. His decorations for valor included five Bronze Stars with the Combat V and the Purple Heart.

William "Bill" Edwards
(LSU Archives)

William "Bill" Edwards, Commander, U.S. Navy: A native of Little Rock, Arkansas, Edwards was a three-year letterman in football in 1939, 1940, and 1941 and was the starting offensive left tackle on the 1941 LSU team. The newspapers at that time called the team the Ole Lou Bengals.[15] The Tigers compiled a 4–4–2 record, and although they finished seventh in the Southeastern Conference, their season highlight was a 0–0 tie played against Mississippi State, which had won sixteen straight games. The game was watched by thirty thousand. The Tigers' triple threat tailback was Leo Bird, who ran through holes in the Mississippi State line made by Edwards and the other offensive line blockers. Bird's accurate punting kept the Maroons deep in their own territory.

After Pearl Harbor, Edwards joined the navy and went to submarine school, and Bird became a naval aviator. Their paths would cross during the war, most unexpectedly. On August 16, 1944, USS *Finback,* a Gato-class diesel-electric U.S. submarine departed from Majuro in the Marshall Islands on her tenth war patrol. She was a formidable weapon, carrying twenty-four torpedoes and a 3-inch deck gun. She was under the command of Lieutenant Commander Robert Russell Williams Jr. and had a crew of six officers and fifty-four enlisted. The first lieutenant of *Finback* was Ensign William "Bill" Edwards.

Finback had a stellar combat record. She would receive thirteen battle stars and be credited with sinking seventy thousand tons of enemy shipping, but as she set to sea on her tenth war patrol, her primary duty was neither to sink more Japanese shipping nor to track enemy warships, but to take up a lifeguard station in the western Pacific and be in position to rescue downed American airman, affectionately known as "*zoomies.*" Returning from bombing missions, many pilots found themselves flying battle-damaged aircraft and had to bail out for survival. Many were then lost at sea for lack of rescue. *Finback* had been ordered to this critical lifeguarding duty near the Bonin Islands, close to Iwo Jima, Chi Jima, and Haha Jima.

On September 1, 1944, a crippled torpedo bomber from the aircraft carrier USS *Enterprise* radioed its position as it streamed smoke from its engine. Constantly losing altitude, the aircraft's crew prepared to bail out. The plane had been hit by antiaircraft fire during a strike on Iwo Jima, and fighter aircraft patrolling the area radioed *Finback* with the location of the bail out, and the sub raced to the site. The lookouts finally spotted three downed airmen bobbing on the surface in their bright yellow flotation jackets. The submarine maneuvered close, and the deck crew heaved out floatation lifelines and pulled them to safety. This initial rescue included Ensign Tom Keene, the pilot, and crewmen Stovall and Dougherty. They became happy members of an elite club: the 504 airmen rescued by lifeguarding submarines.[16]

The following day, September 2, another air strike against Japanese forces in the Bonin Islands was ready to launch from the USS *San Jacinto* (CVL-30), a light carrier deployed in the North Pacific.[17] The target was a radio station on Chi Chi Jima, 620 miles south of Japan and 150 miles north of Iwo Jima. It had been intercepting U.S. radio transmissions concerning aerial attacks and sending early warnings to the targeted areas. With this vital intelligence, the Japanese gunners had plenty of time to prepare for the American strikes, with predictable results. It had to be destroyed, and the aircraft of *San Jacinto* prepared for the mission.

Leo Bird
(LSU Archives)

Flying high fighter cover that day for the bombers was **Ensign Leo Bird (USNR),** LSU's great halfback from the 1941 football team and former teammate of Ensign Bill Edwards, patrolling below on USS *Finback.* Bird's proficiency in the *Hellcat* was every bit as good as his running had been on the football field, and he would eventually be awarded the Distinguish Flying Cross for valor. Leading the strike force of four *Avenger* torpedo bombers of Squadron 51 (VT-51) was Lieutenant (junior grade) George H. W. Bush (USNR). Each plane was armed with four 500-pound bombs, and beefing up the four *Avengers* were eight *Helldiver* bombers from Squadron VB-20—all escorted by twelve F6F *Hellcat* fighters.

Time over the target was scheduled for 0830, and in Lieutenant Bush's final approach his bomber was struck by antiaircraft ground fire. The plane shuddered from the impact, but he shrugged it off and pressed this attack home and released his bombs. Pulling out of the bomb run with his engine on fire and trailing a black stream of smoke, he flew his mortally wounded aircraft out to sea in an effort to get as far away from the island as possible, knowing that capture by the Japanese meant torture and certain death.[18] But smoke soon filled the cockpit and he ordered a bail out. Choking and gasping for air, Bush abandoned the bomber, still close to the island, and bailed out at fifteen

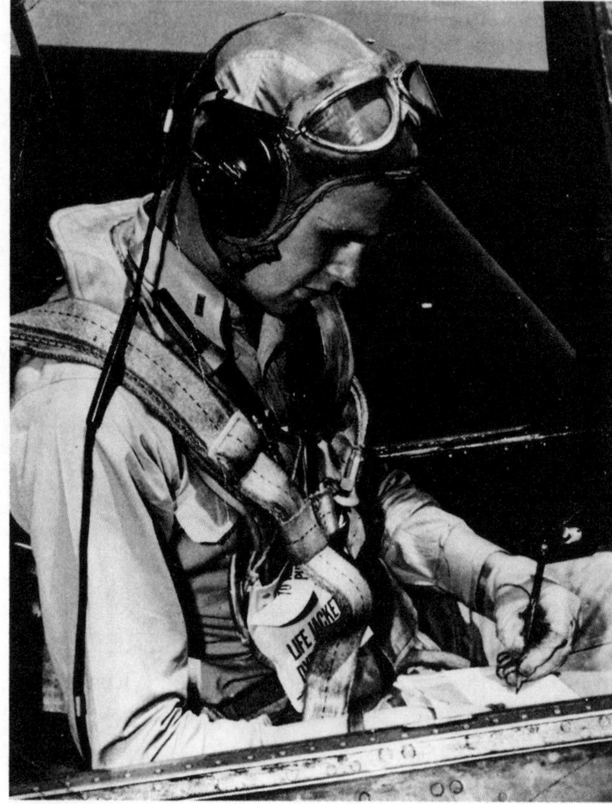

George H. W. Bush
(Courtesy George H. W. Bush Presidential Library)

hundred feet. As he floated to the water, he looked anxiously around for his crewmen and spotted one other descending figure. But the parachute never deployed, and there was no sign of a third parachute.

Finback was on station nearby, and at 0933 received a signal that an aircraft was down nine miles northeast of the island of Minami-Jima. Two F6F fighters flew overhead to escort the sub in, and *Finback* maneuvered around the southern end of Chi Chi Jima. It made sure to remain seven and a half miles from the shoreline, out of Japanese artillery range. At 1130, two hours after the alert, the submarine arrived in the rescue area. Lookouts on their high perches scanned the surface with binoculars and finally spotted a bobbing yellow raft. Lieutenant Bush had inflated his one-man life raft and crawled in, and now drifted at the mercy of the current.

In the distance Bush could see a Japanese boat putting out from the shore and it soon headed for him. Flying overhead was Lieutenant Doug West in an attacking *Avenger*. He promptly strafed the boat and drove it off.

Bush saw *Finback* gliding toward his position and said, "Jeez, I hope it's one of ours."[19] *Finback*'s captain, Lieutenant Commander Williams, maneuvered the 312-foot submarine close to the small raft and flooded the ballast tanks to sink the sub's deck to just a few feet off the waterline. Five of the crew tossed Bush a lifeline and hauled him aboard.

From the conning tower, Bill Edwards filmed the entire rescue with his 8 mm camera. After Bush was aboard the sub, Edwards asked him what carrier he was from. "The *San Jacinto*," Bush replied. "Do you know Leo Bird?" Edwards asked. "I played football with him at L.S.U." Bush pointed to one of the *Hellcat* fighters circling above,[20] then was hustled below, and *Finback* continued the search. The rescue had been completed at 1156, and the search was now for the other missing airmen. At 1236 the sub's radioman announced another rescue signal. This time a fighter had spotted another yellow raft. The position given was confusing. It plotted out to be in the hills of the island of *Haha Jima* to the south. Despite the confusion, *Finback* pressed forward in that direction and continued to request new information on the sighting, but there was none. Finally, an unknown plane broke in and reported that he was circling over the raft, and then another transmitted that he was "west of *HAHA*."[21]

Finback took up a position nine miles west of the island. The report now was that the raft was only a mile and a half off the beach and was being shelled by Japanese shore gunners. Edwards and his fellow officers knew that there was no way that they could go in for the rescue under Japanese fire. At 1505, *Finback* submerged to fifty-five feet—periscope depth. The captain was able to observe the scene and recorded, "Planes in sight zooming a spot in the water 1 mile WSW of MEGANE IWA [island]."[22]

Twenty-five minutes later, the sub spotted the raft. *Finback* surged past the downed flyer, who could only observe her visible periscope. Maneuvering in the confined waters was most difficult, but the sub's log reported, "We twisted around and started stalking him. The pilot hooked on, and we headed out away from the beach. Tried to make two-thirds speed. The pilot had one arm around the periscope and the other around the raft with a bailing bucket bringing up the rear."[23]

All this maneuvering proved too much for the clinging pilot, and he was soon adrift again. Ten minutes later he was back in his raft, hugging the peri-

***Finback* rescue**
(U.S. Navy photo)

***Finback* and rescued pilots (*kneeling, from left:* Ensign Tom Keene, Lt. George Bush, Ensign James Beckman, Crewman Stovall, and Crewman Dougherty; *standing behind George Bush:* Ensign William E. Edwards)** (U.S. Navy photo)

scope, and *Finback* went ahead at two-thirds speed. But the wake from the periscope and the choppy seas filled the raft and again dumped the airman back in the ocean. "Finally, [we] came up to 38 feet to keep him out of the water until at a range of 5 miles away from the beach, we planed up and opened the hatch."[24] It had been four hours since the original report, and the waterlogged flyer was dragged aboard. He was Ensign James W. Beckman, an F6F *Hellcat* pilot from the carrier *Enterprise.*

The rescued airmen happily posed for a photograph with their rescuers on the deck of *Finback.* They now became part of the ship's company and stood watch, and lookout, and manned battle stations. They even participated in two torpedo attacks to sink Japanese shipping and endured the ensuing, terrifying depth charge counterattacks. Fellow rescued pilot Thomas Keene said, "It must have seemed like a dream to George Bush. One minute he was all alone on the ocean, and the next he was on board a submarine being served food in a red-lighted compartment that had music playing."

And the aviators gained the utmost respect for the submariners. "I thought I was scared at times flying into combat, but in a submarine, you couldn't do anything, except sit there." Lieutenant Bush summed it up. "There was a certain helpless feeling when the depth charges went off. I didn't experience that when flying my plane against AA."[25]

⚜ ⚜ ⚜

Commander Bill Edwards retired from the U.S. Navy after a twenty-six-year career. He was uniquely qualified in both submarines and naval aviation, and proudly wore both the Silver Dolphins and the Wings of Gold. He later commanded Attack Fighter Squadron VA-196 and was operations officer on the aircraft carrier USS *Hancock.* He received the Distinguished Flying Cross. He later became Commandant of the U.S. Navy Marine Training Facility in New Orleans.

On January 20, 1989, Bill Edwards and thirty former shipmates of the USS *Finback* attended the swearing-in of President George H. W. Bush as the 41st President of the United States.[26]

Paul Dietzel
(U.S. Air Force photo)

Paul Franklin Dietzel, First Lieutenant, USAAC Aviator: Dietzel was born on September 5, 1924, in Fremont, Ohio. The family moved to Mansfield, where he excelled in athletics and was highly recruited. In 1942 he accepted a football scholarship from Duke University but answered the call to arms and enlisted in the United States Army Air Corps. He and was ordered to Biloxi, Mississippi, to begin training that would lead to his qualification as an aviator.

He received his wings on August 4, 1944, as a nineteen-year-old in Seymour, Indiana, and continued training on the B-24 *Liberator* at Smyrna Army Air Base in Tennessee. But at that time, the new B-29 *Superfortress* was coming off Boeing's assembly lines to replace the *Liberators* as America's heavy bomber. It was a huge aircraft that pilots had neither seen before nor ever imagined. It was thirty-three feet longer, ten feet taller, and sported a 141-foot wingspan that was thirty-one feet longer than anything else flying. Dietzel was in awe of the bomber. "I couldn't believe it. You mean to tell me we're going to fly this thing? It's so sleek. It's beautiful. It's like moving from a Model T to a Rolls Royce!"[27]

He was assigned to the 6th Bomb Group, part of the 313th Bomb Wing, and was deployed to the Mariana Islands on the tiny island of Tinian. The big aircraft began arriving on Tinian in January 1945, just five months after the Marines had seized it from the Japanese defending forces. Navy Seabees immediately began building four eighty-five hundred-foot runways on North Field utilizing crushed coral and converted this remote island into what became the busiest airport in the world. As a twenty-year-old, Dietzel became an aircraft commander in the 40th Squadron of the 6th Bomb Group. His bomber was named *The Banana Boat.*

During the first few months, flying was limited to training missions, but the all-out air assault against Japan began on the evening of March 9, and it would be the tenth mission flown by the 6th Bomb Group. Thirty-two bombers were assigned to the mission, and *The Banana Boat* was one of them. But this mission, code named Enkindle 5, was to be like none previously flown. This was an aerial assault to firebomb Japanese urban areas, and the first target was Tokyo.

General Curtis LeMay, the newly appointed Commander of the 20th Air Force, personally announced the plan to the pilots in their briefing rooms.

The Banana Boat (U.S. Air Force photo)

Dietzel and aircraft crew
(LSU Archives)

The air group's log recorded the event. "A SERIES OF MAXIMUM EFFORT HIGH INCENDIARY ATTACKS WERE TO BE MADE ON MAJOR JAP INDUSTRIAL CITIES. BOMBING ALTITUDES WOULD BE 5,000 TO 8,000 FEET. NO ARMAMENT OR AMMUNITION WOULD BE CARRIED AND THE SIZE OF THE CREW WOULD BE REDUCED."[28]

Dietzel was shocked. "We dropped bombs from 20,000 to 21,000 feet, because that's what the B-29 was designed to do." But at that height, many bombs fell well off of their target due high jet stream winds. "We were bombing the hell out of all the fish at sea," said Dietzel, "but we weren't hitting our targets. General Curtis LeMay said, 'that's nonsense, we're not doing that anymore; we're going in at 5,000 feet,' and we all said, 'you've got to be kidding.'"[29]

At 1853 on March 9, 1945, thirty-two bombers of the 6th Bomb Group roared off of Tinian's runways for the fifteen-hundred-mile flight to Tokyo. They carried 989 five-hundred-pound incendiary clusters, and were over the target area at 0200 on the 10th. "What a scary way to start your tour of duty in the Army Air Corps," said Dietzel. "You know that every B-29 is on the same mission, at the same altitude, and in total darkness. You have no idea where they are, and you're all on the same path. I don't know how everyone missed everyone else. But we were strictly focused on that bomb run."[30] The bombing runs concluded at 0310. Expected enemy opposition was surprisingly light, and no planes were lost. The pilots, after pulling out of their bomb runs, looked down at the resulting conflagration, and one said, "Tokyo was burning like a forest of pine trees."[31]

> Tokyo was made of bamboo [and paper] houses. You talk about burning . . . we came back two nights later and did the same thing—more people were killed by those fire raids than were ever killed by the atomic bombs. It was a terrible thing.
>
> I think we felt sorry that we were killing so many people, but we were in a war. You aren't going to win a war without killing people. War is war. War is hell.[32]

But Dietzel felt that his most dangerous mission was a mining operation around Yawata, at the Shimonoseki Straits separating Kyushu from the main island of Honshu. Eighty percent of the Japanese merchant fleet passed through the straits, which provided the only western exit from the Inland Sea. Admiral Chester Nimitz had pressed the Air Corps to initiate this mining offensive and asked for 150 sorties a month. The entire military brain trust, with the singular exception of General George C. Marshall, the Army Chief of Staff, was in favor of this operation, sensing that it would bring about the downfall of Japan faster than any other operation and it would eliminate the need for a dreaded ground invasion.[33]

It was code named Operation Starvation and was designed to prevent the transfer of food and materiel to Japan, to prevent supply and movement of military forces, and to disrupt shipping in the Inland Sea. It was planned in three phases from March 27 to April 30. Mining Mission No. 1, for Dietzel's squadron, was scheduled for the night of March 27, and two minefield areas had been designated in the Shimonoseki Straits: MIKE, on the western side of the straits, and LOVE on the eastern side.

The entire wing of 105 bombers was scrambled for this operation. Dietzel's *Banana Boat* was among the thirty aircraft of the 6th Bomb Group chosen to lay a string of mines in the target LOVE area.[34] "The most dangerous mission was to mine the waters around the heavily defended Yawata naval base in the strait between the Japanese islands of Honshu and Kyushu," said Dietzel. An earlier attempt to attack Yawata had been a disaster, and veterans of that mission were now part of this mining mission and freely talked about the previous debacle. "We went in to get our briefing," he said. "A hush fell over the audience. They had told us so much about Yawata that we were almost as scared as they were."[35]

The aircraft took off for the twenty-eight-hundred-mile round trip. Each proceeded independently to the target area and arrived in total darkness. But the Japanese searchlights lit the sky and crisscrossed it with intense anti-aircraft fire. Dietzel ignored warnings to keep his speed down so the wind wouldn't rip the parachutes from the mines and bore in for his run. Holding a steady course, he dropped his twelve-thousand-pound string. It consisted of an assortment of both two thousand- and one thousand-pound mines. The parachutes held, and at the end of the low-altitude run Dietzel pulled out and turned his B-29 over Kyushu for the long return flight.[36] For the next thirty days, *The Banana Boat* flew dangerous, low-altitude mining runs. In July, Lieutenant Dietzel and other select pilots were ordered back to the States to train as squadron leaders. While they were there, Japan surrendered after the surprise of the atomic bombings.

The mine laying operation proved to be everything that the planners had hoped it would be. It accounted for 1.25 million tons of shipping during only the last five months of the war and helped bring about Japan's downfall. Its effectiveness was testified to by the Japanese themselves after the war. A Captain Tamura noted at a military conference: "The mine warfare coupled with the bombing raids, prevented our utilizing our war strength, and nullified our plans to the extent of forcing us to abandon them."[37]

After the war, Paul Dietzel pursued a career in college football. He arrived on LSU's campus in 1955 to become the head coach. During his first three years, his coaching record was lackluster. None of those first teams even managed to compile a winning record. But in 1958, he and his team rose to a level that few athletic teams achieve. They became the darlings of college football—amassing a perfect 11–0 season, winning a national championship, and having Billy Cannon named the Heisman Trophy recipient as the best player in the land.

But Dietzel had fielded more than just a great football team in 1958. His leadership skills, carried from his military experience, led him to create a legend. In an era of limited substitution, he created three eleven-man squads to deal with the restrictions of the substitution rules. But so had other coaches. Dietzel was different in that he elevated his three squads to a special uniqueness, and he instilled in each a special sense of pride. When the time came for each squad to take its turn on the playing field, they ran out to the thunderous sound of blaring music, special songs, raucous applause, and ear-shattering screams.

Dietzel gave each squad its own separate identity, and its own source of pride—much like his B-29's special identity. It was not just Plane #64 among the rest of the planes of the 40th Squadron. It was *The Banana Boat!* Dietzel's three squads bore similar, scintillating names: The White Team, the Go Team, and the darlings of them all, the Chinese Bandits. As the third string, the Bandits were, in reality, the least talented of the entire team; but Dietzel told them that they were much more than that. He told them, "Chinese bandits are the meanest, most vicious characters in the world. From now on, that's what you'll be called." And they lived up to his challenge and became part of LSU's lore and captured national attention. *Life* magazine wrote, "No one suggests they are third stringers anymore. Being a Bandit . . . has become even more of an honor than making the starting lineup."[38]

Dietzel continued coaching at LSU through 1961 before departing to coach at Army and the Long Gray Line. But he had created in his 1958 team a legend beloved by every generation of LSU fans. It would take sixty-two years for another LSU team to post another perfect season (13–0), win the national championship, and have the Heisman Trophy recipient, Joe Burrow. In 2023, LSU did not win the national championship, nor was the team undefeated (9–3), but quarterback Jayden Daniels became LSU's third Heisman Trophy recipient.

⚜ ⚜ ⚜

In 1995, while the world celebrated the fiftieth anniversary of the end of the war, the ugly head of revisionism rose to challenge and besmirch the heroism of the American Bomber Command and its heroic pilots who flew the B-29 missions that ended the war against Japan. That end had come without invasion, and without the anticipated 1 million American casualties. For the previous fifty years, Americans had no second thoughts. They thanked God for deliverance from the bloodbath that awaited them on the battlefields of Japan against an enemy determined to die to the man.

Now came the revisionists, ignoring facts, including Japan's long history of barbarism and atrocities against anyone not Japanese. Japan had officially stated that it would fight down to the honorable death of every last person ("100,000,000 people, as one man, will be in active resistance to the enemy").[39]

Leading the leftist revisionists was the head of the Smithsonian Museum, who portrayed the Japanese as victims rather than as the ones that started the war and pursued it with the cruelest of vigor. In the exhibit featuring the B-29, *Enola Gay,* the American flyers were presented as the actual evil doers, and it besmirched the courage, sacrifice, and valor of B-29 pilots like Paul Dietzel. And ABC network applauded this insult. News anchor Peter Jennings, surrounded by what the *Washington Post* called "a stacked deck of revisionist historians,"[40] wrote and starred in a ninety-minute television special. In it he labeled the veterans of the 20th Air Force Association as "bullies" for publicly rejecting the Smithsonian revisionists, and then proclaimed that the estimates of a million American casualties saved by the bombings was simply "enduring fiction."[41] But Peter Jennings and this type of revisionist behavior was nothing new. His ilk has always shuffled across the pages of history, hand-wringing and second-guessing from the comfort and security of quiet offices, salons, and libraries.

The destruction of Japan's ability to wage war with the firebombings and the atomic bombs decisively brought World War II to an end. But the images of destruction were indeed horrific, and the photographed gruesomeness of those attacks set the revisionists off to mindless proclaiming. But Supreme Court Justice Felix Frankfurter recognized this folly and accurately wrote of those who criticize from afar: "The language of the picket line is very different from the language of the drawing room." And California's Judge Ralph Nutter was even more succinct, writing of the bombings and armchair generals:

> I would say the problems facing a combat commander are very different from those of scholars and philosophers in the comfort of a library.
>
> It was a good faith decision on [General] LeMay's part when faced with one million American casualties![42]

Charles C. "Hondo" Campbell
(U.S. Army photo)

Charles C. "Hondo" Campbell, General, U.S. Army: Born in Shreveport, Louisiana, on August 24, 1948, Campbell enrolled at LSU and received his second lieutenant commission through the Army ROTC. His initial assignment was as an instructor at the Infantry Training Command. This began a forty-year army career that can only be described as "the essence of leadership."

Deployed to Vietnam, he served with the Army Advisor Group and as A-Detachment Commanding Officer. Throughout his long career, Campbell commanded at all levels: 2nd Armored Division at Fort Hood, Texas; the 3rd Armored Division in Europe; 2nd Infantry Division in South Korea; 7th Infantry Division at Fort Carson, Colorado; and Eighth Army in South Korea.

In 2007, he took command of Forces Command (FORSCOM), responsible for 750,000 soldiers. His dedication to the individual soldier was second to none. He once said of those serving difficult tours in Iraq and Afghanistan, "I think as a nation we are blessed with young men and women in our ranks who made a choice to serve our nation at a seminal point in our history. They are doing the dangerous and difficult work of a free society, and we are blessed to have them."[43]

General Campbell's legendary leadership would be called to the fore for a difficult decision after the catastrophic Battle of Wanat in the rugged mountains of Afghanistan. On July 13, 2008, nine soldiers were killed and twenty-seven wounded at a patrol base defended by forty-nine soldiers of 2nd Platoon, *Chosen* Company, of the 503rd Infantry Regiment. They came under an intense attack from a 150-man force of well-armed and well-equipped insurgents. In the months leading up to the attack, these insurgents had completely infiltrated and compromised the local police force, and had coerced support from the local population. This allowed the entire town, including the mosque and surrounding buildings, to become both a command post for a surprise attack and a massive, secret arsenal crammed with arms, ammunition, and explosives. The insurgents were, in reality, on the outpost's doorstep.

When the attack came, it was overwhelming. By the time Apache helicopters arrived on the battle scene, 75 percent of 2nd Platoon were either dead or wounded. A following investigation concluded that three senior officers—the company commander, the battalion commander, and the brigade commander—had been derelict in their duties. The army's investigation cited that the ill-fated soldiers had no heavy construction supplies with which to build a base, no potable water, and that no officer had visited the defenders at Wanat until the day before the attack. General David Petraeus, U.S. Central Command, approved the reprimands and even defended them before a Senate hearing.

But General Campbell felt that these reprimands were unfairly flawed. He stepped in and rescinded the letters, clearing the officers' records. It was a tough decision and one that stunned the grieving families of the soldiers killed at Wanat. But having reviewed the findings from the U.S. Central Command-directed investigation, and after interviewing the three reprimanded officers, he could not take the easy way out and sit idly by. He particularly noted some obvious mitigating circumstances that challenged the dereliction charges: the size of the area of responsibility, the rugged terrain, the limited roads and air assets, and particularly the demand that soldiers constantly engage with the local population—always a nest of spies and assassins. These were factors beyond any commander's control, and were a tremendous advantage to an attacking enemy, and a detriment to the defenders. Campbell would write:

> One must understand there is no such thing as a perfect decision in war, where complexity, friction, uncertainty . . . the actions of independent individuals, and the enemy, all affect the outcome of events.
>
> That U.S. casualties occurred at Wanat is true; however, they did not occur as a result of deficient decisions, planning, and actions of the chain of command. The U.S. casualties occurred because the enemy decided to attack the [combat outpost] at Wanat.

Campbell retired from the army in 2010. His decorations include the Army Distinguished Service Medal, Defense Superior Service Medal, (4) Legion of Merit, and Bronze Star Medal.

St. Clair Bienvenu
(U.S. Navy photo)

Lieutenant St. Clair Bienvenu, USNR: When the Japanese attacked Pearl Harbor, Bienvenu left his studies at LSU to join the navy and became an aircraft carrier fighter pilot, flying the F6F *Hellcat* fighter off USS *Langley.* He conducted air strikes all across the Pacific: in the Marshall Islands, New Guinea, Truk, and the Mariana Islands. He was in the sky for the climactic "Marianas Turkey Shoot," where more than three hundred Japanese aircraft were shot down. In a later attack against the Japanese stronghold in the Philippine Islands, Lieutenant

Bienvenu's fighter was shot down, and he floated in the Pacific with a broken back until rescued and evacuated.

After the war he returned to Baton Rouge and continued his life as a successful businessman. But patriotism and history were never far from his mind, and when the call came he was front and center. In 1982, the navy designated three World War II destroyers for use as memorials, and Louisiana congressman William Henson Moore selected the USS *Kidd* to serve as a memorial for Louisiana's World War II veterans. St. Clair Bienvenu served on the USS *Kidd* Commission, then as its President, and as President of the USS *Kidd* Foundation.[44]

George S. Bowman Jr. (U.S. Army photo)

Major General George S. Bowman Jr., USMC: Bowman graduated from LSU in 1936 after an active college career in sports where he was a five-letter man in football and track. He was commissioned in the army through the ROTC but resigned his Army Reserve commission to accept appointment as a Marine second lieutenant on July 10, 1936, beginning a thirty-five-year career in the Corps.[45]

His heroic actions during the Korean War merited the Distinguished Flying Cross. On March 26, 1953, Colonel Bowman served as Tactical Air Coordinator in a massed aerial assault on enemy supply installations in the vicinity of Chinnampo. Flying two attack bombers, he and his wingman reached the objective prior to the arrival of the strike force to conduct a preliminary reconnaissance. Suddenly and without warning they were attacked from above by four enemy jet fighters. Colonel Bowman led his wingman in countering and beating off this attack, and then proceeded to direct the Marines' striking force of jet and propeller-driven fighter bombers in a closely coordinated and highly destructive attack on the enemy installations.[46]

Ensign Rodney Shelton Foss, USN: Born in Monticello, Arkansas, on August 17, 1919, Rodney Foss graduated from Pine Bluff High School. He attended the University of Arkansas and Louisiana State University before enlisting in the navy on September 5, 1940. He was commissioned as an ensign.

In 1941, Ensign Foss was stationed at the Naval Air Station, Kaneohe Bay,

Rodney Shelton Foss
(U.S. Navy photo)

Hawaii, and was assigned to patrol squadron VP-11 on Ford Island. On the night of December 6, 1941, the squadron's crews worked late to install self-sealing fuel cells in the wings of the PBY-5 seaplane aircraft. Foss was the squadron's graveyard shift duty officer that night, and was in the hangar when the Japanese attack struck at 0752 on December 7, 1941. He had been scheduled to be relieved at 0800.

The Japanese began their attack at Ford Island, and the first bombs and strafing fell on the flight line jammed with the seaplanes of VP-11 and VP-12. The PBYs were lined up in two neat rows along the ramp. The attack was devastating. None of the aircraft survived, and the hangar was bombed and riddled with 20 mm bullets and set aflame. Ensign Foss was struck and killed instantly by a 20 mm bullet during the first strafing run, most probably making him the first U.S. casualty in the Pacific Theater, and the first U.S. combat death of the entire war. When the Sunday duty officer, Ensign Joe Smartt, arrived in the hangar, Foss was already dead.[47]

Whitney A. Langlois
(U.S. Army photo)

Second Lieutenant Whitney A. Langlois, U.S. Army (Bataan Death March): Whitney Langlois was born on June 4, 1916, in New Roads, Louisiana. He was commissioned a second lieutenant of Field Artillery through the Army ROTC program at Louisiana State University in May 1940 and went on active duty on August 15, 1940. He completed Field Artillery School at Fort Sill, Oklahoma, and a year later was ordered to the 23rd Field Artillery Regiment of the Philippine Scouts at Fort Stotsenburg in the Philippine Islands, fifty miles north of Manila. He arrived in August 1941.[48]

Four months later, Langlois was among the twelve thousand American and sixty-six thousand Filipino defenders desperately formed into a defensive line to try to hold off the Japanese attack to seize the Philippine Islands after Pearl Harbor. But within a month, the Japanese had captured Manila, the capital, and the American and Filipino defenders had retreated to the adjacent Bataan Peninsula for a last-ditch stand—finally fighting with their backs to the ocean. All was not gloomy, however. They reached the southern coast and were now under the protection of the big guns firing from the Island of Corregidor that lobbed large-caliber shells into the pursuing Japanese. But the Japanese air forces continually pummeled the island.

Finally, on April 9, 1942, with his forces crippled by starvation and disease, Major General Edward King Jr. decided to surrender. Many of the men were disappointed, including fellow Louisianan James Bollich from Lafayette. "We were flabbergasted by the news," he said. "We were surrendering. I could not believe it; all these men still with arms, and now under the protection of Corregidor's big guns, we were giving up. But it was true."[49] The men were told to get rid of their weapons in any way possible.

The Japanese column arrived and soon rounded up the seventy-five thousand-man American/Filipino force at Marivales, on the extreme southern coast. With rifle butts and flashing bayonets, they prodded the men forward to walk in a long column. The Japanese along the road quickly began to taunt and abuse the prisoners. They took watches, rings, wallets, anything they thought might be worth something. More importantly, they quickly took the canteens, emptied them of water, and knocked the hats of the men off, exposing them to the heat of the sun.

"We endured constant beatings, beatings, beatings every time we ran into a new group down the road," said Bollich. In that shuffling column, Lieutenant Whitney Langlois also trudged along. "They hit us with rifles, sabers, sticks and anything else they could get their hands on. We were half starved and many so weak they could hardly walk, but they showed us no mercy."[50]

The men shuffled on for two days without water. Beatings and shootings became routine, but the torture of thirst for water became unbearable. They had passed several artisan wells along the way, and the Japanese splashed and drank from them freely while laughing and joking, but any prisoner attempting to steal a quick sip met instant death. One soldier said, "Many of the men were rushing every well that we came to, despite being shot as they did. There were one or two guards close enough to do the shooting; if ten or twelve went at once, the chances were six to eight would get back without being hit."[51]

After five days of torture, stumbling along in the choking dust, at the point of a bayonet, without food or water, cruelly beaten at the whim of barbaric guards, the Death March survivors collapsed in a place called San Fernando. They had walked over sixty miles. Their final destination was still seven miles away, although that was unknown.

Suddenly some of the exhausted column was herded into rail cars for the

Bataan
(National Archives)

Bataan death march
(National Archives)

final run to Camp O'Donnell. The rest would continue the agonizing march, including Langlois and Bollich. Those who thought the train ride was a blessing quickly learned it was not; and found themselves stuffed like sardines, eighty to one hundred men in a rail car meant to carry thirty. Everyone had to stand, shoulder to shoulder, back to front, without any possibility to sit. The cars were then locked and left in the blistering sun for hours. Many died from suffocation, gasping for air, their lifeless bodies propped up by the sheer mass of the other prisoners.

Camp O'Donnell was straight out of hell. A few spoons of rice were finally given to eat, and a single spigot dripped water—a quart every fifteen minutes—the supply for the entire camp. Thus began the eternal line that inched along, twenty-four hours a day.

Death was the order of every day. Disease became the relentless killer. Dysentery, malaria, and diphtheria turned Camp O'Donnell into a camp of horrors. Desperately sick men shed their excrement-encrusted clothes to crawl outside, naked, to be close to the filthy latrines. The flies blackened the air, and their staggering numbers actually bent the small branches of the trees down. They hung by the thousands on the edges of the thatched roofs because there was no more room in the trees.[52] And the stench was ghastly. But the greatest swarming was around the latrines. "To die of dysentery was the most horrible way to go," said James Bollich. "They had to stay outside near the latrines because with dysentery, you pass blood and mucous every twenty minutes. There is no way to keep from it."[53] Those who died at the latrine wore a shroud of flies the following morning, to cover their nakedness. A feeble attempt was made to isolate these desperately ill men from the rest. It was called a hospital, but that was just a name since it had no ability to treat the patients or diseases, and it soon became known a "Zero Ward," because that was your chance if you checked in.[54]

Those still physically able were put to work, and in short order the Japanese began transporting them to the far reaches of their empire as slaves in their war industry, a procedure completely condemned in international law. This movement was done by merchant vessels, and the barbarity conducted on these ships quickly earned them the title of "hell ships." "The holds were floating dungeons, where inmates were denied air, space, light, bathroom facilities, and adequate food and water—especially water. Thirst and heat claimed many lives in the end, as did summary executions and beatings."[55] Fifteen hundred prisoners died in these dungeons.

By the end of the war, Japan was using 134 of these floating torture chambers, and they were lucrative targets for British and American air forces hunting them as merchantmen, unaware that they contained Allied prisoners of war. These attacks resulted in the additional deaths of nineteen thousand Allied prisoners, either killed in the attack, or shot attempting to escape, or drowned chained in the holds.[56]

Such was the case of the *Oryoku Maru,* which left Manila on December 13, 1944, with 1,620 prisoners of war, including 1,556 Americans packed in the three cargo holds. Lieutenant Whitney Langlois was in the bowels of the forward cargo hold.[57] As the ship neared the naval base at Subic Bay, U.S. Navy attack aircraft from the carrier USS *Hornet* swooped down on the unmarked *Oryoku Maru* furiously attempting to evade. On December 15, after two days of relentless attack, the ship sank; and 270 died. Some died from suffocation or dehydration; others not killed in the attack drowned or were shot while escaping.

Lieutenant Langlois survived to, somehow, slip over the side and keep his head above water several hundred yards off the shoreline. He was eventually rounded up with other survivors and held under guard on open tennis courts. Again, there were no sanitary conditions, no water, and they were subjected to severe mistreatment. More men died.

They were then sent to San Fernando for further movement, but while in San Fernando fifteen of the weakest men were loaded onto a truck with the promise of medical treatment. The medical treatment turned out to be a quick ride to a nearby cemetery, fifteen beheadings, and a dump into a common grave.[58]

On December 21, the ever-diminishing numbers of American prisoners were transported from Subic Bay back to San Fernando and confined in a jail. On December 27, Langlois with a thousand other survivors were reloaded onto the Japanese ship *Enoura Maru* for the continued voyage to Japan. The other three hundred survivors boarded the smaller *Brazil Maru*. Both ships sailed to Taiwan and arrived on New Year's Day 1945—minus twenty-one more men who died at sea during the trip and were thrown overboard.[59]

On January 8, the two ships underwent repair while at anchor, the prisoners still confined, and suddenly the American aircraft attacked again, like lions on two wounded zebras. "Planks, hatches, and other debris flew through the air. Some men were pinned by the hatches. One floor gave way, dropping prisoners thirty or forty feet below. When it was over, only three doctors remained to care for seventy-five injured. Thirty-five dead were placed in a pile."[60] But there was more. The main damage was in the forward hold, where a bomb fragment left a gaping hole. "There were mangled Americans, some 300 of them, piled three deep and pinned down with large steel girders and hatch covers."[61]

On January 14, the survivors of this second sinking, including Lieutenant Langlois, were crammed yet again into the holds of the surviving *Brazil Maru* and sailed for Moji, Japan. It arrived two weeks later, and the prisoners finally disembarked, little more than walking skeletons. Of the 1,556 Americans that had boarded *Oryoku Maru* in Manila six week earlier, fewer than 500 were alive.

Whitney Langlois was trucked off to Fukuoka POW Camp #1, in Fukuoka city, on the southern mainland of Kyushu. He arrived in February to continue his now almost three-year fight to survive in a world of utter barbarism. But Divine Providence smiled down upon him sometime in April, and on April 27 he arrived at the Jinsen POW Camp in Korea. This camp was first occupied by the British and the Australians in 1942, and now 140 Americans arrived from Fukuoka, Japan. Most had been captured on Bataan and Corregidor. If ever a prisoner could think that he had stepped out of hell and into paradise, this was it.

Conditions at Jinsen were by far the best of any camp the Americans had ever been confined in. The camp consisted of several one-story Japanese army barracks. The prisoners slept on raised wooden platforms, about eighteen inches off the floor, on straw mats. The latrines were located fifteen feet from the living quarters, with eight cubicles in each latrine, and each cubicle had a straddle hole. The bathhouse was a separate building, and the prisoners were allowed to bathe frequently. Hot water was available and there was a cold-water faucet. Wooden buckets were provided to douse the individual who soaped up.[62]

Buckets of food were brought out of the kitchen to the barracks. The basic food rations consisted of a bowl of soup and a small bowl of rice in the morning and evening, a bun, and a bowl of soup at noon. Most of the vegetables came from the camp garden. Occasionally a small portion of fish would be added to the diet. The prisoners were issued one Red Cross food parcel for two men every other Sunday.[63]

And one can only imagine Langlois's astonishment when he actually got paid a prisoner's wage: 70 yen a day, with deductions of course for rations and quarters. No prisoner was permitted to have in his possession more than 50 yen at any one time, and the excess was placed in a postal savings account to be redeemed at the end of hostilities.[64]

On September 8, 1945, Camp Jinsen was liberated. Langlois arrived back in the United States on October 16, 1945, and was hospitalized to recover from his injuries before being honorably discharged a year later on September 24, 1946. In 1947, he returned to Louisiana with his family and raised cattle and chickens and grew an enormous garden. He said that neither he nor his family would ever be hungry or thirsty again.[65]

His decorations include the Bronze Star Medal, Purple Heart, Prisoner of War Medal, Prudential Unit Citation (3), and Combat Infantryman Badge.

Pontoon bridge over the Rhine
(National Archives)

Charles William Hair Jr.
(LSU Archives)

Colonel Charles William Hair Jr., U.S. Army: Charles Hair joined the 88th Engineers as a second lieutenant in July 1941, shortly after graduating from LSU, where he had completed ROTC. The battalion was formed at Camp Beauregard, near Alexandria. On August 1, 1944, he commanded General George S. Patton's Third Army's 88th Engineer Heavy Pontoon Battalion and made fifteen river crossings as Third Army rolled through France, Luxembourg, and Germany. Hair's battalion built its first combat bridge on August 25, 1944, crossing the Seine River at Montreau, France.

In late March 1945 they were on the west bank of the Rhine River, and it was Hair's 88th Engineers that threw the pontoon bridge across it in record time. On March 24, Patton crossed into Germany.[66] Colonel Hair was active in the Army Reserve until his retirement in 1971. He was instrumental in bringing the USS *Kidd* to Baton Rouge and establishing the Louisiana War Memorial, for which he served as chairman.[67]

Colonel Gordon L. Jenkins, USAF: At LSU, Jenkins served as Cadet Wing Commander of the two-thousand-man Air Force Reserve Officer Training Corps. He graduated with a degree in economics because it seemed the easiest way to pursue his dream and that was to fly jets. He graduated and was commissioned in January 1965, and received his wings in 1966 and completed

Gordon Jenkins
(U.S. Air Force photo)

advanced training at the Tactical Fighter School. He was deployed to Thailand at Takhli Royal Thai Air Force Base [RTAFB] as an F-105D *Thunderchief* fighter pilot with the 501st Tactical Fighter Squadron. His standard mission was to attack targets in North Vietnam employing his 50,000-pound, Mach-2-fast, fighter/bomber with its six 750-pound bombs. He became one of the most decorated first lieutenants in the air force and was awarded two Silver Stars, three Distinguished Flying Crosses, and fifteen Air Medals.[68] He flew one hundred combat missions over North Vietnam in Operation Rolling Thunder, thirty-seven of them over Hanoi and its fierce anti-aircraft fire. He spent twenty-one years in the service, retiring in 1986.

Marvin J. Roberts
(LSU Archives)

Captain Marvin J. Roberts, U.S. Army: Roberts briefly attended LSU before enlisting in 1961. He was selected for Officer Candidate School at Fort Benning. After commissioning, he was sent to Vietnam with the 101st Airborne Division, and on June 20, 1966, was awarded the Silver Star for heroism under fire. In 1969 he was the Commanding Officer of Company B, 1st Battalion, 11th Infantry Regiment, 5th Mechanized Infantry Division. On March 27, 1969, his company engaged a strong enemy force at Cam Lo in Quang Tri Province.

Captain Roberts was wounded while leading his men against the entrenched enemy but still managed to direct artillery and tactical air strikes as he moved through the ranks shouting encouragement and giving directions. As the attack neared the crest of the hill, an enemy grenade bounded into the company command group. Roberts fielded it and hurled it back. But concealed enemy machine guns raked his assaulting force and inflicted numerous casualties. Realizing the deadly threat of those guns, Roberts drew his pistol and charged up the hill and hurled a grenade into the position, silencing the enemy weapons. He was mortally wounded and died the next day on the hospital ship *Repose.* He received the army's highest award, the Distinguished Service Cross.[69]

Walter S. Savage Jr.
(U.S. Navy photo)

Ensign Walter S. Savage Jr., USN (SC): Savage received his LSU degree in commerce in 1938 and received a commission in the Supply Corps of the U.S. Navy. He was serving as Assistant Supply Officer on board the battleship *Arizona* at Pearl Harbor on December 7, 1941, and was killed in action on the first day of the war when a Japanese bomb penetrated into the battleship's magazine and *Arizona* exploded. Later, the destroyer USS *Savage* was named in his honor.

Wiltz Paul "Flash" Segura
(Courtesy Air Museum)

General Wiltz Paul "Flash" Segura, USAAC: Segura graduated from New Iberia High School and attended LSU in 1940 before entering the Air Corps in 1942. He joined the famed 23rd Fighter Group of the 14th Air Force, which had been organized with members of the American Volunteer Group known as Chennault's *Flying Tigers*. During his World War II tour of duty in China, Segura flew 102 combat missions, destroyed one Japanese bomber and five fighter aircraft, and was credited with damaging three more. He was shot down twice by ground fire, but each time parachuted to safety and successfully evaded enemy capture behind the lines.

Twenty years later he was in Vietnam at Bien Hoa Air Base. Beginning in November 1965, he assumed command of the 3d Tactical Fighter Wing and flew more than 125 combat missions in the F-100 *Super Sabre* and F-5 *Freedom Fighter* aircraft. He was awarded the Legion of Merit, three Distinguished Flying Crosses, eight Air Medals, and the Purple Heart. He logged more than 6,500 flight hours and retired after twenty-eight years of service.

William Preston Johnston, Colonel, CSA: Johnston was born in Louisville, Kentucky, in 1831 and graduated from Yale in 1852. On April 19, 1862, he was deemed unfit for field service in the Confederate Army because of a long, severe illness. Just thirteen days previously, the Confederate Army of Tennessee had fought a significant battle at Shiloh under the command of his father, General Albert Sidney Johnston, who, at the decisive moment of the battle, was struck down and killed.

William Preston Johnston
(Public domain)

Now President Jefferson Davis, presumably with great respect for his father, offered William Johnston a commission as aide-de-camp on his staff, with the rank of colonel of cavalry. He would spend the rest of the war with Davis and, after Lee's surrender at Appomattox, was captured with Davis at Irwinville, Georgia—and imprisoned for several months at Fort Delaware. In 1880, Colonel William Preston Johnston became the fifth President of LSU, serving a four-year term until 1883. He then resigned to become President of the newly designated Tulane University in New Orleans.

Ronald J. Drez
(U.S. Navy photo)

Captain Ronald J. Drez, USMC (the author): A multi-book published LSU Press author, he graduated from Tulane University in 1962 and commissioned a second lieutenant in the United States Marine Corps through the Navy ROTC.

On December 6, 1968, he was the Commanding Officer of Company H, 2nd Battalion, 5th Marines, during Operation Meade River in northern Quang Nam Province, Vietnam, southwest of Danang. It was the largest cordon and search, helicopter-borne operation ever conducted during the Vietnam War. The cordon encircled fourteen square miles of a notorious enemy hangout, appropriately called Dodge City.[70]

It began on November 20, 1968, and Company H had little to do except establish several blocking positions along a road and a small river, twenty yards wide, to ensure that the enemy trapped on the other side did not escape. By early December, the battle was in its final stages with the trapped enemy fighting for its life. No Marines in the blocking position were complaining. The sounds of battle were all on the other side: the artillery and naval gunfire bombardments, the air strikes, and the machine gun fire were relentless. And all that was to support two battalions of the 9th and 26th Marines slugging it out with the encircled 36th NVA Regiment. Drez's Marines could even hear loudspeakers from the Psychological Operations teams broadcasting to encourage the trapped enemy to give up; there was no way out. "North, south, east, and west, you are completely surrounded, and the circle is getting smaller. Today you cannot go a thousand meters in any direction. Tomorrow, will you be killed in your bunker? Tomorrow, will your legs be blown from your body and will you die in a hole in the ground far from your home?"[71] The broadcast prattled on, but no one surrendered. "The enemy troops were unimpressed." The 1st Marines reported, "They chose to fight."[72]

But all of that had nothing to do with Company H. The operation was coming to an end. The 2nd Battalion, 5th Marines had already been ordered to prepare for a helo lift to extract it from its blocking positions and return its companies to the base at An Hoa.[73] "Captain Ronald J. Drez' Company H, on the battalion's southern flank, waited for the lift. After 15 days of what had been for them, a very unexciting operation [the Marines] were anxious to return to base. They sat eating C-Rations and idling away the time until the helicopters arrived."[74]

But within moments, everything changed. The battalion commander informed Drez that the arriving helicopters would not be lifting his company back to the base with the rest of the battalion (Companies, E, F G), but directly into the assault line just eight hundred meters on the other side of the river. The choppers were shortly there and lifted the company for what "Drez characterized it as 'the shortest tactical airlift in history.'"[75]

Minutes after takeoff, the Marines were again down and scrambling to extend the left flank of the 3rd Battalion, 26th Marines. Moments later, another satellite Marine company plucked out of the cordon blocking force also choppered in to further extend the left flank, and it immediately came under a hail of ground fire. Within fifteen minutes, the new battle line had become a formidable force of five rifle companies; a thousand Marines abreast in a continuous, regimental-size attacking line. The battalion call sign was "Barkeep" and Company H became "Barkeep Hotel."

On December 7 the frontal attack began, and Barkeep Hotel was quickly engaged by a determined enemy fighting from dug-in positions in a cemetery. The official Marine Corps history recorded: "North Vietnamese troops fighting from two pagodas laid down heavy fire. Much of the ground was under water, forming a quagmire through which the Marines were unable to maneuver. . . . Under the intense fire, the attacking companies recovered their wounded only with great difficulty."[76] Early in the morning, the wounded and dead were evacuated, the enemy had fallen back, and Company H pressed on.

At 1120 on December 8, the bulk of the enemy, the 3rd Battalion of the 36th NVA Regiment, defended from a final line of fortified positions yet unseen by the approaching Marines. It would become known as the "Northern Bunker Complex." "Captain Drez surged his 180-man company forward in a classic assault line, but the enemy was waiting for him. When Company H reached a rice paddy a few hundred meters from their starting point, Communist troops hidden in a tree line suddenly opened fire, trapping Marines in the paddy. For 30 minutes, the Marines returned fire individually, then began moving in small groups toward a large bunker which appeared to be the lynchpin of the Communist defenses."[77]

Drez called in air support, and multiple flights of navy, Marine, and air force aircraft, along with a low-flying army gunship, provided close air support. But "because of the proximity of [other] Marines, the aircraft had difficulty attacking targets without endangering friendly troops. In one instance a napalm bomb impacted directly on Company H [twenty yards from Drez] but miraculously bounced away before detonating . . . but neither the airstrikes nor mortar and 3.5″ rocket fire overcame the enemy resistance."[78]

When the smoke cleared, the bunker still stood menacingly in front of him, blasting his Marines as they attempted to surround it. Drez requested support from a South Vietnamese Armored Personnel Carrier that was armed with two 75 mm recoilless rifles that could make short work of the bunker by firing a round into its side-facing openings. The APC rumbled into the battle area but then did nothing but sit and watch. No amount of waving or cursing or cajoling served to budge it from its "spectator's" position. Whether they were out of ammunition or just didn't want to get involved was never known. What is known: it never fired its weapons.

Finally, at wit's end, Drez summoned his combat engineer, PFC Michael A. Emmons, to crawl to him, and they began to collect every form of explosive from the Marines pinned down around them. Quarter-pound sticks of C-4 and hand grenades were tossed to them, and Emmons molded the C-4 explosive into a large silly-putty ball. Fifteen hand grenades were then embedded into the putty-like mass. To top off this explosive concoction, Drez and Emmons stuffed the warheads of two 3.5" rockets into their satchel with "a five-second fuse,"[79] wrote Marine Corps historians. The 3.5" warheads alone each had the hitting power of a 105 mm howitzer.

"With the assistance of another Marine, they [dragged] the satchel charge to the top of the bunker where Drez lit the fuse and Emmons flipped the charge through the embrasure. The others ran, but Emmons momentarily remained."[80] When he finally took off, he almost cleared the top of the bunker on a dead run when the massive blast went off. It tossed him into the air like a rag doll and dropped him into the slop of the rice paddy. The charge ripped

Company H—Drez standing third from right. (Courtesy USMC Barry Broman)

the bunker apart, projecting the bunker's reinforcing railroad crossties and sections of steel rails fifty feet into the air.

"The blast smashed the bunker killing all but one of the five NVA inside," reported Marine historians. The fighting all around that bunker had been furious. "The Marines reported 39 enemy dead and 1 prisoner. The combat on December 8 was so intense that some senior Marines said that it was 'the fiercest fighting they had ever seen.'"[81] Strewn in the wake of the Company H attack from December 6 through December 9 were fifty-five dead NVA and forty-three of their weapons. Emmons was awarded the Silver Star.

But Company H did not escape the fierce fighting in Dodge City unscathed. Seven of its Marines were dead and another twenty-one wounded and evacuated. But the enemy's defenses were shattered, the key positions blown up, and the Northern Bunker Complex was no more. The operation ended the next day.

Part of Captain Drez's citation for bravery on December 8 read: "Disregarding his own safety, Captain Drez skillfully maneuvered his unit to an advantageous position and directed the fire of his men against the enemy. . . .

Early in the morning, Captain Drez organized a small reaction group and led an aggressive assault against a heavily fortified enemy bunker. Ignoring the enemy rounds impacting nearby, he continued to advance until his men overran the hostile position and silenced the enemy fire." His awards include two Bronze Stars with Combat V, the Vietnamese Cross of Gallantry with Silver Star, Combat Action Ribbon, and the Presidential Unit Citation.

Epilogue

There is a moral to every meaningful story. This story of military service by Louisiana State University dating from 1860 continues to the present day. The moral of the story is that the history of the world has always shown that a nation must be prepared to defend itself and its way of life, as detailed by events presented in this book. LSU has done its part and continues to do so today in a manner that few universities can equal. The agenda of university President William F. Tate IV has a five-point academic program, one of which is "Defense." That means LSU will continue to develop top-tier basic research programs, build its new cybersecurity department into a world-class program, and, of course, continue and improve its military through education and development of leaders for the future. The Long Purple Line will continue.

APPENDIX

LSU Students Killed in Action

CIVIL WAR

Seth Andrus
Alexandre Arcenaux
Wiley M. Barrow
Albert Bringhurst
George Brown
Charles L. Comes
George Compton
Philogene H. Couvillion
Jacques Dupre
Samuel Ruffin Gray
Francis Marion King J. W. Mundy
William Hickman Ringgold
Lawrence Sexneider
William G. Swilley A. M. Tauzin
Robert Wilkerson
John D. Workman

WORLD WAR I

Milton W. Adams
Leslie Philip Backes
Lawrence Edward Brogan
David Jenkins Ewing Jr.
John F. Goodrich
Ike Hahn Gottlieb
James Oliphant Hall
Henry N. Huck
Leslie Carl Hunt
John Seymour Joseph
Alan Louis Labbe
David Thompson Land Jr.
Ireanus J. Lietemeyer
Philip John McMahon
Lewis Hypolite Martin
Allen Loughery Melton
William Digby Morgan
Cecil Anthony Neuhauser
Jasper Joseph Neyland
David J. Ory
Walter Asbury Phillips
Maurice Joseph Picheloup Jr.
Thomas James Powell Jr.
Duane Horton Rutledge
Julian Bowles Sanford
Wilburn Edward Scott
Stuart D. Simonton
Charles N. Singletary
Henry Ras Thames
Charles P. Willis

WORLD WAR II

George M Abraham
Richard H. Alexander Jr.
Jules B. Allain
Leroy J. Allain Jr.
Edward Mark Allenberg
Jesse Carl Allmand Jr.
Joseph Maxwell Anderson
Wilson Preston Arnaune
Leslie R. Arnold
Louis Wilfred Arpin
Whitney P. Aucoin Jr.
Richard S. Averitt
Harold J. Ayme
E. Watson Ayres
Aubrey Bacon
Al D. Baggett
Charles Adolph Bahn Jr.
Venoy Baker
John Ainsley Ball
Samuel A. Barkoff
Harold Eugene Barlow
Robert Lawrence Barnes
William Preston Barnes
Bob Wade Barnett
Charles L. Barnett
James R. Barrow Jr.
Bennett B. Battaile
Charles E. Beal
Thomas D. Becker
Carlh Becnel
William Douglas Beene
Arthur Behrens
Bernard G. Bel
Marion F. Bell
William A. Bell
Billy Benjamin
P. O. Benjamin Jr.
Arvis Reldon Bennett
Louis Ignanus Bentz
Justin Bergeron
Merrill Bernard Jr.
Oswald L. Bernich
Bernard Francis Berry
Richard A. Blach
Jean Blackburn
Lal C. Blanchard Jr.
Leslie P. Bordelon
Stephen Andrew Borne
Joseph C. Bostick
Lewis Burton Bostwick
Rudolph Boudreaux
James Elstner Bourgeois
Simeone Alex Box
James B. Breathwit
Alex R. Breyer
John M. Brizzard
Frank Brocato
Edward T. Brodnax Jr.
William Earle Brookins
Kermit Johnson Brooks
George O. Broussard
Kirby M. Brown
Leon H. Brown
Hal L. Bruner
Walter Clyde Buaas
Melvin Buckner
Ernest J. Burandt
Austin W. Burdick
Marion E. Burnett
Walter Burnett
Leonard E. Burton
Young Bussey

Robert Dabney Calhoun Jr.
Irving E. Campbell
Neri Philip Cangelosi
William Henry Canty
Ernest E. Cardona
John Mario Carimi
John Eathan Carmichael Jr.
Warren B. Carriere
Adrian Phillips Carruth
Charles A. Carruth
William Cazort
Leslie C. Cazzell
Bert L. Cenac
Alvin W. Chaudoir
John Newton Chambers
Claude Howell Champagne
Edward Chase
Thomas A. Chastant
David Chin-Bing
Francis Xavier Clark
Wilton G. Clay
George W. Cline
H. R. Clugston
Henry L. Cockerham
Loy Grady Coffee
James H. Cooney Jr.
Jack E. Copeland
Eldridge Courrege
Milton M. Coverdale
Kendall H. Cram
James Alex Crawford
James S. Crawford
Milton Hatcher Crawford
William B. Crowson Jr.
Thomas W. Croxton Jr.
Paul Montgomery Daily
Bruce Darsey
Justin Darsey
John Warren Daspit
Percival R. Davidson Jr.
Thomas Decker
Hubert Francis Delhaye
Lloyd Kenneth Demoss
Amos J. Derouen Jr.
Pierce Didier
William Hawkins Dillon Jr.
Allen Edward Dimmick
John Allen Dixon
Francis C. Dolese
Stanislaus A. Donaldson
David Doughty
George H. Douglas Jr.
Aubrey E. Drake Jr.
Fortune Anthony Dugan
Harry Peck Dugas
Elbert L. Dukate
John D. Duplantis
Roy Samuel Dupree
Albert P. Dyer
Wilbur L. Edgerton
John Clarpha Ellender Jr.
Leo Elliot
Cyril Marcus Ellzey
John Charles Eskew Jr.
Frank E. Evans
Lawless Falcon
Nolan A. Fanguy
David Samuel Felsenthal
Hillard S. Fenlaw
Bowman Fetzer
Arthur Edward Fielder
Raymond H. Fleming Jr.
William P. Foil
Thomas Walton Fort
Luther Sexton Fortenberry
Rodney Shelton Foss

Charles A. Fourrier
Bert Fowler Jr.
Alonzo P. Francis Jr.
Julian B. Francis Jr.
John W. Franklin
Charles H. Freeman
Morgan S. Freeman
Philo French
Chester Sebastion Fresh Jr.
W. W. Freshwater
Cleitus R. Garrett
Edward T. Gatto
James D. Gauthier Jr.
James Waldron Geagan
Benjamin D. Gibbens
James Irving Gibbs
Richard C. Gibbs Jr.
Dewey Larry Gilbert
Harry Gilbert
Henry L. Gilbert
Gus N. Girlinghouse
John Golden
Charles Oswald Golson
John Blanton Goode
Franklin C. Gough Jr.
Howard Herman Gragon
Boyd Sutcliff Grant
George A. Gray
John Leslie Green
Wingate Green Jr.
Harold Greenberg
Marvin Greenberg
Archie J. Grefer
A.W. Grillet Jr.
Kenneth Russell Grundman
Levy A. Guidry Jr.
Marion Guthrie Jr.
Arthur Cramer Haas
John B. Hammatt
L. D. Hammonds
Vic R. Hammonds
John D. Hammons
Selser R. Harmanson III
R. S. Harper Jr.
Allen C. Harris
Austin C. Harris
James A. Harris
John E. Harris
James Edward Harris
Robert Allen Harris
Charles R. Harrison
Arthur C. Harvell
Travis H. Heard
Willie Stokes Heard Jr.
Andrew Cleve Heath
Otis Charles Hebert
Thomas Eads Hebert
Earl Heckler
Alva Eugene Hemming
Alvin C. Henderson
David Joseph Henderson
John Morton Henderson
William W. Heyerdale
Gerald Hightower
Frank H. Hilderbrand
S. Clarence Hixon
Woodrow Wilson Holland
Marvin P. Hollander
James Drayton Holston
Emmett J. Holt
John Benjamin Holton
James Bernard Hopper
Jeff C. Hornsby
Albert Taylor Hughes Jr.
Sidney W. Hugston
Frank Bateman Hulsizer

Compton R. Hummel Jr.
John E. Hunley
Robert Rand Hunter
Stephen S. Hunter
Donald William Hurd
George C. Hurd
Linwood P. Ingram
Nelson H. Irwin
Harold C. Jackson Jr.
Sidney Jackson
Sidney W. Jacobson
Charles E. Jenkins
George W. Johnson
James E. Johnson
Louis Albin Johnson
William Johnston
James Garland Jones
Roger Claude Jones
Thomas Ray Jones
William R. Jones
Clyde Jordan
Paul S Julienne
Andrew Keenan Jr.
Thomas David Kennedy
Joseph Adam Kernan Jr.
Ben H. Kerr Jr.
Charles Willis Kessler
Bryan L. Kethley Jr.
Douglas M. Kilpatrick
G. B. Kimbrow
William Jack Kircheval
Saul Kokotovich
Robert L. Knox
Joseph L. Lacombe
Edward Lacour
Lewis L. Landon III
Larry Daniel Landry Jr.
Harold Edgar Lane
Cyril I. Lankford
Germaine Catheríne Laville
William Philip Lay
Joseph Howell Laycock
Dabney Hull Lea
Noel Fairfax Learned
William Percy Leeke Jr.
Arthur Andre Lejeune Jr.
Francis Lemoine
Odie Lawrence Lemoine
Hugh L. Lesher
Melvin Erwin Levin
Marcy Hartman Levy
John Robert Lewis
Ray Lewis
David M. Lewy
John Milton Lindholm
Henry W. Liner Jr.
Warren Russell Lobdell
John Alfred Lofland
Louis Albert Lorio
Edward V. Loustalot
Don Sterling Loyd
Patrick D. Lyons
George M. Macnichol
Ford Raines Maddux
Earl Clifford Mallette
James Hubert Marmande
Robert Marshall
Ivan L. Martin
James Bicom Martin
Robert W. Martin
Albert Mason
Fred Joseph Masset
Varner Erwin Mathews
Cleveland Charles Matt
Robert D. Maxent
William Mabry Mayfield

Samuel C. McArthur
Hugh McCall Jr.
Burton Grover McCallum
Martin L. McCoy
Marvin Roy McCoy
Ray G. McDonnell
Percy Edward McKay
Robert G. McKay
Charles Felder McMichael
John C. Melsheimer
Ralph A. Metz
Jack Davis Middleton
Robert O. Middleton
Antoine J. Miller
James H. Miller
John C. Miller Jr.
Marshal H. Miller
Donald W. Milliren
Walter A. Minsch
Ralph R. Mix
Alvin Joseph Monlezun
Gibbs Allen Monrose
Edmond Montagano
Dan Woodrow Moore
Clinton W. Mornhinveg
Marshal W. Mortimer Jr.
Abram Hugh Moss
Louis O. Moss
Floyd John Mouton
Bluford B. Mullins
James G Murchison Jr.
Raymond Thomas Murphy
William Paul Nabours
Kenneth T. Neibaier
Eugene. M. Nettles
Dorothy Mae Nichols
Ralph Nix Jr.
W. O. Noble
Herschel Norton
John W. O'Bryan Jr.
Lester Cecil O'Neal
Henry N. Oates O. T. Oden
Wiley G. Odom
James C. Oliver
Wallace K. Olmstead Jr.
Ralph Papizan
Robert W. Parker
George Kenneth Pavy
Allen J. Perkins
Jules F. Peytral III
Jack O. Philley Sr.
King W. Pipes
Charles S. Pitcher
Joseph L. Pittari
Leslie Earle Poe
Charles E. Polette
William A. Pondrom
Newby H. Pope Jr.
George Shelby Powel
James L. Powell
E. L. Pratt
August C. Prechter Jr.
Norman Price
Theron Dan Price
Edgar Harrs Pritchard
Basil George Rains
Louis T. Ranson
Houison Griffith Reed
Billy Register
Frank Andrew Reid
Frank Burrus Reid
John A. Reinken
Madison E. Reisor
Henry Philip Repak
Ellis J. Resweber
Charles Warren Reynolds

Maurice H. Rhae
Lyman Francis Rhodes IV
Laymon A. Rice Jr.
John A. Richardson
Charles Thomas Rieder
Lloyd Riff
James E. Risher
Merwyn Reginald Robert
Ray M. Roberts
Howard S. Robinson
James L. Robinson
Woodrow G. Robinson
Floyd Rockhold
Peter B. Rodrigue
Jules S. Rosenthal
Morris D. Rosenthal
Beamon Dewitt Ross
Francis B. Rolbique Jr.
Eugene Bertrand Roux
Carroll St. Amant
Joseph T. Salemi
William A. Sample
L. Cary Saurage
Carroll Conway Saux
Walter Samuel Savage Jr.
Carl John Savoie
Raymond F. Schroeder
Anthony Sconza
T. Aubrey Scott Jr.
William K. Seago
William Sellers
Roy E. Sentilles
Arthur Darrel Shamp Jr.
George Sharpe
David Earl Shaver
John Gilbert Shea
Carl J. Shetler
Frederick Shelton
Thomas C. Shields
Raymond T. Siegel
Eugene E. Simpson
Sidney Lee Sims Jr.
Stewart Slack Jr.
Torbert Slack Jr.
Joseph E. Slattery
Edwin Bomer Smith
Harry B. Smith
Jack Hughes Smith
Other Forrest Smith
Paul Matthew Smith Jr.
Sidney Forsyth Smith
Robert E. Smitherman
M. G. Snell
Glenn William Soap
Walter T. Staffort
Harry J. Stahl
Robert P. Stallings
Maurice D. Stark
Willard F. Starns
Frank Julius Stehr Jr.
Kenneth Steward Jr.
Marshall Neil Stickel Jr.
Channing Stowell Jr.
Raylon L. Stokes
Reilly Stonecipher
H. Wallace Stopher Jr.
Edward M. Sturges
Victor J. Sutker
Max. Tatum
Joseph P. Texada
Marshal A. Tharpe
Robert Donald Thaxton
Charles S. Thirlkeld Jr.
Norman Karlson Thomas
Glenn Taylor Thompson
Henry Ellison Thompson

Hilton O. Thompson
Howard Thompson Jr.
Mason O. Thompson
Willam Thornall
W. I. Thurman Jr.
Roberta L. Treme
Harry Lorraine Trowbridge
Artemus W. Turner
Billy A. Turner
Garland B. Turner
Harry Thomas Van Fossen Jr.
Frank Von Sprecken Jr.
Harry J. Walkey
Kenneth E. Walsh
Roy F. Wanamaker
James E. Ward
Roy E. Warren
Carl Zachary Webb
I. Barron Webb
William Keith Webb
Charles Paul Weeks
Edgar Ray Weiland Jr.
William Weill
Thomas Henry Wells
Marion Armond Westbrook
Bennie White
Richard Franklin White
Theodore Elroy White
Willis A. White
Leroy Ellis Whitehead
Wesley T. Whitfield
Joserh Clarence Whitlock
Adrian Ingalls Wilcombe Jr.
Wilbur Eugene Wilder
Adrian Delton Willams
Harves Dean Willams
Oscar Marion Weliams
Turner Williamson
Jewell Gordon Wilson
John Wilson
Thomas Edward Wilson
Thomas Woodrow Wilson
William Hannamen Wilson
John Edward Winkler
James J. Winn
William E. Worsham Jr.
Henry Wyche
Earl W. Yancey
Bernard "Ben" R. Yglesias
Edward M. Young
John Rayner Young
David N. Zoller

KOREAN WAR

Leon Andrews
William Robert Ball Jr.
Curtis Donald Baskin
William Murphy Bolton
Price Brannon
Bascom A. Brooks
Neilson V. Brouillette
Al Brownfield
Clifton Z. Couch Jr.
A. J. Cronin
Lionel J. Delcambre
James D. Falgoust
Leonard P. Garcia Jr.
Herbert H. Gielen
Frank S. Hagan
Emile Oscar Hempel Jr.
Stanly Milam Jones
William Harold Kellum

Davis A. King
Bruce Lloyd
G. J. Mandina
Douglas Emmitt McInnis
Thomas Foster Murphy
Carl W. Plitt
Charles William Poe
Maurice H. Rahe
Carroll Anthony St. Martin
C. Edward Schiele Jr.
Carroll Edward Schmidt
John Henry Shepherd
Jim Shoaf
Edward S. Simmons
Frank Stephens
Clement Norwood Street
Monte Strong
David Tatum
Lote Thistlethwaite
Hereford Percy Thompson
Charles Raiford Tircuit
Lear P. Titard
John Charles Trent
Joe A. Tully

VIETNAM WAR

Thomas Dalton Babin Jr.
Robert Arthur Belcher
Aubrey R. Boswell
Michael Allen Clause
Robert Franklin Coady
Donald J. Cook
Terry Cordell
Louis David Davis Jr.
Arthur J. Delahoussaye Jr.
Carl W. Drake
Michael Emmett
Jack Graham Farr
Arthur Few
George Edward Flynn
Donald F. Ginart
James Hansen
Kenneth Jude Hibbets
James Marshall Hill
Carroll Genehogeman
Phillip Holmes
Deryl Ramon Kirkwood
Joseph Starns Kopfler
Gregory L. Lafleur
Anthony Lavite III
Gerald Thomas Leblanc
Marvin Lindsey
James Edward Miremont
Clifton P. Moak
William Joor Morgan
Henry Lee Prather III
Severo James Primm III
William Archibald Richard
Marvin James Roberts
Layne Joseph Romagosa
Harry W. Smith
Woodrow W. Snyder Jr.
Walter A. Souther
Virgil Grant Stewart
Auldon K. White
Robert Logan Wickliffe

SEPTEMBER 11, 2001

Michael Scott Lamana

IRAQ: OPERATION IRAQI FREEDOM

Christopher William Barnett
Lee Hamilton Deal
Brandon Ryan Dronet

AFGHANISTAN: OPERATION ENDURING FREEDOM

Severin West Summers
Robert James Reeves
Nathan Brock Carse
Aaron Dale Istre

NOTES

INTRODUCTION

1. Paul E. Hoffman, *Louisiana State University and Agricultural and Mechanical College 1860–1919* (Baton Rouge: Louisiana State Univ. Press, 2020), 1.

2. https://www.gilderlehrman.org/history-resources/teaching-resource/statistics-education-america-1860-1950 (accessed 2/19/2021).

3. Walter Lynwood Fleming, *Louisiana State University, 1860–1896* (Baton Rouge: Louisiana State Univ. Press, 1936), 21.

4. Hoffman, *Louisiana State University,* 2.

5. Benjamin L. Price, "Origins of the Ole War Skule: The Creation of the Louisiana Seminary of Learning and Military Institute," *Louisiana History* 52, no. 1 (2011): 41.

6. Fleming, *Louisiana State University,* 29.

7. Price, "Origins of the Ole War Skule," 51.

8. J. A. Turner, "What Are We to Do?" *DeBow's Review* 29 (1860): 71.

9. Price, "Origins of the Ole War Skule," 50.

10. Walter L. Fleming, ed., *General W. T. Sherman as College President,* part 1 (Cleveland, Ohio: Clark, 1912), http://www2.latech.edu/~bmagee/louisiana_anthology/texts/sherman/sherman--lsu.html#CshermantocN02.

1. THE OLE WAR SKULE

1. Fleming, *Louisiana State University,* 37.

2. In the 1840s, Sherman had been a lieutenant in Captain Bragg's artillery company.

3. Fleming, ed., *General W. T. Sherman as College President,* Bragg to WTS, Baton Rouge, February 13, 1860.

4. George Mason Graham, "Autobiography of George Mason Graham," *Louisiana Historical Quarterly* 20 (January 1937): 43.

5. James L. Barnidge, "George Mason Graham: The Father of Louisiana State University," *Louisiana History* 10, no. 3 (summer 1969): 225–26.

6. Ibid., 230.

7. Fleming, *Louisiana State University,* 23–24.

8. John Duffy, *The Sword of Pestilence: The New Orleans Yellow Fever Epidemic of 1853* (Baton Rouge: Louisiana State Univ. Press, 1966), 167.

9. Fleming, *Louisiana State University,* 36.

10. Ibid.

11. Barnidge, "George Mason Graham," 227, 228n6.

12. Fleming, *Louisiana State University*, 25–26.

13. W. T. Sherman, *Personal Memoirs of Gen. W. T. Sherman* (New York: Appleton, 1889), 1:172–193.

14. Fleming, ed., *General W. T. Sherman as College President*, WTS to Brother-in-Law, Alexandria, January 21, 1860.

15. Ibid., G. Mason Graham to S. A. Smith, Alexandria, January 21, 1860.

16. Fleming, *Sherman, Madison Democrat*, November 1859.

17. William Tecumseh Sherman, *Personal Memoirs of Gen. W.T. Sherman* (New York: Charles L. Webster & Co., 1890), 1:17.

18. Fleming, ed., *General W. T. Sherman as College President*, editorial in *Louisiana Democrat* [Alexandria, La.], July 20, 1859.

19. Fleming, *Louisiana State University*, 45.

20. Ibid., 31–33.

21. Ibid., 46.

22. Ibid.

23. Ibid., 48.

24. Sherman, *Personal Memoirs of Gen. W.T. Sherman*, 1:172–193.

25. Robert C. Wood, ed., *Confederate Handbook* (Falls Church, Va.: Sterling Press, 1982), 57.

26. John D. Winters, *The Civil War in Louisiana* (Baton Rouge: Louisiana State Univ. Press, 1963), xii.

27. Act of July 15, 1870, https://history.army.mil/html/faq/5star.html.

28. https://history.army.mil/html/faq/5star.html (accessed April 2021).

29. Ibid.

30. William A. "Bud" Davis, *LSU's Military History: 141 Years of the "Ole War Skule"* (Baton Rouge: Cadets of the Ole War Skule Memorial Tower, 2001, 8.

31. Ibid.

2. AWKWARD AND UNTUTORED

1. Fleming, *Louisiana State University*, 49. In addition to this early chronicling of the early history of LSU, Walter Fleming was a prolific historian, publishing ten books and 166 articles and reviews. He was born in Alabama during Reconstruction and became an authority on that period. Fleming taught history at Louisiana State University from 1907 to 1917.

2. Ibid., 49.

3. Ibid.

4. Fleming, ed., *General W. T. Sherman as College President*, WTS to G. Mason Graham, February 10, 1860.

5. Ibid.

6. Ibid., part 4, "Student Troubles."

7. Ibid.

8. Ibid.

9. Ibid.

10. Ibid., WTS to GMG, February 6, 1860.

11. Ibid., WTS to GMG, February 10, 1860.

12. Ibid., GMG to PTR—R, February 10, 1860
13. Ibid., GMG to PTR—R, February 10, 1860.
14. Ibid., WTS to Mrs. WTS, February 10, 1860.
15. Quoted in Fleming, *Louisiana State University*, 67–68.
16. Ibid., 68.
17. Fleming, ed., *General W. T. Sherman as College President*, WTS to Mrs. WTS, February 10, 1860.
18. Quoted in Hoffman, *Louisiana State University*, 15.
19. Fleming, *Louisiana State University*, 50.
20. Ibid.
21. Ibid.
22. Quoted in ibid., 80.
23. Ibid.
24. Ibid., 82.
25. Quoted in ibid., 83.
26. Ibid., 86.
27. Ibid., 87.
28. Ibid., 88.
29. Fleming, ed., *General W. T. Sherman as College President*, WTS to John Sherman, June 1860.
30. Quoted in Lloyd Lewis, *Sherman: Fighting Prophet* (Lincoln: Univ. of Nebraska Press, 1993), 133.
31. Fleming, ed., *General W. T. Sherman as College President*, WTS to GMG, January 5, 1861.
32. Ibid., WTS to MRS. Sherman, January 5, 1861.
33. Lewis, *Sherman: Fighting Prophet*, 135.
34. Ibid., 136.
35. Fleming, *Louisiana State University*, 96.
36. Ibid.
37. Quoted in Lewis, *Sherman: Fighting Prophet*, 136.
38. Ibid., 138.
39. Fleming, ed., *General W. T. Sherman as College President*, WTS to his wife, January 8, 1861.
40. Ibid., January 13, 1861.
41. Lewis, *Sherman: Fighting Prophet*, 100–101.
42. Ibid., 101.

3. REUNION AT MANASSAS

1. Fleming, ed., *General W. T. Sherman as College President*, WTS to Mrs. WTS, February 22, 1861.
2. Ibid.
3. Ibid.
4. Alcée Fortier, *A History of Louisiana* (New York: Goupil, 1904), 4:4.
5. Ibid., 4:5.
6. Fleming, ed., *General W. T. Sherman as College President*, WTS to Mrs. WTS, February 22, 1861.
7. Ibid., WTS to DFB, April 21, 1861.

8. Ibid., WTS to Montgomery Blair, April 1861.

9. Ibid., WTS to D. F. Boyd, May 13, 1861.

10. Lewis, *Sherman: Fighting Prophet,* 161.

11. David French Boyd Papers, Mss. 40, 90, 794, etc., Louisiana and Lower Mississippi Valley Collections, Louisiana State University Libraries, Baton Rouge, Louisiana.

12. Fleming, *Louisiana State University,* 104.

13. Fortier, *A History of Louisiana,* 4:6.

14. Fleming, *Louisiana State University,* 108

15. Ron Soodalter, "The Saga of the Original Louisiana Tiger: Chatham Roberdeau Wheat," https://www.myneworleans.com/the-saga-of-the-original-louisiana-tiger/ (accessed May 5, 2021).

16. Ibid.

17. Ibid.

18. Ibid.

19. Ibid.

20. Ibid.

21. Ibid.

22. Robert Underwood Johnson and Clarence Clough Buel, eds., *Battles and Leaders of the Civil War,* vol. 1 (New York: Century, 1984–1987), 150n.

23. Soodalter, "The Saga of the Original Louisiana Tiger."

24. Ibid.

25. Arthur W. Bergeron Jr., *Guide to Louisiana Confederate Military Units 1861–1865* (Baton Rouge: Louisiana State Univ. Press, 1989), 149–50.

26. The name "Louisiana Tigers" lived on long after Major Wheat's exploits in the Civil War. At one point they became known as "Lee's Tigers," since the Louisianans fought many battles with Lee's Army of Northern Virginia. In the early 20th century, as LSU developed its athletic programs, especially football, using a tiger as the school mascot was most likely suggested because of the tradition of the fighting battalion, and perhaps because of the memory of former President David F. Boyd, who had been a major in the Confederate Army and had fought with Wheat, and later with Robert E. Lee. (David G. Baker and W. Sheldon Bivin, *Mike the Tiger: The Roar of LSU* [Baton Rouge: Louisiana State Univ. Press, 2003], 1–6).

27. "Funeral of Charles Dreux, 1st Louisiana Battalion," *New Orleans Picayune,* July 16, 1861, http://www.civilwarlouisiana.com/2012/01/funeral-of-charles-dreux-1st-louisiana.html (accessed April 30, 2021).

28. Fortier, *A History of Louisiana,* 4:7.

29. Fleming, *Louisiana State University,* 105.

30. Fleming, ed., *General W. T. Sherman as College President,* epilogue.

31. Quoted in: Samuel M. Bowman and Richard B. Irwin, *Sherman and His Campaigns* (New York, Charles Richardson, 1865), 25.

32. Lewis, *Sherman: Fighting Prophet,* 169.

33. V. C. Jones, *First Manassas: The Bull Run Campaign* (Harrisburg, Pa.: Eastern Acorn Press, 1980), 18.

34. Ibid.

35. Lewis, *Sherman: Fighting Prophet,* 170.

36. Ibid.

37. Ibid., 171.

38. Jones, *First Manassas,* 9.

39. Clifford Dowdy, *A History of the Confederacy: 1832–1865* (New York: Barnes and Noble Books, 1955), 117.

40. Terry L. Jones, *Lee's Tigers: The Louisiana Infantry in the Army of Northern Virginia* (Baton Rouge: Louisiana State Univ. Press, 1987), 19.

41. Ibid., 19–20.

42. Francis F. Wilshin, *Manassas: National Battlefield Park Virginia,* Historical Handbook Series No. 15 (Washington, D.C.: Government Printing Office, 1953).

43. Dowdy, *A History of the Confederacy,* 120.

44. Douglas Southall Freeman, *Lee's Lieutenants: A Study in Command* (New York: Charles Scribner's Sons, 1942), 1:42.

45. Ibid., 1:48.

46. Later it was named the Army of Northern Virginia.

47. Johnson and Buel, eds., *Battles and Leaders of the Civil War,* 1:203.

48. http://www.firstbullrun.co.uk/Potomac/Sixth%20Brigade/7th-louisiana-infantry.html (accessed April 2021). The 7th Infantry Regiment had been organized in May 1861 at Camp Moore from men from New Orleans, Baton Rouge, Donaldsonville, and Livingston. The 944 men were mustered under the command of Colonel Harry T. Hays. On June 22–23, they arrived at Manassas Junction on the Orange & Alexandria Railroad and were assigned to General Jubal Early's Sixth Brigade.

49. Lewis, *Sherman: Fighting Prophet,* 171.

50. Ibid., 172.

51. Gary Schreckengost, "1st Louisiana Special Battalion at the First Battle of Manassas," *America's Civil War,* May 1999.

52. Lewis, *Sherman: Fighting Prophet,* 173.

53. Jones, *First Manassas,* 26.

54. Ibid., 26–27.

55. Johnson and Buel, eds., *Battles and Leaders of the Civil War,* 1:206.

56. Jones, *First Manassas,* 26–27.

57. Ibid.

58. Johnson and Buel, eds., *Battles and Leaders of the Civil War,* 1:206.

59. Jones, *First Manassas,* 28–29.

60. Johnson and Buel, eds., *Battles and Leaders of the Civil War,* 1:206.

61. Bergeron, *Guide to Louisiana Confederate Military Units,* 150.

62. Jones, *First Manassas,* 32.

63. Freeman, *Lee's Lieutenants,* 1:66–67.

64. Ibid., 1:67.

65. Ibid., 1:68–72.

66. Report of Col. Samuel P. Heintzelman, 1st Battle of Manassas, dated 31 July 1861, http://www.thomaslegion.net/colonelsamuelpheintzelmansofficialreportforthebattle offirstmanassas.html (accessed November 24, 2024).

67. Freeman, *Lee's Lieutenants,* 1:68–72.

68. William Parker Snow, *Lee and His Generals* (New York: Richardson, 1867), 232.

69. Ibid.

70. Ibid., 232–33.

71. Freeman, *Lee's Lieutenants,* 1:68–72.

72. Report of Col. Samuel P. Heintzelman, 1st Battle of Manassas, dated 31 July 1861, http://www.thomaslegion.net/colonelsamuelpheintzelmansofficialreportforthebattleof firstmanassas.html (accessed November 24, 2024).

73. Report of Brig. Gen. Irvin McDowell, Commanding U.S. Forces at the Battle of 1st Manassas (Bull Run), dated August 4, 1861, https://spirit61.info/primary-sources/irvin-mcdowell-3/ (accessed May 19, 2021).

4. THE FLUSH OF VICTORY AND THE SEMINARY

1. Jones, *Lee's Tigers,* 253; http://www.firstbullrun.co.uk/Potomac/Fifth%20Brigade/schaeffers-battalion-infantry.html (accessed August 2021).

2. "Second South Carolina Infantry," http://www.firstbullrun.co.uk/Potomac/First%20Brigade/2nd-south-carolina-infantry.html (accessed August 2021).

3. Jones, *First Manassas,* 43; Wilshin, *Manassas.*

4. Ibid.

5. Ibid.

6. Freeman, *Lee's Lieutenants,* 1:72.

7. Johnson and Buel, eds., *Battles and Leaders of the Civil War,* 1:252.

8. James Longstreet, *From Manassas to Appomattox* (Bloomington: Indiana Univ. Press, 1960), 51–52.

9. Ibid., 52.

10. Ibid.

11. Ibid., 53.

12. Richard Taylor, Lieutenant General in the Confederate Army, *Destruction and Reconstruction: Personal Experiences of the Late War* (New York: Appleton, 1879), 19–20.

13. Wilshin, *Manassas.*

14. "Report of Col. William T. Sherman, Thirteenth U. S. Infantry, Commanding Third Brigade, First Division. At The Battle of Bull Run (1st Manassas)," dated July 25, 1861, http://www.thomaslegion.net/colonelwilliamtshermansofficialreportforthebattleoffirstmanassas.html (accessed May 19, 2021).

15. Johnson and Buel, eds., *Battles and Leaders of the Civil War,* 1:159.

16. Ibid.

17. Fleming, *Louisiana State University,* 106.

18. Ibid.

19. Ibid., 109, 123.

20. Ibid., 108–10.

21. Ibid., 112.

22. Ibid., 113.

23. Ibid., 114.

24. Ibid.

25. Jones, *Lee's Tigers,* 54.

26. Ibid., 55.

27. Ibid.

28. Taylor, *Destruction and Reconstruction,* 16–17.

29. Ibid.

5. TIGERS IN THE VALLEY

1. Johnson and Buel, eds., *Battles and Leaders of the Civil War,* 2:282.

2. Freeman, *Lee's Lieutenants,* 1:201.

3. Ibid., 1:203.

4. Johnson and Buel, eds., *Battles and Leaders of the Civil War,* 2:284n.

5. Ibid., 2:285.

6. Bergeron, *Guide to Louisiana Confederate Military Units,* 92–93.

7. Fleming, *Louisiana State University,* 121.

8. Freeman, *Lee's Lieutenants,* 1:321.

9. Henry Kyd Douglas, *I Rode with Stonewall Jackson* (St. Simons Island, Ga.: Mockingbird Books, 1974), 38.

10. Ibid., 56.

11. Taylor, *Destruction and Reconstruction,* 44–45.

12. Ibid., 46.

13. Ibid., 24–25.

14. Ibid., 47.

15. Ibid., 47–48.

16. Ibid., 47.

17. Ibid., 49.

18. Jones, *Lee's Tigers,* 68.

19. Taylor, *Destruction and Reconstruction,* 50.

20. Jones, *Lee's Tigers,* 71.

21. Taylor, *Destruction and Reconstruction,* 38.

22. Freeman, *Lee's Lieutenants,* 1:363.

23. Douglas, *I Rode with Stonewall Jackson,* 335.

24. Ibid., 59.

25. Taylor, *Destruction and Reconstruction,* 51.

26. Belle Boyd, *Belle Boyd, in Camp and Prison* (London: Saunders, Otley, 1865), 1:60.

27. Ibid., 1:63.

28. Ibid., 1:65.

29. Ibid., 1:65–66.

30. Ibid., 1:66–67.

31. Ibid., 1:69–70.

32. Ibid., 1:95.

33. Ibid., 1:105.

34. Ibid., 1:127–28.

35. Ibid., 1:129–30.

36. Ibid., 1:131–33.

37. Taylor, *Destruction and Reconstruction,* 52–53.

38. Jones, *Lee's Tigers,* 74.

39. Ibid., 53.

40. Jones, *Lee's Tigers,* 74.

41. Ibid., 75.

42. Taylor, *Destruction and Reconstruction,* 54.

43. Ibid., 55.

44. Douglas, *I Rode with Stonewall Jackson,* 64.

45. Taylor, *Destruction and Reconstruction,* 57.

46. Ibid.

47. Ibid., 58–59.

48. Ibid., 59.

49. Quoted in Freeman, *Lee's Lieutenants,* 1:402.

50. Douglas, *I Rode with Stonewall Jackson,* 65–66.

51. Quoted in Freeman, *Lee's Lieutenants,* 1:407.
52. Jones, *Lee's Tigers,* 102.
53. Ibid.
54. Ibid., 103.
55. Ibid.
56. Ibid., 104.
57. Ibid., 110–11.
58. Fleming, *Louisiana State University,* 123, 361.
59. Tony O'Bryan, "Jayhawkers," Civil War on the Western Border: The Missouri-Kansas Conflict, 1854–1865, Kansas City Public Library, https://civilwaronthewesternborder.org/encyclopedia/jayhawkers (accessed Friday, July 9, 2021).
60. Fleming, *Louisiana State University,* 124.
61. Ibid.
62. Ibid.
63. Ibid., 125.
64. Ibid., 127.

6. RAPHAEL SEMMES

1. Raphael Semmes, *The Cruise of the Alabama and the Sumter* (New York: Carlton Publishers, 1864), 8.
2. Ibid.
3. Admiral Raphael Semmes, *Memoirs of Service Afloat During the War Between the States* (Baltimore: Kelly, Piet, 1869), 83.
4. Ibid., 86.
5. Ibid., 86.
6. Ibid., 88.
7. "Secession," Library of Virginia, https://www.encyclopediavirginia.org/entries/virginia-convention-of-1861/ (accessed November 25, 2024).
8. Semmes, *Memoirs,* 92–93.
9. Semmes, *The Cruise of the Alabama and the Sumter,* 7–8.
10. Semmes, *Memoirs,* 92–93.
11. Ibid., 93–94.
12. Ibid., 94.
13. Ibid.
14. Ibid., 97.
15. Semmes, *The Cruise of the Alabama and the Sumter,* 11.
16. Semmes, *Memoirs,* 98.
17. Semmes, *The Cruise of the Alabama and the Sumter,* 10; Semmes, *Memoirs,* 98–99.
18. Semmes, *Memoirs,* 99–100.
19. James G. Randall, *Constitutional Problems Under Lincoln* (New York: Appleton, 1926), 61, 67.
20. Quoted in Gordon H. Warren, *Fountain of Discontent: The Trent Affair and Freedom of the Seas* (Boston: Northeastern Univ. Press, 1981), 107.
21. Semmes, *Memoirs,* 100.
22. Semmes, *The Cruise of the Alabama and the Sumter,* 13.
23. Ibid., 15–16.
24. Semmes, *Memoirs,* 117.

25. Ibid.
26. Ibid., 118.
27. Ibid.
28. Ibid.
29. Ibid., 126.
30. Ibid.
31. Ibid., 127.
32. Semmes, *The Cruise of the Alabama and the Sumter,* 17.
33. Semmes, *Memoirs,* 127–28.
34. Ibid.
35. Ibid., 128–29.
36. Ibid., 129.
37. Ibid., 130.
38. Ibid., 131.
39. Ibid., 132–33.
40. Ibid., 139.
41. Ibid., 140.
42. Ibid., 139–40.
43. Ibid., 141.
44. Ibid., 386.
45. Ibid., 180.
46. Ibid., 181.
47. Ibid., 302–3.
48. Ibid., 329.
49. Ibid., 350.
50. Ibid.
51. Ibid., 351.
52. Semmes, *The Cruise of the Alabama and the Sumter,* 97–100.
53. Ibid.
54. Ibid.
55. Ibid., 107.
56. Ibid.
57. Semmes, *Memoirs,* 424.
58. Semmes, *The Cruise of the Alabama and the Sumter,* 115.
59. Semmes, *Memoirs,* 428.
60. Ibid., 445.
61. Ibid., 449.
62. Ibid.
63. Ibid., 450.
64. Ibid.
65. Ibid., 458.
66. Ibid.
67. Ibid.
68. Ibid., 525.
69. Ibid.
70. Ibid., 525–26.
71. Ibid.
72. Ibid., 526.

73. Ibid., 526–27.

74. Semmes, *The Cruise of the Alabama and the Sumter,* 144.

75. Semmes, *Memoirs,* 519.

76. Ibid., 520.

77. Ibid.

78. January 1, 1863.

79. Semmes, *Memoirs,* 543.

80. Ibid.

81. Ibid.

82. Ibid.

83. Semmes, *The Cruise of the Alabama and the Sumter,* 153–54.

84. Ibid., appendix IV, "The Engagement with the Hatteras," 273.

85. Semmes, *Memoirs,* 545–46.

86. Semmes, *The Cruise of the Alabama and the Sumter,* 228–29.

87. Ibid., appendix I, "Prizes of Alabama," 239–48.

88. Semmes, *Memoirs,* 750–51.

89. Ibid., 752.

90. Ibid., 753.

91. Ibid., 753–54.

92. Ibid.

93. Semmes, *The Cruise of the Alabama and the Sumter,* 233.

94. Ibid., 233–34.

95. Semmes, *Memoirs,* 761.

96. Ibid., 761–62.

97. Ibid., 762.

98. Ibid., 764–65.

99. Ibid., 786.

100. Hoffmann, *Louisiana State University,* 288–89.

7. JOHN ARCHER LEJEUNE

1. Fleming, *Louisiana State University,* 172.

2. Ibid., 174.

3. Ibid., 175.

4. Ibid., 177.

5. Ibid., 193.

6. Fortier, *A History of Louisiana,* 4:193

7. Fleming, *Louisiana State University,* 316.

8. John A. Lejeune, *The Reminiscences of a Marine* (Washington, D.C.: Marine Corps Association, 1990), 12.

9. Ibid., 13.

10. Ibid., 15.

11. Ibid., 16.

12. Ibid.

13. Peter S. Michie, *The Life and Letters of Emory Upton, Colonel of the Fourth Regiment of Artillery, and Brevet Major-General, U.S. Army* (New York: Appleton, 1885), 108.

14. Ibid.

15. Ibid., 109.

16. Ibid., 95–96.
17. Ibid., 96.
18. Ibid., 97.
19. Ibid.
20. Ibid.
21. Ibid., 97–98.
22. Ibid., 98.
23. Ibid., 99–100.
24. Ibid., 115.
25. Brian Mast, "Emory Upton changes U.S. Army Tactics" (May 21, 2010), https://www.army.mil/article/39572/emory_upton_changes_u_s_army_tactics (accessed November 14, 2021).
26. Lejeune, *The Reminiscences of a Marine,* 17.
27. Ibid.
28. Ibid.
29. Ibid., 23.
30. Ibid.
31. Ibid., 27–41.
32. Ibid., 42.
33. Ibid., 43.
34. Ibid.
35. Ibid.
36. Orlando Crowcraft, "Armistice Day: WWI Was Meant to Be the War That Ended All Wars. It Wasn't," https://www.euronews.com/2021/11/11/armistice-day-wwi-was-meant-to-be-the-war-that-ended-all-wars-it-wasn-t (accessed November 21, 2021).
37. Mitchell's diary, quoted in *Colonel Billy Mitchell and His Aerial Armada,* http://www.homeofheroes.com/wings/part1/5_mitchell.html (accessed November 21, 2021).
38. Justin Jurek, "Top 10 Deadliest Battles of World War I," https://www.toptenz.net/top-10-bloodiest-battles-of-world-war-i.php (accessed November 21, 2021).
39. National Army Museum, https://www.nam.ac.uk/explore/battle-somme (accessed November 22, 2021).
40. Quoted in ibid.
41. Ibid.
42. Anthony Bruce, *An Illustrated Companion to the First World War* (London: Penguin, 1989), 405.
43. Ibid., 406.
44. The damning Zimmermann telegram read: "Most secretly: We intend to begin on the first of February unrestricted submarine warfare. We shall endeavor in spite of this to keep the United States of America neutral. In the event of this not succeeding, we make Mexico a proposal of alliance on the following basis: make war together; make peace together; generous financial support and an understanding on our part that Mexico is to reconquer the lost territory in Texas, New Mexico, and Arizona. The settlement in detail is left to you. You will inform the President of the above most secretly as soon as the outbreak of war with the United States of America is certain and add the suggestion that he should, on his own initiative, invite Japan to immediate adherence and at the same time mediate between Japan and ourselves. Please call the president's attention to the fact that the ruthless employment of our submarines now offers the prospect of compelling England in a few months to make peace." See *Collier's Encyclopedia* (1969), 761.

45. Lejeune, *The Reminiscences of a Marine,* 107.

46. Quoted in Edward Howard Simmons and Joseph H. Alexander, *Through the Wheat: The U.S. Marines in World War I* (Annapolis, Md.: Naval Institute Press, 2008), 10.

47. Ibid., 11.

48. A. W. Catlin, *With the Help of God and a Few Marines* (New York: Doubleday, Page, 1919), 8.

49. Lejeune, *The Reminiscences of a Marine,* 107.

50. Quoted in David Bonk, *Château Thierry and Belleau Wood 1918: America's Baptism of Fire on the Marne* (London: Osprey Publishing, 2007), 24.

51. Lejeune, *The Reminiscences of a Marine,* 107.

52. Ibid., 107–8.

53. Ibid., 108.

54. Ibid.

55. Simmons and Alexander, *Through the Wheat,* 15.

56. Ibid., 9.

57. Catlin, *With the Help of God and a Few Marines,* 17.

58. Ibid., 17, 25.

59. Simmons and Alexander, *Through the Wheat,* 34.

60. Ibid.

61. Ibid., 55.

62. Quoted in Bonk, *Château Thierry and Belleau Wood 1918,* 27.

63. Mike Phifer, "WWI's Massive German Spring Offensive of 1918," https://warfarehistorynetwork.com/2021/07/23/wwis-massive-german-spring-offensive-of-1918/ (accessed December 7, 2021).

64. Ibid.

65. Ibid.

66. Ibid.

67. Ibid.

68. Ibid.

69. Lejeune, *The Reminiscences of a Marine,* 111.

70. Ibid., 112.

71. Ibid.

72. Ibid., 113

73. Ibid.

74. Ibid., 115.

75. Ibid., 116.

76. Ibid.

77. Ibid., 117.

78. Ibid.

79. Ibid.

80. Ibid., 123.

81. Ibid., 126.

82. Ibid., 127.

83. Ibid., 128.

84. Ibid.

85. Ibid., 128–29.

86. Simmons and Alexander, *Through the Wheat,* 178.

87. Ibid., 179.

88. Lejeune, *The Reminiscences of a Marine,* 137.
89. Simmons and Alexander, *Through the Wheat,* 188.
90. Ibid., 192.
91. Simmons and Alexander, *Through the Wheat,* 198.
92. Lejeune, *The Reminiscences of a Marine,* 154.
93. Simmons and Alexander, *Through the Wheat,* 200.
94. Lejeune, *The Reminiscences of a Marine,* 154–55.
95. Ibid., 155.
96. Ibid.
97. Ibid., 157.
98. Ibid., 159.
99. Simmons and Alexander, *Through the Wheat,* 217.
100. Ibid., 219.
101. Ibid., 225; Lejeune, *The Reminiscences of a Marine,* 172.
102. Lejeune, *The Reminiscences of a Marine,* 175.
103. Robert B. Asprey, "John A. Lejeune: True Soldier," *Marine Corps Gazette,* April 1962.

8. GRAVES BLANCHARD ERSKINE

1. General Graves Blanchard Erskine interview, "Oral History Transcript, General Graves Blanchard Erskine, U. S. Marine Corps (Retired)," interviewed by Benis M. Frank, History and Museums Division, Headquarters, U.S. Marine Corps, Washington, D.C., 1975, 5, hereafter cited as Erskine interview.
2. Ibid., 2.
3. Ibid., 4.
4. Ibid., 15.
5. Ibid., 5–6.
6. Ibid., 6.
7. Ibid., 8.
8. Ibid., 9.
9. Ibid.
10. Ibid., 12.
11. Ibid., 14–15.
12. Ibid., 15.
13. Ibid., 15–16.
14. Catlin, *With the Help of God and a Few Marines,* 15.
15. Simmons and Alexander, *Through the Wheat,* 7.
16. Catlin, *With the Help of God and a Few Marines,* 255.
17. Ibid., 256.
18. Catlin, *With the Help of God and a Few Marines,* 57–63.
19. Ibid.
20. Ibid., 59.
21. Ibid.
22. Ibid., 60.
23. Ibid., 66–67.
24. David Bonk, *Château Thierry and Belleau Wood 1918: America's Baptism of Fire on the Marne* (Oxford: Bloomsbury Publishing, 2012), 40
25. Ibid.

26. Catlin, *With the Help of God and a Few Marines,* 68.
27. Bonk, *Château Thierry and Belleau Wood 1918,* 41.
28. Ibid.
29. Ibid.
30. Catlin, *With the Help of God and a Few Marines,* 74.
31. Ibid.
32. Ibid., 70.
33. Erskine interview, 33.
34. Catlin, *With the Help of God and a Few Márines,* 70.
35. Ibid., 84.
36. Ibid., 86.
37. Bonk, *Château Thierry and Belleau Wood 1918,* 46.
38. Ibid., 49.
39. Simmons and Alexander, *Through the Wheat,* 94.
40. Catlin, *With the Help of God and a Few Marines,* 87.
41. Simmons and Alexander, *Through the Wheat,* 95.
42. Catlin, *With the Help of God and a Few Marines,* 92.
43. Ibid., 93.
44. Bonk, *Château Thierry and Belleau Wood 1918,* 66.
45. Erskine interview, 35–36.
46. Ibid.
47. Ibid., 37
48. Ibid.
49. Simmons and Alexander, *Through the Wheat,* 115.
50. Ibid.
51. Ibid., 54.
52. Bonk, *Château Thierry and Belleau Wood 1918,* 88.
53. Simmons and Alexander, *Through the Wheat,* 167, 194; Erskine interview, 66–68.
54. B. C. Mossman and M. W. Stark, *The Last Salute: Civil and Military Funerals, 1921–1969* (Washington, D.C.: Department of the Army, Government Printing Office, 1991), 3.
55. Erskine interview, 69.
56. Ibid.
57. Ibid.
58. Mossman and Stark, *The Last Salute,* 4–8.
59. Ibid.
60. Ibid., 10.
61. Ibid., 9.
62. J. R. Neubeiser, "With the Hand of God He Will Be Delivered Home," September 22, 2020, 7–8, https://tombguard.org/assets/images/news/Olympia-Marines-With-the-Hand-of-God.pdf (accessed January 10, 2022).
63. Erskine interview, 70.
64. Neubeiser, "With the Hand of God He Will Be Delivered Home," 8.
65. Ibid., 9.
66. Erskine interview, 70–71.
67. Colonel Dennis D. Nicholson, "In Good Hands," *Marine Corps Gazette* 48, no. 11 (1964), 57.
68. Mossman and Stark, *The Last Salute,* 10–11.
69. Michael Robert Patterson, "Graves Blanchard Erskine—General, United States Marine Corps," http://arlingtoncemetery.net/gberskin.htm (accessed January 11, 2022).

70. Erskine interview, 295.

71. Ibid., 296.

72. Patterson, "Graves Blanchard Erskine."

73. Erskine interview, 353.

74. Ibid., 356.

75. Ibid.

76. Joseph H. Alexander, *Closing In: Marines in the Seizure of Iwo Jima* (Washington, D.C.: Marine Corps Historical Center, 1994), 3.

77. Quoted in Robert Leckie, *The Battle for Iwo Jima* (New York: ibooks, 2004), 11.

78. Yoshitaka Horie, "Fighting Spirit—Iwo Jima," memoir, manuscript copy, 1965, Rebstock/Drez archive, 17.

79. Ibid., 75–77.

80. Ibid., 83.

81. Lieutenant Colonel Whitman S. Bartley, USMC, *Iwo Jima: Amphibious Epic* (Washington, D.C.: Historical Section, Division of Public Information, Headquarters, U.S. Marine Corps, 1954), 39.

82. Horie, "Fighting Spirit—Iwo Jima," 83.

83. Alexander, *Closing In*, 5.

84. Major General G. B. Erskine, "3d Marine Division Reinforced Iwo Jima Action Report, 31 October 1944–16 March 1945," April 30, 1945, 1, hereafter cited as 3d Division Report.

85. Colonel Joseph H. Alexander, USMC (Ret.), "Iwo Jima: Hell with the Fire Out," *Leatherneck* 78, no. 2 (February 1995).

86. Charles W. Tatum, *Iwo Jima: Red Blood Black Sand* (California: Chuck Tatum Productions, 2002), 156–57; Charles Tatum telephone interview by Ronald Drez, 2006.

87. Erskine interview, 356–57.

88. Richard F. Newcomb, *Iwo Jima* (New York: Signet Books, 1965), 122.

89. Captain Raymond Henri, *Iwo Jima, Springboard to Victory* (New York: U.S. Camera Publishing Corporation, 1945), 78.

90. 3d Division Report, 1.

91. Ibid., 5

92. Bill D. Ross, *Iwo Jima: Legacy of Valor* (New York: Vintage Books, 1986), 219.

93. Lieutenant Colonel Whitman S. Bartley, USMC, *Iwo Jima: Amphibious Epic* (Washington, D.C.: Historical Section, Division of Public Information, Headquarters, U.S. Marine Corps, 1954), 122.

94. Erskine interview, 358.

95. Ibid.

96. Ibid., 359

97. Ibid., 359–60.

98. Ibid., 363.

9. ARTHUR ABRAMSON

1. Publius Terentius Afer, Roman playwright (195–159 BC).

2. Arthur Abramson interview by Ronald Drez, New Orleans, Louisiana, January 19, 1999, hereafter cited as Abramson interview.

3. Ibid.

4. Ibid.

5. Ibid

6. E. B. Potter and Chester W. Nimitz, *The Great Sea War* (New York: Bramhall House, 1960), 348–54.

7. Clark G. Reynolds, *The Carrier War* (Alexandria, Va., Time-Life Books, 1982), 140.

8. Louis la Vache, "The 'Jeep' Carriers of WWII: Unsung Heroes of the WWII Navy Fleet: The CVEs, Escort Carriers," https://56packardman.blog/2016/07/10/steamship-sunday-the-jeep-carriers-of-wwii/ (accessed January 25, 2022).

9. "Escort Carrier," https://en.wikipedia.org/wiki/Escort_carrier (accessed January 24, 2022).

10. Abramson interview.

11. Joseph H. Alexander, *Utmost Savagery: The Three Days of Tarawa* (Annapolis, Md.: Naval Institute Press, 1995)

12. Abramson interview.

13. Ibid.

14. Ibid.

15. Ibid.

16. Ibid.

17. Reynolds, *Carrier War,* 142

18. Abramson interview.

19. Ibid.

20. Ibid.

21. Potter and Nimitz, *The Great Sea War,* 357.

22. Ibid., 358.

23. Barrett Tillman, *Hellcat: The F6F in World War II* (Washington, D.C.: Naval Institute Press, 1979), 87.

24. Potter and Nimitz, *The Great Sea War,* 358; Abramson interview.

25. Tillman, *Hellcat,* 87.

26. Ibid.

27. Potter and Nimitz, *The Great Sea War,* 359.

28. Tillman, *Hellcat,* 89.

29. Abramson interview.

30. Potter & Nimitz, *The Great Sea War,* 359.

10. MAJOR GENERAL TROY MIDDLETON

1. Stephen E. Ambrose, *Eisenhower: Soldier and President* (New York: Simon and Schuster, 1990), 165.

2. Ibid., 166–68.

3. Ibid., 161.

4. Ibid., 170.

5. Quoted in: Cesare Salmaggi and Alfredo Pallavisini, eds., *2194 Days of War* (New York: Gallery Books, 1977), 633.

6. Hugh M. Cole, *The Ardennes: Battle of the Bulge* (Washington, D.C.: Office of the Chief of Military History, Department of the Army, 1965), 56.

7. Ibid., 55

8. Quoted in Frank James Price, *Troy H. Middleton: A Biography* (Baton Rouge: Louisiana State Univ. Press, 1974), 210.

9. Ibid., 214–15.

10. Ibid., 117–18.

11. Ibid., 175.
12. Ibid., 89–91.
13. Ibid., 212.
14. Ibid., 221.
15. Ibid., 212–13.
16. Ibid., 269.
17. Ibid., 213.
18. Cole, *The Ardennes*, 71.
19. Ibid., 56.
20. Eleanor Beardsley, "At Versailles Palace, An American Looks for Clues to Family History," https://www.npr.org/sections/parallels/2017/04/08/522047155/at-versailles-palace-an-american-looks-for-clues-to-family-history (accessed February 2022).
21. Forrest C. Pogue, *The Supreme Command: United States Army in World War II: The European Theater of Operations* (Washington, D.C.: U.S. Army Center of Military History, 1954), 67–71.
22. Ambrose, *Eisenhower*, 171.
23. Ibid.
24. Ibid.
25. Stephen E. Ambrose, *Citizen Soldiers: The US Army from the Normandy Beaches to the Bulge to the Surrender of Germany* (New York: Simon and Schuster, 1997), 200–201; Stephen E. Ambrose, *Band of Brothers* (New York: Simon and Schuster, 1992), 178.
26. Ambrose, *Band of Brothers*, 179.
27. Price, *Troy H. Middleton*, 215.
28. Cole, *The Ardennes*, 176.
29. Price, *Troy H. Middleton*, 219.
30. Cole, *The Ardennes*, 190.
31. Price, *Troy H. Middleton*, 117.
32. Cole, *The Ardennes*, 294.
33. Charles Gutierrez, "Blood for Time: 9th Armored at Bastogne and the Battle of the Bulge," https://warfarehistorynetwork.com/2016/08/11/blood-for-time-9th-armored-at-bastogne-and-the-battle-of-the-bulge/ (accessed February 2022).
34. Price, *Troy H. Middleton*, 90.
35. Ibid., 262.
36. Ibid.
37. Ibid., 271.
38. Cole, *The Ardennes*, 125.
39. Gutierrez, "Blood for Time."
40. Ibid.
41. Ibid.
42. Gutierrez, "Blood for Time."
43. Ibid.
44. Cole, *The Ardennes*, 176.
45. Ibid., 221–22.
46. Ibid., 211.
47. Ibid., 305–6.
48. Price, *Troy H. Middleton*, 270.
49. Ibid., 231.
50. Cole, *The Ardennes*, 307.

51. Ibid.
52. Ibid., 309.
53. Ibid., 307.
54. Price, *Troy H. Middleton*, 235.
55. Ibid., 241.
56. Cole, *The Ardennes*, 453.
57. Price, *Troy H. Middleton*, 242.
58. Ibid.
59. Ibid., 243.
60. Ibid., 270–71.
61. Cole, *The Ardennes*, 449.
62. Ibid., 454.
63. Ibid.
64. Ibid.
65. Ibid.
66. Ibid., 554; Price, *Troy H. Middleton*, 264.
67. Price, *Troy H. Middleton*, 276–78.
68. Ibid., 271.
69. Ibid., 391.
70. Ibid., 274.
71. Dwight D. Eisenhower, *Crusade in Europe* (New York: Doubleday, 1948), 408–9.
72. Price, *Troy H. Middleton*, 306.
73. Ibid., 346.
74. Ibid., 345.
75. Quoted in ibid., 368–70.
76. Ibid., 374.
77. Ibid., 375–76.

11. CLAIRE LEE CHENNAULT

1. Edwin Holt, *Japan's War* (New York: McGraw Hill, 1986), 60.
2. Ibid., 57–60.
3. Ibid., 85.
4. Ibid.
5. Quoted in ibid., 91.
6. Ibid., 145.
7. Ibid., 153.
8. *Japan Times*, September 10, 1937; Holt, *Japan's War*, 158.
9. Holt, *Japan's War*, 172.
10. Ibid., 156.
11. Ibid., 473n7. Until 1983, most Japanese had never heard of the Rape of Nanking, proof of Japan's continued and long-standing posture of denial, suppression, and exaggeration of facts concerning its actions during the war. In 1983 this diary was revealed at an exhibition sponsored by Kyoto University.

In February 2014 the art of Japanese denial was alive and well. Japan's NHK broadcasting company featured one of its reporters who denied the Rape of Nanking and the murder of three hundred thousand. Naoki Hyakuta said on air that it never happened and that the war crimes trials of Japanese leaders had been staged. Jamie Seidel, "Japan's

Rising, Worrisome Rhetoric Denying Its World War II Crimes Has China and the West Worried about Its New Ultranationalism," https://www.news.com.au/world/japans-rising-worrisome-rhetoric-denying-its-world-war-ii-crimes-has-china-and-the-west-worried-about-its-new-ultranationalism/news-story/3651abcffbcd9102110182bbd47ce043 (accessed 6/20/2022).

12. Quoted in Holt, *Japan's War*, 167.

13. Ibid.

14. Quoted in Iris Chang, *The Rape of Nanking: The Forgotten Holocaust of World War II* (New York: Perseus, 1997), 154.

15. Ibid., 168.

16. Daniel Arnstein, "An Ex-Taxi Driver Checks Up on the Famous Burma Road," *Life*, October 6, 1941.

17. Claire Lee Chennault, *Way of a Fighter: The Memoirs of Claire Lee Chennault, Major General, U.S. Army (Ret.)* (New York: G. B. Putnam Sons, 1949), 3.

18. Ibid., 18.

19. Ronald J. Drez, *Predicting Pearl Harbor: Billy Mitchell and the Path to War* (Gretna, La.: Pelican, 2017), 151.

20. Chennault, *Way of a Fighter*, 20.

21. Ibid.

22. Ibid., 26.

23. Drez, *Predicting Pearl Harbor*, 160.

24. Chennault, *Way of a Fighter*, 16.

25. Ibid.

26. Ibid., 21.

27. Ibid., 29.

28. Ibid., 25.

29. Ibid., 26.

30. Ibid., 30.

31. Ibid., 3.

32. Ibid.

33. Ibid., 30.

34. Ibid.

35. Ibid.

36. LSU did not accept students until they were sixteen. To get in, Chennault listed his birth year as 1890 instead of 1893.

37. Chennault, *Way of a Fighter*, 6.

38. Ibid.

39. Ibid.

40. Ibid., 6–7.

41. Ibid., 7.

42. Ibid., 31.

43. Ibid.

44. Ibid., 32.

45. Ibid., 32–33.

46. Ibid., 34.

47. Ibid., 36.

48. Ibid., 37.

49. Ibid., 38.

50. Ibid.
51. Ibid., 42.
52. Ibid., 51.
53. Ibid., 52–53.
54. Ibid., 59.
55. Ibid., 61.
56. Ibid., 63.
57. Ibid., 64–65.
58. Ibid., 65.
59. Ibid.
60. Ibid., 73.
61. Ibid., 80.
62. Ibid., 88.
63. Ibid.
64. Ibid., 90.
65. Ibid., 92.
66. Ibid.
67. Daniel Ford, *Flying Tigers* (Washington, D.C.: Smithsonian Institution Press, 1991), 43.
68. Chennault, *Way of a Fighter,* 96.
69. Ford, *Flying Tigers,* 44.
70. Ibid., 47–48.
71. Ibid., 52–54.
72. Chennault, *Way of a Fighter,* 102.
73. Navy Department to Nimitz, August 7, 1941. https://www.warbirdforum.com/recruit.htm (accessed July 16, 2022).
74. Ibid.
75. Ibid.
76. Statement of Bruce Leighton (CAMCO), https://www.warbirdforum.com/leighton.htm (accessed July 16, 2022).
77. R. T. Smith, "History of the American Volunteer Group," http://www.usshawkbill.com/tigers/ (accessed July 16, 2022).
78. Ibid.
79. Ibid.
80. Ken Jernstedt interview, January 16, 1999, http://www.usshawkbill.com/tigers/ken.htm (accessed July 17, 2022), hereafter cited has Jernstedt interview.
81. Richard Darling, "The 'Flying Tigers' First American Volunteer Group (AVG), Corgi Aviation Archive Update and Diorama New Announcements," https://www.flying-tigers.co.uk/2019/the-flying-tigers-first-american-volunteer-group-avg-corgi-aviation-archive-update/ (accessed August 22, 2022).
82. Charles H. Older, "Hammerhead Stalls and Snap Rolls," circa 1985, http://www.usshawkbill.com/tigers/chuck.htm (accessed, July 17, 2022).
83. Chennault, *Way of a Fighter,* 104.
84. *Wings Over China: The Story of the AVG Flying Tigers,* written and produced by Susan Yu (Max Media Asia, circa 1999), https://www.youtube.com/watch?v=MhKhBo4WQog (accessed July 18, 2022).
85. Smith, "History of the American Volunteer Group."
86. Ford, *Flying Tigers,* 52.
87. Ibid., 56–58.

88. Byron Glover, "Assembling and Testing P-40s in Burma," *Aviation Magazine*, December 1942, https://www.warbirdforum.com/glover.htm (accessed July 21, 2022).

89. Chennault, *Way of a Fighter*, 109.

90. Ibid., 111.

91. Ibid., 112.

92. Ibid.

93. Ronald V. Regan. "American Volunteer Group: Claire L. Chennault and the Flying Tigers," *Aviation History*, November 2000.

94. Chennault, *Way of a Fighter*, 113.

95. Ibid.

96. *Wings Over China.*

97. Ford, *Flying Tigers*, 89.

98. Chennault, *Way of a Fighter*, 121.

99. Ibid., 123.

100. Ibid., 127.

101. Ibid., 128.

102. Ibid.

103. *Wings Over China.*

104. Chennault, *Way of a Fighter*, 130.

105. Ford, *Flying Tigers*, 117.

106. Chennault, *Way of a Fighter*, 130.

107. Ibid., 133.

108. Ibid., 133–34.

109. Ibid., 135.

110. Ibid.

111. "Battle of China: Blood for the Tigers," *Time*, December 29, 1941.

112. Quoted in *Wings Over China.*

113. Chennault, *Way of a Fighter*, 137.

114. Ibid.

115. Ibid.

116. Ibid., 138.

117. Quoted in ibid., 144.

118. Ibid.

119. Jernstedt interview.

120. Ibid.

121. Chennault, *Way of a Fighter*, 146.

122. Ibid., 147.

123. Ibid.

124. Ibid.

125. Charles Bond and Terry Anderson, *A Flying Tiger's Diary* (College Station: Texas A&M Univ. Press, 1984), 134.

126. Ibid.

127. Quoted in Ford, *Flying Tigers*, 273.

128. Ibid., 275.

129. Holt, *Japan's War*, 428.

130. Chennault, *Way of a Fighter*, 148.

131. Ibid., 149.

132. Ibid.

133. Ibid., 150.
134. Ibid., 161.
135. Ibid., 164–65.
136. Ibid., 164.
137. Ibid., 163.
138. Ibid., 165.
139. Ibid., 166.
140. Ibid.
141. Ibid.
142. Ibid.
143. Quoted in *Wings Over China*.
144. Ibid.
145. Ibid.
146. Chennault, *Way of a Fighter*, 172.
147. Quoted in *Wings Over China*.
148. Chennault, *Way of a Fighter*, 178.
149. CNAC was China National Aviation Corporation.
150. Colonel Ward Boyce, USAF (Ret.), and Tripp Alyn, comps., "WWII Fighter Aces Who Served in the AVG Flying Tigers," https://flyingtigersavg.com/avg-aces/ (accessed August 23, 2022).
151. https://web.archive.org/web/20110615072415/http://www.tecom.usmc.mil/HD/Whos_Who/Boyington_G.htm (accessed August 23, 2022).
152. Chennault, *Way of a Fighter*, 174.
153. Ibid., 174–75.
154. Ibid., 177.
155. https://flyingtigersavg.com/avg-history/ (accessed August 23, 2022).
156. Chennault, *Way of a Fighter*, 175.
157. Ibid., 192.
158. Ibid., 142.
159. Ford, *Flying Tigers*, 245.
160. Chennault, *Way of a Fighter*, 171.
161. Ford, *Flying Tigers*, 246.
162. Chennault, *Way of a Fighter*, 201.
163. Ibid., 203.
164. Ibid., 205.
165. Ibid., 206.
166. Ibid., 211.
167. Ibid., 212.
168. Ibid., 212–13.
169. Ibid., 218.
170. Ibid., 220.
171. Ibid., 224.
172. Ibid., 226.
173. Ibid., 350.
174. Ibid., 204.
175. Ford, *Flying Tigers* (1991), 370.
176. Ibid., (2007), xi–xii.
177. Ibid., (1991), 369.
178. Ibid., (2007), xi.

179. Ibid.

180. Ibid., xii–xiii.

181. Daniel Ford, "About Those AVG Victory Credits," https://www.warbirdforum.com/vics.htm (accessed September 8, 2022).

182. Ibid.

183. Ibid.

184. "Dr. Frank John Olynyk," https://www.sscfuneralhomes.com/obituary/DrFrank-Olynyk (accessed September 8, 2022).

185. Ibid.

12. ROBERT HILLIARD BARROW

1. Robert Barrow interview by Peter Soderbergh, Oral History Center at Louisiana State University, March 24, 1992, 25, hereafter cited as Barrow interview, March 24, 1992.

2. Ibid., 26.

3. Ibid., 27.

4. Ibid., 7.

5. Ibid., 5.

6. Ibid., 21.

7. Ibid., 22.

8. Ibid., 23.

9. Ibid., 32–33.

10. Ibid., 36.

11. Ibid.

12. Ibid., 27.

13. Ibid., 29.

14. Gregory J. W. Urwin, "The Battle of Wake Island: Nation's Morale Lifted in 1941," https://www.nationalww2museum.org/war/articles/battle-of-wake-island-1941 (accessed October 29, 2022).

15. Ibid.

16. Ibid.

17. Ibid.

18. Barrow interview, March 24, 1992, 29.

19. Ibid., 33.

20. Ibid., 47–48.

21. Ibid., 62.

22. Ibid., 63–66.

23. Ibid., 68.

24. Ibid., 72.

25. Ibid., 72–73.

26. Ibid., 75.

27. Ibid., 9, 78.

28. Ibid.

29. Ibid., 80.

30. Ibid.

31. General Robert H. Barrow interview by Brigadier General Edwin H. Simmons, January 27, 1986, 47–48, https://www.usmcu.edu/Portals/218/Robert%20H_%20Barrow.pdf (accessed October 29, 2022), hereafter cited as Barrow interview, January 27, 1986.

32. Barrow interview, March 24, 1992, 83.

33. Barrow interview, January 27, 1986, 51.

34. Barrow interview, April 1, 1992, 6–7.

35. Barrow interview, March 24, 1992. 83–84.

36. Barrow interview, April 1, 1992, 28–28.

37. Ibid., 29.

38. Roy E. Appleman, *South to the Naktong, North to the Yalu: June–November 1950* (Washington, D.C.: Center of Military History U.S. Army, 1992), viii.

39. Ibid., 37.

40. Ibid.

41. Barrow interview, April 1, 1992, 39.

42. Ibid., 40–41.

43. Barrow interview, January 27, 1986, 96–97.

44. Ibid.

45. General O. P. Smith interview by Benis M. Frank, June 1969, 194, Historical Division, U.S. Marine Corps, Washington, D.C., https://www.usmcu.edu/Portals/218/Smith%2C%20Oliver%20P_.pdf (accessed November 2, 2022), hereafter cited as Smith interview.

46. Ibid.

47. Brigadier General Edwin H. Simmons, *Over the Seawall: U.S. Marines at Inchon* (Washington, D.C.: History and Museums Division, Headquarters, U.S. Marine Corps, 2000), 2–3.

48. Quoted in ibid., 3.

49. Ibid.

50. Ibid., 5.

51. Ibid., 6.

52. Alan Rems, *Naval History Magazine* 33, no. 3 (June 2019), https://www.usni.org/magazines/naval-history-magazine/2019/june/propaganda-machine-stalins (accessed December 5, 2022).

53. Simmons, *Over the Seawall*, 6.

54. Quoted in Appleman, *South to the Naktong, North to the Yalu*, 496.

55. Lynn Montross and Nicholas Canzona, *The Inchon-Seoul Operation*, vol. 2, *U.S. Marine Operations in Korea, 1950–1953* (Washington, D.C.: Headquarters, U.S. Marine Corps, 1955), 41.

56. Appleman, *South to the Naktong, North to the Yalu*, 498–99.

57. Barrow interview, April 1, 1992, 41.

58. Ibid.

59. Simmons, *Over the Seawall*, 24.

60. Barrow interview, January 27, 1986, 102.

61. Quoted in Simmons, *Over the Seawall*, 26.

62. Ibid.

63. Appleman, *South to the Naktong, North to the Yalu*, 506.

64. Ibid.

65. Simmons, *Over the Seawall*, 31.

66. Barrow interview, January 27, 1986, 103.

67. Montross and Canzona, *The Inchon-Seoul Operation*, 2:212–13.

68. Ibid., 2:216–18.

69. Ibid., 2:226–27; Appleman, *South to the Naktong, North to the Yalu*, 518.

70. Barrow interview, January 27, 1986.

71. Ibid.

72. Quoted in Appleman, *South to the Naktong, North to the Yalu*, 535.

73. Appleman, *South to the Naktong, North to the Yalu*, 535–36.

74. Ibid., 538.

75. Ibid., 539.

76. *New York Times*, October 29, 1950.

77. Appleman, *South to the Naktong, North to the Yalu*, 573.

78. Ibid., 669.

79. Ibid., 608–9.

80. Ibid., 609.

81. Ibid.

82. Ibid., 615–16.

83. Ibid., 621.

84. General Robert H. Barrow interview by Peter Soderbergh, Baton Rouge, Louisiana, May 1, 1992, 3.

85. Quoted in Lynn Montross and Nicholas Canzona, *The Chosin Reservoir Campaign*, vol. 3, *U.S. Marine Operations in Korea, 1950–1953* (Washington, D.C.: Headquarters, U.S. Marine Corps, 1957), 62n3.

86. Ibid., 3:69.

87. Barrow interview, May 1, 1992, 4.

88. Ibid.

89. Ibid.

90. Barrow interview, January 27, 1986, 134.

91. Barrow interview, May 1, 1992, 6; Montross and Canzona, *The Chosin Reservoir Campaign*, 3:70.

92. Barrow interview, January 27, 1986, 135.

93. Barrow interview, May 1, 1992, 9.

94. Appleman, *South to the Naktong, North to the Yalu*, 742.

95. Smith interview, 216.

96. Appleman, *South to the Naktong, North to the Yalu*, 674–75.

97. Montross and Canzona, *The Chosin Reservoir Campaign*, 3:81–82.

98. Ibid., 3:82.

99. Appleman, *South to the Naktong, North to the Yalu*, 676.

100. Ibid.

101. Ibid., 678.

102. Montross and Canzona, *The Chosin Reservoir Campaign*, 3:106.

103. Ibid., 3:119.

104. Ibid., 3:124.

105. Appleman, *South to the Naktong, North to the Yalu*, 769.

106. Ibid., 770.

107. Roy Appleman, *Disaster in Korea: The Chinese Confront MacArthur* (College Station: Texas A&M Univ. Press, 1989), 37–38.

108. Montross and Canzona, *The Chosin Reservoir Campaign*, 3:146.

109. Appleman, *Disaster in Korea*, 37

110. Ibid., 63–64.

111. Joseph C. Goulden, *Korea: The Untold Story of the War* (New York: Times Books, 1982), 338.

112. Appleman, *Disaster in Korea*, 63–64.

113. Goulden, *Korea*, 342.

114. Montross and Canzona, *The Chosin Reservoir Campaign,* 3:178.

115. Edward Simmons, *Frozen Chosin: The U.S. Marines at the Changjin Reservoir* (Washington, D.C., 2013), 48.

116. Ibid., 77.

117. Montross and Canzona, *The Chosin Reservoir Campaign,* 3:243–44.

118. Smith interview, 222.

119. Montross and Canzona, *The Chosin Reservoir Campaign,* 3:239.

120. Ibid.

121. Smith interview, 238.

122. *Frozen Chosin,* 87.

123. Ibid.

124. Montross and Canzona, *The Chosin Reservoir Campaign,* 3:275.

125. *Frozen Chosin,* 88.

126. Montross and Canzona, *The Chosin Reservoir Campaign,* 3:280.

127. Ibid., 3:281.

128. Ibid., 3:282.

129. Ibid., 3:301–2.

130. Ibid., 3:302–3.

131. Ibid., 3:222.

132. Barrow interview, May 1, 1992, 10.

133. Montross and Canzona, *The Chosin Reservoir Campaign,* 3:309–11.

134. Barrow interview, May 1, 1992, 14–15.

135. Barrow interview, January 27, 1986, 150–51.

136. Barrow interview, May 1, 1992, 16.

137. Barrow interview, January 27, 1986, 153.

138. Barrow interview, May 1, 1992, 16.

139. Barrow interview, January 27, 1986, 152.

140. Ibid., 152–53.

141. Montross and Canzona, *The Chosin Reservoir Campaign,* 3:320.

142. Barrow interview, January 27, 1986, 156.

143. Montross and Canzona, *The Chosin Reservoir Campaign,* 3:351.

144. Serving also with that 1957–1960 NROTC unit, as Executive Officer, was Commander William E. Edwards, USN. See chapter 13, "Conspicuous Gallantry and Heroism and the Ole War Skule."

13. CONSPICUOUS GALLANTRY AND HEROISM OF THE OLE WAR SKULE

1. "Capt A J DeLaHoussaye USMC," https://airandspace.si.edu/support/wall-of-honor/capt-j-delahoussaye-usmc (accessed January 27, 2023).

2. Drew Miller, "Honoring Our Heroes: Captain Arthur Joseph delaHoussaye, United Sates Marine Corps," https://www.houmatimes.com/lifestyles/point-of-vue/honoring-our-heroes-captain-arthur-joseph-delahoussaye-united-sates-marine-corps/ (accessed January 27, 2023).

3. Lawrence G. Byrnes, *History of the 94th Infantry Division in World War II* (Washington, D.C.: Infantry Journal Press, 1948), 506, https://archive.org/stream/HistoryOfThe94thInfantryWWII/HistoryOfThe94thInfantryWWII_djvu.txt (accessed January 22, 2023); "Ralph T. Brown," https://valor.militarytimes.com/hero/21927 (accessed January 22, 2023).

4. Jefferson DeBlanc interview, https://www.youtube.com/watch?v=ZWrrYtlgMao (accessed December 2022).

5. The fighters were in fact Nakajima Ki-43 *Oscars*. The Ki-43, often mistaken for the Mitsubishi A6M, was sometimes called the Japanese army's *Zero*.

6. Quoted in https://chsarkansasgreatwar.weebly.com/jarman-sanderford.html (accessed January 25, 2023).

7. Stanley A. Frankel, "Battle of Bougainville: 37th Infantry Division's Battle for Hill 700," *World War II*, September 1997, https://www.historynet.com/battle-of-bougainville-37th-infantry-divisions-battle-for-hill-700/ (accessed December 2022).

8. Ibid.

9. Ibid.

10. Quoted in David Chrisinger, "Fighting to Go Home: Operation Desert Storm, 30 Years Later," https://thewarhorse.org/fighting-to-go-home-operation-desert-storm-30-years-later/ (accessed January 25, 2023).

11. Eliot Brenner, "Iraqi Soldiers Buried Alive," https://www.upi.com/Archives/1991/09/12/Iraqi-soldiers-buried-alive/8831684648000/ (accessed January 25, 2023).

12. Ibid.

13. "Robert J. Reeves," http://www.veterantributes.org/TributeDetail.php?recordID=2013 (accessed January 25, 2023).

14. "Robert J. Reeves: A Navy SEAL Who Died in Extortion 17," https://special-ops.org/robert-j-reeves-navy-seal/ (accessed January 25, 2023).

15. Norman Walker, "Ole Lou Bengals Battle Maroons of Mississippi State to 0-0 Draw," https://www.newspapers.com/clip/36093575/ole-lou-bengals-battle-maroons-of/ (accessed January 27, 2023).

16. Nathaniel S. Patch, "Mission Lifeguard: American Submarines in the Pacific Recovered Downed Pilots," *Prologue: The Journal of the National Archives* 46 (fall 2014): 12–21.

17. Timothy J. Christmann, "Vice President Bush Calls World War II Experience 'Sobering,'" *Naval Aviation News* 67 (March–April 1985): 12–15.

18. Ibid.

19. Quoted in ibid.

20. Quoted in Bud Johnson, "A Memoir: Bush 41 and LSU Tigers," https://www.lsualumni.org/blog/bush41 (accessed January 27, 2023). Also see William Edwards interview by Ronald Drez, January 19, 1992.

21. Ship's log in Patch, "Mission Lifeguard."

22. Ibid.

23. Ibid.

24. Ibid.

25. Timothy J. Christmann, "Vice President Bush Calls World War II Experience 'Sobering,'" https://www.history.navy.mil/research/histories/biographies-list/bios-b/bush-george-h-w/ltjg-george-bush-in-world-war-ii.html (accessed January 27, 2023).

26. https://apnews.com/article/96c11cc442ad8013c05b2be81a5fe216 (accessed January 27, 2023).

27. George Morris, "Before His Football Fame, World War II Shaped Paul Dietzel," https://ww2thebigone.com/2016/09/09/before-his-football-fame-world-war-ii-shaped-paul-dietzel/ (accessed January 27, 2023).

28. "Mission 10—Tokyo Urban Area (Mar 9)," https://6thbombgroup.com/mission-10-tokyo-urban-area-mar-9/ (accessed January 27, 2023).

29. "Paul Dietzel, 6th Bombardment Group," https://www.youtube.com/watch?v=Z9YO-k2dV64 (accessed January 27, 2023).

30. Ibid.

31. "Mission 10—Tokyo Urban Area (Mar 9)," https://6thbombgroup.com/mission-10-tokyo-urban-area-mar-9/ (accessed January 27, 2023).

32. Morris, "Before His Football Fame."

33. Gerald R. Mason, "Operation Starvation," an essay submitted to the Air War College, Air University, Maxwell Air Force Base, Alabama, 6.

34. Ibid., 11.

35. Morris, "Before His Football Fame."

36. Mason, "Operation Starvation," 12.

37. Ibid., 16.

38. "Chinese Bandits of L.S.U.," *Life* 57 (October 12, 1959).

39. Edwin P. Hoyt, *Japan's War: The Great Pacific Conflict* (New York: Da Capo, 1986), 402–3.

40. Ken Ringle, "History Through a Mushroom Cloud," *Washington Post*, July 27, 1995.

41. Peter Jennings, *Hiroshima: Why the Bomb Was Dropped* (ABC Special Report), 1995, https://www.youtube.com/watch?v=9-WnLNLe3sk (accessed February 2, 2023).

42. "The Bombing of Tokyo—The Wrath of the B-29s," November 25, 2014, www.historyonthenet/bombing-of-tokyo.

43. Michelle Tan, "'Hondo' Campbell, Former FORSCOM Boss, Vietnam Vet, Dies," https://www.armytimes.com/news/your-army/2016/02/09/hondo-campbell-former-forscom-boss-vietnam-vet-dies/ (accessed January 27, 2023).

44. "St. Clair Bienvenu Sr.," https://obits.theadvocate.com/us/obituaries/theadvocate/name/st-clair-bienvenu-obituary?id=20761122 (accessed February 11, 2023).

45. Michael Robert Patterson, "George S. Bowman, Jr.—Major General, United States Marine Corps," https://www.arlingtoncemetery.net/gsbowmanjr.htm (accessed February 15, 2023).

46. "George Shepard Bowman," https://valor.militarytimes.com/hero/47978#80354 (accessed February 15, 2023).

47. "Rodney Shelton Foss," https://military-history.fandom.com/wiki/Rodney_Shelton_Foss#Early_life (accessed February 11, 2023).

48. "Whitney A. Langlois," http://veterantributes.org/TributeDetail.php?recordID=1233 (accessed May 1, 2023).

49. James Bollic, *Bataan Death March: A Soldier's Story* (Gretna, La.: Pelican, 1993), 68.

50. Ibid., 73–74.

51. Ibid., 81.

52. Ibid., 90–92.

53. Ibid.

54. Ibid.

55. Gregory F. Michno, *Death on the Hellships: Prisoners at Sea in the Pacific War* (Annapolis, Md.: Naval Institute Press, 2001), 292.

56. Ibid.

57. *Oryoku Maru* roster, https://www.west-point.org/family/japanese-pow/Erickson_OM.htm (accessed May 2, 2023).

58. "*Ōryoku Maru,*" https://en.wikipedia.org/wiki/%C5%8Cryoku_Maru#Experience_of_the_survivors (accessed May 2, 2023).

59. Ibid.; Lee A. Gladwin, "American POWs on Japanese Ships Take a Voyage into Hell,"

Prologue Magazine 35, no. 4. (winter 2003), https://www.archives.gov/publications/prologue/2003/winter/hell-ships (accessed May 2, 2023).

60. Ibid.

61. Ibid.

62. Captain James I. Norwood and Captain Emily L. Shek, "Prisoner of War Camps in Areas Other Than the Four Principal Islands of Japan," Liaison and Research Branch, American Prisoner of War Information Bureau, 1946, https://www.axpow.org/medsearch/PAC-JINSEN%20CAMP.pdf (accessed May 7, 2023).

63. Ibid.

64. Ibid.

65. "Whitney A. Langlois," http://veterantributes.org/TributeDetail.php?recordID=1233/ (accessed May 1, 2023).

66. George Morris, "Building Patton's Bridges," https://ww2thebigone.com/2016/03/31/building-pattons-bridges/comment-page-1/ (accessed February 20, 2023).

67. https://www.dignitymemorial.com/obituaries/baton-rouge-la/charles-hair-9882534.

68. https://www.youtube.com%watch?V=-1LJkws6Zbw/ (accessed February 24, 2023).

69. "Marvin James Roberts," https://www.virtualwall.org/dr/0RobertsMJ01a.htm.

70. Jack Shulimson, Lieutenant Colonel Leonard A. Blasiol, Charles R. Smith, and Captain David A. Dawsonl, *U.S. Marines in Vietnam: The Defining Year, 1968* (Washington, D.C.: History and Museums Division, Headquarters, U.S. Marine Corps, 1997), 425–27.

71. Ibid., 431.

72. Ibid., 432.

73. Ibid., 433.

74. Ibid.

75. Ibid.

76. Ibid., 434.

77. Ibid., 435.

78. Ibid.

79. Ibid.

80. Ibid.

81. Ibid.

INDEX